FP

WITNESSING PALACE, POWER AND POLITICS

MEMOIR OF A MILITARY SECRETARY

Vivek Kumar Shah

FP

Published by FinePrint Books
FinePrint Pvt. Ltd
Bishalnagar Basti Ka, Bishalnagar,
Kathmandu, Nepal
Phone No.: +977-1-4443263
E-mail: info@fineprint.com.np
www.fineprint.com.np

First Published by FinePrint March, 2018

Translated by : Ghanashyam Raj Kafle
Edited by : Peter J. Karthak

ISBN : 978-9937-665-42-1

This book is an English translation of *Maile Dekheko Durbar*, a bestseller and one of the most commented and quoted books published in the Nepali Language.

REVIEWS

The book is informative. At a time when mostly rumors, rather than facts, about the Royal Palace Massacre, came to public, the book has brought truths to light for the people to learn about it. A few human flaws came to my notice while reading the book which I, however, don't consider deliberate.

Lokendra Bahadur Chand
former Prime Minister of Nepal

I thoroughly read the much talked-about book, 'Maile Dekheko Durbar', by Vivek Kumar Shah. Keeping the bulk of the book aside, I found the contents interesting, the language lucid and arresting while reading it. The book will be of use to and help for readers to learn about, and make an opinion of the time and the personalities who operated the country's security organs, the government, political parties, and other matters. The book will also prove useful to those students willing to make a study of the convergence and divergence of opinions of the immediate past two Kings, Gyanendra, and the late Birendra, in their patterns of rule amidst the international events of their times. The book minutely analyses the political events of the times with its far-reaching implications. Because the author in his book is the Military Secretary to the King to brief the ruling monarch on the current state of affairs of importance and the challenges of the times – namely, the three events: the rebellion against the political system of the Panchayat decades, the Royal Palace

Shooting, and the developments thereafter, leading to King Gyanendra – this volume carries much importance. I consider it to be worth reading as it succinctly describes the matters it deals with.

Dr. Suresh Raj Sharma,
Vice Chancellor, Kathmandu University

The book has brought hitherto unknown facts to everyone's notice to generate overwhelming responses from people at home and abroad. Except a few deceased ones, most of the characters described in the book are alive as sources of the knowledge of the events. Since there has been no sound refutation against the facts and information by those individuals, save for a few informal comments, the book enjoys authenticity in the references made therein. Another revelation made by the book is the way it informs the readers on how Nepal's constitutional monarchy controlled and operated even popularly elected and valid governments at times.

Sharad Chandra Wosti,
Editor, *Annapurna Post*

The book, 'Maile Dekheko Durbar' by former Military Secretary to the King, Lt. Gen Vivek Kumar Shah, is an attempt to bring to light the truths and facts hidden in the deepest corners of the Narayanhiti Raj Durbar (Royal Palace). The book valiantly and unhesitatingly unearths events, such as the Royal Palace Massacre, King Gyanendra's October 1 Move, the Maoists' Armed Rebellion and its genesis and the forces backing it, the Comprehensive Peace Agreement (CPA) between Nepal's mainstream political parties and the Maoists, the end of monarchy in Nepal, and the former kingdom's new avatar as a federal democratic and secular republic. It's understandable that some of the most confidential and sensitive information would never be revealed, taking into consideration the author's professional ethics and personal conscience. However, whatsoever has come down to us is more that what we could expect.

Madan Mani Adhikari,
Annapurna Post

Most importantly, the success of the book lies in bringing out the mythlike mysteries of the Royal Palace Massacre to the public. Even then, Nepalis in general still aren't willing to accept Prince Dipendra as the main persona to have caused the massacre. Shah, in his book, doesn't just describe the dual character of Crown Prince Dipendra, he also goes on to explain his sadistic nature. Although Dipendra emerges as the sole factor behind the Royal Palace Massacre, the book considers elements from home and abroad to have incited him to unleash the massacre. The book often directs its fingers to India. The facts presented by Lieutenant General Shah give a new direction to the analysis of the Royal Palace Massacre.

The book also documents the reasons to have caused the long-entrenched monarchy's end, and the arrival of republicanism in Nepal: Disillusionment with political parties, and the Royal Palace's intention to rise to power again. But the Royal Palace chose those as its very confidants who had lost their mass base, and those very elements always encouraged the Royal Palace to resort to autocracy.

Gunaraj Luitel
Nagarik

The author of the book, the Military Secretary but dismissed by the King prior to the completion of his service tenure, has done justice to Nepali history by singling out Dipendra as the perpetrator of the Royal Palace Massacre. By reading the book, one may have a change of impression: that the monarchy that was supposed to remain constitutional in the post-1990 changeover was ever so. The Royal Palace, while it continued taking an interest even in such petty matters as the transfers, promotions, and placements of security officers and civil servants, saw the end of its own existence in 2006.

Babita Basnet
Ghatana Ra Bichar

Those very self-anointed revolutionaries emerge as anti-revolution in the book. In it, the CPN (UML) General Secretary, Madhav Kumar Nepal, is revealed inciting the Royal Palace against the then Prime Minister [Girija Prasad Koirala]. The reference to his meeting the Military Secretary [the author of this book] on Monday, January 28, 2002 speaks of one such machination where listing out the faulty functions of the incumbent Prime Minister, the General Secretary offers the Palace his advice to opt for an alternative in Sher Bahadur Deuba as the next PM of Nepal. The book also reveals that the UML leader Bam Dev Gautam was on the list of the Royal Palace's favorites and that it concentrated its efforts to have Gautam win the elections of 1999, giving him the cloak of a 'nationalistic leader' to wear. Readers also discover through the book that the Royal Palace played its hand to cause one candidate to win and another to lose in the elections. Particularly illuminating is the cajoling of the Members of Parliament to have Sher Bahadur Deuba elected as the leader of the Parliament from the Nepali Congress in 2002 [because] the Royal Palace had a hardened attitude to the Nepali Congress leader, Girija Prasad Koirala.

Hari Bahadur Thapa
Kantipur

The crux of the Royal Palace Massacre is what Vivek has brought out in the book. He audaciously writes who in the Palace were hardliners and who else were softies. We learn from his book that the late King Birendra sincerely wished to go with the political parties by bringing the Maoists into the mainstream politics to address the problems facing Nepal. Under constant pressures from such royal hardliners as Gyanendra, Dipendra, Aishwarya, and Ratna, with some hardcore military and also foreign elements' interests, it caused the murder of King Birendra.

Keshari Bandhu in an email

Other factors notwithstanding, Nepal doesn't lack books to boggle the mind. There are books with such force of thought and language. *Gaddhar Pushpalal* (Sly Pushpalal) by Mohan Bikram [Singh] enjoyed its sway within the Communist circle and its movement. Prithvi Narayan Shah's 'Divyopadesh' (Great Teachings) and 'Jung Bahadurko Belayat Yatra' (Jung Bahadur's Visit to Britain) left their influence on Nepali readers. The latest books 'Raktakunda' (Pool of Blood) by Krishna Abiral, and 'Maile Dekheko Durbar' by Vivek Shah created ripples in many sections of the Nepali society.

Kusum Bhattarai
Annapurna Post

The fresh disclosure by a previous king's military aide and secretary suggests that India was behind promoting the Maoist insurgency against the monarchy in Nepal. He has revealed that India provided military training to Nepal's anti-monarchist Maoist insurgents. In his memoir, 'Maile Dekheko Darbur', meaning, 'The Palace, as I Saw', the former Military Secretary to the King, Lt. General Vivek Kumar Shah, writes, 'The Indian military trained Nepali Maoist insurgents at a military training camp at Chakrata, near Dehradun.'

In his 599-page [Nepali edition] memoir, General Shah even suggests that India could have possibly incited the killing of King Birendra Bir Bikram Shah Dev and his family members in the Narayanhiti Royal Palace Massacre, on Friday, June 1, 2001.

Further, according to Shah, though it was undoubtedly Dipendra who pulled the trigger and caused the carnage in which nine other royals died, he could've been incited by foreign power/s. Shah adds that not only his political moves but also his plan to break the hold of India as the sole supplier of weapons to Nepal sealed King Birendra's fate. General Shah writes as much, 'The politics of weapons is, in one way or the other, linked to King Birendra's assassination.'

Deepak Gajurel
Shri Lanka Guardian

The writer has revealed the Palace's weaknesses at considerable risks to himself. It's an expose of the inner working of the most hallowed of institutions in the country, the monarchy. More than on himself, it's the daring revelations of the institutions, their characters, and the events of contemporary Nepal that border on the explosive. Given Nepali conditions, one is fearful for Shah. But the service he has rendered is in hard print, and that's that.

People's Review
English weekly from Kathmandu

Another point related to the Royal Palace Massacre was Shah's own role during the crisis. What were he and his security system doing when the King, the Queen, and other members of the royal family inside the dining hall of the Narayanhiti Royal Palace were being shot dead one by one? Whom can Shah blame when the four-tier security system surrounding the palace premises was breached?

Through several pages, Shah explains that the hall where the royals had gathered for a private dinner wasn't accessible to security personnel. Nobody could've checked the Crown Prince for carrying weapons. Security cordons alone cannot offer foolproof security to anyone, says Shah, citing such assassinations as those of the former US President, John F. Kennedy and the former Indian Prime Minister, Indira Gandhi.

Dhruba Adhikary
Asia Times Online

Dedicated to

King Birendra and Queen Aishwarya,
who died in the horrific royal palace massacre of June 1, 2001

CONTENTS

Prologue XV
Preface XXI

Chapter: One
AN UNFINISHED JOURNEY

The Context of My Resignation 1
The Royal Palace Massacre 13
June 1: A Dreadful Night 14
The Funeral Processions, Followed by Last Rites 21
Coronation of Gyanendra Bir Bikram Shah Dev 23
The Funeral Procession of the Late King Dipendra Amidst a Curfew 26
The Hospital as (the King's) Office 27
Dipendra and Royal Palace Massacre: The Undercurrents and the Enigmas 29
Some Unanswered Questions 39
King Birendra's Plans for the Immediate Future 41
The Maoist Rebellion: Its Genesis and Growth 43
The Royal Palace's Perceptions of the Maoist Problem 45
Nepal's Political Parties and the Maoist Problem 50
The Maoist Problem and the Royal Nepalese Army (RNA) 52
The Maoist Problem and the Nepal Police 54
The Maoist Problem and the Armed Police Force (AFP) 55
The Maoist Problem and the National Investigation Department (NID) 57
The Maoist Problem and India 58
Maoist Problem and China 64
The Politics of Conspiracies: The Kismat of Nepal 65
The Downfall of the Panchayat Polity and the Restoration of Democracy 66
The First Communist Government in Nepal 70
Six Governments in Five Years! 72
Koirala's Resignation: Holeri a Mere Excuse 76

The Late Night Dissolution of the Parliament 79
The October 4 Move 81
A Bracing Peace Endeavor 85

Chapter: Two
PAGES OF REALITY: THE GENERAL'S JOURNALS

The Undercurrents that Culminated in the Present States of Affairs in Nepal 97

Chapter: Three
THE END OF MY JOURNEY

Scenarios after I left the Royal Palace 383
Peace Efforts in Nepal, and the Carter Center 385
The Royal Move of February 1, 2005 391
The Republic of Nepal, the Ground Realities, and the Nepal Army 397
Some Unforgettable Episodes 400
Arms Diplomacy vis-à-vis Nepal 400
The Social Services National Coordination Council, to Me 405
Handing the ADC's Office over to Me 407
The Sudden Retirement of Colonel Shashi Pratap KC 407
My Emergency Telephone Call to the Press Secretary 408
The Plan to Rig the Ballot Box 409
Queen Aishwarya's Rage 410
My Consultations with the King's Principal Private Secretary 411
The Creation of Two More Vacancies of Lieutenant Generals 413
Undesirable to the RNA Headquarters but Recognized by the Royal Palace 415
Two Men of the Royal Palace on My Memories 416
My Journey, from Rukum to the Royal Palace 418
My Interest in Music 429
Ups and Downs 432
The End of the Road 436

PROLOGUE

For centuries, Nepal has been imagined as heaven on earth, a Shangri-La between two big neighbors – India and China. The country's total landmass of 147,181sq km is geographically varied, unevenly populated with vast linguistic and cultural diversities, and stretches to a length and breadth of 885kms at the broadest and 193km at the narrowest, all endowed with subtropical climate from 60 meters to an arctic climate of 8,848 meters from the sea level. This perpendicular land commands eight of the 14 highest mountains of 8,000 meters above the sea level in the world, with many peaks above 5,000 meters. The snowcapped high mountains are the sources of many perennial rivers, with a collective hydropower potential of more than 80,000MW, which is deemed second only to that of Brazil.

Nepal is one of the most sought-after tourist destinations in the world for mountaineering and high-altitude adventure trekking, with nature and safari tours, rafting and kayaking trips. It houses over 800 species of flora and fauna. The tropical Tarai lowlands are home to the one-horned rhino, the Royal Bengal tiger, musk deer, barking deer, stripped hyena; and snow leopard in the alpine regions, to name only a few of its unique fauna. Its human population stands almost 30 million today.

Archeological remains point to the evidence of human settlements as long ago as around 2,000-3,000 BC while Nepal's recorded history dates from the 5th century AD. Nepal as the birthplace of Lord Buddha and the land of the brave Gurkhas are some facts the Nepalis take pride in. Until recently, Nepal was the only Hindu Kingdom where the monarch was revered as an incarnation of Lord Vishnu. Throughout its history, the country has seen the rise and fall of many royal dynasties, some of them ending up in bloodshed.

There was a time, until recently, when only the Kathmandu Valley was called Nepal. The idea of oneness and togetherness grew into a nationality under the same identities of Nepal and Nepali, regardless of one's geographical location, community, religion, and mother language in the land.

Until 1796 AD, Nepal had been ruled by several dynasties for over 14 centuries, existing in many principalities whose sole aim was to extend their territory and expand influence over the immediate neighbors either by war or intrigue. This continued until King Prithvi Narayan Shah, the ruler of Gorkha, emerged with a pioneering desire for conquest to stitch together the shaky and hostile principalities into a single country which he would call Nepal. Born and raised in Gorkha, he envisioned a stronger, richer, and better Nepal where people could live in peace, prosperity and social well-being.

That would not happen just by wishing, however. It would require a series of engagements for a good cause, or one day some elements from the outside would rip off the territories and annex them to their own. Some readers of the history of the conquests may smell bloodshed and hear the echo of brutality, but carving a country doesn't happen without paying some price at times.

Prithvi Narayan Shah established the Gorkha Dynasty, founded by his ancestor, Drabya Shah, revered as the lord of Gorkha, as the ruling dynasty of what was to become Nepal.

However, the Shah Dynasty and its rule suffered its ill fortunes through palace intrigues, acts of cruelty, bouts of insanity, and queens vying to place their own eldest sons on the throne – because in those days, a king married more than one queen – no matter what their own precedence prevailed.

It was not surprising, therefore, that the House of Gorkha was pushed to the background by Jung Bahadur Kunwar (Rana) who, in 1846, established his Rana family rule of Maharajahs which lasted for 104 years, ending only in 1950 AD.

Jung Bahadur rose through a series of murders, all of which may not well have been committed by him. But the Kot Massacre stands out in which the Kunwars slaughtered a number of powerful Chautarias – the descendants of Drabya Shah – as well as Thapas, Basnyats and Pandes. The survivors suffered even more severe lives; many were driven to exile, chief among them was the Junior Queen, clearly shown having a hand behind the bloodbath. Exiled by Jung, her husband, King Rajendra, followed her, thus affording Jung Bahadur with the opportunity to place Prince Surendra on the throne under Jung's own guardianship. Surendra, people believed, simply stood as an onlooker to let acts of brutality go rampant, leading to untold deaths. In fact, Surendra was somewhat mentally unstable and exhibited his insanity. The Ranas thus rose to power and ruled the roost in Nepal for 104 years.

Based on current political consequences, Nepal has not really come much further and, no doubt, when the current period is over, it will be remembered in similar ways.

After the Rana rule, King Tribhuvan, who is hailed as a kind of savior for siding with the exiled Congress Party of Nepal which stood for democracy – it was a party nurtured by the Indian Congress and a beneficiary of Rana wealth donated by disaffected members of the clan – left for medical treatment in Switzerland, and his son, Crown Prince Mahendra, ascended the throne after his demise. Out of patience with the squabbling of the political parties – the Congress and the opposition, the nascent Communists – King Mahendra took over to introduce his Panchayat governance, calling it a system suited to the soil of Nepal under the active leadership of the King. When King Mahendra passed away in 1971, his son, Crown Prince Birendra, succeeded him to the throne.

Many people, this author included, believe that King Birendra genuinely wished to establish a constitutional monarchy and multiparty democracy. Unfortunately, vested interests, both within and from outside Nepal, with pushes and pulls active within his own family, and the Palace administration's own cliques rendered his goal unattainable. In 1980, there was a nationwide Referendum declared by the King for the people to vote either for a reformed Panchayat system or a multiparty democracy. The Panchayat with Reformation plank won the vote, though with slender margins (55:45%); and many believed that the polls were rigged by wily Panchayat politicians.

In 1990, Prime Minister Krishna Prasad (KP) Bhattarai spoke to the media, admitting that he was impressed by the compassionate nature of the King and his belief in democracy and development. He said he found the King democratic not just in principle but in practice, too. Unfortunately, in the general elections one year later, political infighting and intrigue caused Bhattarai to lose his seat to Girija Prasad (GP) Koirala.

Koirala, in turn, took over a decade of political infighting, leading to the culmination of the Maoists declaring their 'People's War' (Jana Yuddha) by 1996. Despite the obduracy of dodgy politicians and their failure to bring the war to an end, King Birendra however envisioned a prosperous Nepal with a system that could embrace broad expectations and deliver a better life with the participation of the Nepalis.

Unfortunately, that could not happen. The Royal Palace Massacre of Friday, June 1, 2001put an end to that dream.

On this night, ostensibly drunk and taken to his quarters to sleep if off, Crown Prince Dipendra burst in upon the royal family and relatives invited to a traditional Friday dinner in a hall of the royal palace. Miraculously sober, he broke in, firing an automatic upon the party. Dressed like a Special Forces commando, he flattened to death or coma nine of his family members and close relatives, injuring three others and leaving unscathed six others who had hidden behind the furniture.

The general public was left speculating as to how and why such a seemingly calm and cool man could end up killing his own family members and kinsfolk. It was something hard to believe, but that was what indeed happened.

Crown Prince Dipendra suffered from a trauma of unrequited love. He was in love with Devyani Rana. But his parents, particularly his mother, Queen Aishwarya, would not even wish to hear that name. It was even rumored that the Queen had told him that, if he so wished, he could marry whom he wished and would receive a certain inheritance, but the crown would go to his younger brother, Nirajan.

There were also talks of Maoist and foreign plots in what happened. Much of the speculations were fueled by the fact that the Crown Prince had blown out his brains on the left side of his head: whether or not a right-handed marksman would be able to shoot the left half of his head. With his father dead, along with his mother, brother, sister, and three aunts and an aunt's husband – and later to be followed by his youngest uncle, Dhirendra – a brain dead Dipendra was declared King as per the nation's constitution of the day, and who died 48 hours afterwards, to be cremated in the curfewed night, without a full state honor due to a King.

The direct royal line that ended with King Birendra now moved to his middle brother, Gyanendra, who was absent from that ill-fated party in the Royal Palace that evening. Gyanendra had been crowned King before, in 1950, at the age of four, by the Rana Oligarchy.

King Gyanendra ascended the throne at a very difficult time. There was a great deal of political instability, and the Maoists (CPN-Maoist) were waging a so-called 'People's War' throughout the country. By 2005, the King, worried at the political bickering, had maneuvered to take over the state powers himself and assert his authority in the name of good governance and for bringing peace. King Gyanendra resolved to restore 'peace and effective democracy' within three years. But the seven-party alliance, along with the Maoists, did not cooperate with his idea. On the contrary, the two sides – the seven-party alliance and the CPN Maoists – signed the 12-Point Agreement to launch their popular movement against the King.

Most Nepalis believe that this agreement was engineered by India, like the one it mediated in 1951. The revolt was successful in its mission, which included elections to a Constituent Assembly and the automatic abolition of Nepal's monarchy.

Despite these events, however, there is no end in sight to the ongoing political instability. The Peace Process was yet to be concluded, and the writing of the new Constitution was further delayed. The people were becoming increasingly restless.

The facts of what transpired that night are hidden within the walls of the Narayanhiti Royal Palace, just as much of the 'true' history of Jung Bahadur's Kot Massacre is hidden in the old buildings with worn out walls of the Thapathali Palace.

1.Killed in the carnage were King Birendra Bir Bikram Shah Dev, Queen Aishwarya Rajya Laxmi Devi Shah, the Princess Shruti Rajya Laxmi Devi Rana, and the youngest son of the King, Prince Nirajan. The other kinsfolk killed were the King's eldest and middle sisters, Princesses Shanti and Sharada, Princess Sharada's husband, Kumar Khadga Bikram Shah, the King's cousin, Princess Jayanti Shah, and later the Crown Prince Dipendra himself, supposedly by his own hand, and his paternal uncle Dhirendra Shah (formerly Prince, youngest brother of King Birendra)

2. The survivors were Princess Shobha, the King's youngest sister, and his cousin, the former Princess Ketki Chester. Two youngest members of the royal family and their spouses were also unscathed. Princess Komal, Prince

Gyanendra's spouse who later became Queen, was severely injured. Her son Prince Paras and daughter-in-law Princess Himani narrowly escaped death, hiding behind or under some furniture, with Brigadier General Ravi SJB Rana and Kumar Mahesh Singh, brothers-in-law of the late King Mahendra, and the daughter and son-in-law of Princess Prekshya who was Dhirendra's wife. Prekshya was to die in a helicopter crash some weeks later. Kumar Gorakh SJB Rana, the husband of the Princess Royal Shruti, was severely injured.

The survivors stated that Crown Prince Dipendra carried four sophisticated military equipment, such as an MP5, a submachine gun, an M16 A4 Commando Rifle, one Pistol and an automatic 12-bore shotgun.

~~~~~

With this background, I wrote the story of the time I spent serving the House of Gorkha at the Narayanhiti Raj Durbar.

My original story in Nepali, entitled 'Maile Dekheko Durbar', was received with an instant immediacy by people from different walks of life. The book turned out to be the most commented-upon work ever published in Nepali.

As has been said earlier, Nepal abounds in natural beauty and resources. But how could this country be accursed to have political instability for so long, to have nine prime ministers from 1801 to 1846, 12 from 1846 to 1951, eight from 1951 to 1961, 16 from 1990 to 2008.

We cannot, however, say, 'Let the past bury the dead.' Let us instead hope that the facts, whatsoever be they, will come to light one day. The true facts of the Kot Massacre in 1842 and the schemes of Jung Bahadur Rana still lie in the dark and in the shadows of the cobbled enclaves in and around the old Royal Palace of Kathmandu, as the mysteries so remain in the case of the Royal Palace Massacre of 2001 as well.

What I can definitely say, however, is that this book is the history of events witnessed by someone who served in important positions during very critical periods at the Narayanhiti Royal Palace for most of his military career. I am that soldier. I venture to think, therefore, that my story has much to teach and to caution Nepal's politicians, administrators, professionals, intellectuals, and soldiers who nurture their very best intentions for our country.
~~~~~

PREFACE

When I left the Royal Palace in 2003 after 30 years of service, the country was passing through a critical and turbulent period: socially, politically, economically, and in every ominous way, with, among other national ills, the Maoists' so called 'Jana Yuddha' (People's War) at its peak and the mainstream political parties agitating in the streets. At a time when the affairs of the state were fluid and the country was veering to a dreadful uncertainty, the common populace desperately wished for peace.

All of a sudden, to my surprise, and for reasons still unclear to me, King Gyanendra asked for my resignation as his Military Secretary, a post I was serving at for the last three and a half years. As an obedient servant of the King, I immediately submitted my resignation for the monarch's pleasure. I thought the King had a good reason for his decision and in his mind a better and more qualified man for the post. Events that followed after my resignation and the officers who replaced me, however, did not show that kind of light.

Meanwhile, it was but natural for the media to sensationalize my resignation and seek to highlight my side of the story. But I preferred to refrain from going public and speaking in the open, guided by my sense of ethics and conscience.

With my promotion to Major General on April 14, 1998, I took charge of the ADC (Aide-de-Camp) Office at the Royal Palace. Later, on August 17, 1999, after serving about 14 months as ADC, I was posted to the Principal Military Secretariat as Military Secretary. When Lt. Gen Shanta Kumar Malla, the then Principal Military Secretary to the King, took his retirement on April 12, 2000, King Birendra appointed me to the post, effective from April 13, 2000. From then on, I worked closely with, and in direct command of, the King.

By that time, the country had entered the extreme phase of the Maoist insurgency. The successive governments of the time thought of mobilizing the army the way the police force had been deployed to keep the rebels at bay. But instead of doing that the government sought to address the problem by deploying the police force, which, however, was not trained and equipped enough to address the armed rebellion.

I spent 25 years in service as ADC to King Birendra, 14 months as Military Secretary, and more than two years as King Gyanendra's Military Secretary. My experience in the service and my personal integrity might be the reasons I was promoted to the rank of Lieutenant General (Lt. Gen) on April 14, 2003.

It was November 7, 2003 when I walked out of the Royal Palace immediately after my resignation from the very position.

The three and a half or so years of my tenure as Military Secretary thus mark some memorable and crucial periods in the history of Nepal. It is because the years are characterized, among other events, by the Royal Palace Massacre in which the immediate family of King Birendra, including himself, was killed, thus severing the traditional line of succession to the throne and transferring the line to the new King.

Another instance of the period was the failure of the two rounds of roundtable talks with the Maoists. In November 2001, the Maoists, for the first time, attacked the Royal Nepal Army barracks at Dang and took away a large cache of weapons.

In the four and a half years that followed the Parliamentary Elections of 2000, there had been five successive governments, three led by the Nepali Congress and the other two after the dissolution of the Parliament.

Amidst all this, King Gyanendra, invoking Article 127 of the Constitution of Nepal, dismissed Prime Minister Sher Bahadur Deuba from office, repeatedly labeling him 'incompetent', and declared himself as a 'constructive' and 'visible' King, unlike his older brother, King Birendra, whom he dubbed 'a mere onlooker.'

Given the critical situation in the country, with the King and the administration on one side, and the political parties to another, and the Maoists insurgency at its climax, the position of a Military Secretary would obviously be challenging, even more so when he is entrusted with the additional responsibilities of having to regulate the functions of the Secretariat in the Royal Palace and the National Security Council (NSC) in perfect concord with the government and the ministries therein.

Ever since the unification of the country, its monarchy was entrenched as a power centre and the symbol of unity in Nepal in every respect. A combination of destiny, luck and perhaps my dedication to duty, or a mixture of all these factors, placed me in this position at the crucial time in the country's history. The job required of me to see to it that the work at the Home, Defense, and other Ministries was in perfect order on matters of security. In addition, I was to visit intellectuals and politicians, too, and the officials of the security organs, listen to and collect their reports, complaints, and suggestions, and then brief or submit the same to the King. Making arrangements for political and other VIPs' meetings with the King was part of my duty, but it would be the King to give a final nod to grant his audience to the individuals of his choice. The prime duty of the Military Secretary was to study and objectively analyze the situation by studying and reviewing all the information and reports received on security matters, and the documents that came from the Army and offer advice of immediate importance to the King. Because he was the Supreme Commander-In-Chief of the Army as per the Constitution of Nepal, 1990, and in pursuance of the Army Act, 1960, I had to play the role of a bridge between the King and the Chief of the Army Staff (CoAS). I would tactfully maintain 'checks and balances' between the two institutions to the best of my ability. Sometimes misunderstandings surfaced but it would subside soon enough.

Given the nature of my duty, it is quite natural and obvious that I had knowledge of confidential and administration-related information of the country. From the first day of my assuming the high-profile office to the day of my resignation, I wrote notes as and when I went to brief the King on important matters pertaining to the country with regards to all aspects of the prevailing situation, be it intelligence reports or diplomatic notes, under the guidance and orders received from the King. I used these notes that I wrote and the dairies that I kept to write the Nepali version of this book, *Maile Dekheko Durbar*. I have omitted a few events which are of little importance to the readers and certain other entries as well because of my sense of duty, ethics, and conscience that would never allow me to divulge them. I have otherwise chronicled the rest, which I think will satisfy your interest and curiosity as a reader. Those highly confidential and sensitive matters, however, will remain with me to the end of my life.

When I walked out of the office after my resignation, I did not take any documents save my diaries. I didn't write them with a book in mind. But while at it, I have honestly adhered to the diary notes to chart them out in this book in an 'as is where is' manner without deletion or addition of anything betraying my interest. It isn't a research work. I haven't consulted anyone or my colleagues at the Palace. This book simply contains my personal memoirs of the times. Therefore, readers are advised to take this book at its face value and entertain no other romances with the contents.

~~~~~

Sudheer Sharma, the then editor of the weekly *Nepal*, had requested me for an interview following my resignation. I had refused to go public then. That was four years ago. I had told him 'I shall speak up when the time comes, and I'll give my first interview to you.'

He revived his request, and I got ready for the promised interview. It began at the appointed time. An advance notice of the interview was published in the *Kantipur* daily of the previous day. The following day, the interview was published in *Nepal* magazine with a cover story, bringing sensational effects on the public.

It was rumored that the Royal Palace had tried to preempt the interview from being made public, and if it did, it would condemn it outright. It also came to my knowledge that some Generals at the Army Headquarters requested the CoAS to ask me about the interview. When the interview was published, mixed reactions followed, while some Palace people allegedly sounded unhappy about it.

~~~~~

After the interview, many of my friends, intellectuals and well-wishers told me that it would be a pretty good idea to write a book on all that. It occurred to me that as a witness to the most important and sensitive times and periods in Nepal, the book would remain as an important document and be of service as a study of the contemporary history of the country. Thus encouraged and reenergized, I tried my hand in writing this book.

The book has been organized in three parts. The first chapter is called 'An Unfinished Journey' and it includes the episodes of my resignation and other events. The second chapter, 'Pages of Reality', deals with my diary entries as Military Secretary; and the third chapter, called

'The End of My Journey' covers the records, memories and experiences of my service, and the moments of bidding adieu to the Royal Palace.

Sharad Adhikari, a journalist by profession, joined me to assist me and made every effort to bring the book to reality. I wonder if I would ever be able to prepare all those ingredients into a book as it is now without his help.

I am indebted to my friend Sudheer Sharma, the editor of the *Kantipur Daily*, who encouraged me to write this book.

I can never forget Mr. Sharad Chandra Wosti and Mr. Paras Prakash Nepal for editing and proofreading the original Nepali manuscript.

I express my thanks to Ram Krishna Rana of Abriti Media for his deft setup and layout of the book.

Thanks are also due to FinePrint for publishing the book

Similarly, special thanks go to my friend, Devi Ram Sharma, the former Chief of National Investigation Department (NID), mediamen Bhushan Dahal, Manoj Dahal, Kiran Bhandari, and Krishna Jwala Devkota for their support and constructive suggestions.

There is a clear hand of my dear wife, Geeta, with our sons Ajaya Bikram Shah and Dr. Kunda Bikram Shah, behind the book in its present manifestation. I highly value Geeta's constant and priceless support in all the moments throughout the process of writing the book, and of course in all the series of ups and downs in my life.

After *Maile Dekheko Durbar* was published, there was much media reaction in the country and from outside. Also a number of people from eminent walks of life called me with their comments and concerns in the media and via the Internet. My foreign friends called to encourage me to have the book translated into English for international readers. I waited for the opportune time, and now the book in English is in your hands.

I can never forget Ghanashyam Raj Kafle for meticulously translating the book from Nepali to English for an international readership. He was laborious enough to accomplish a challenging task.

I also thank Dr Shashi Pratap Bom Malla for inspiring me to publish this book in English.

My thanks also go to Brig. Gen (Retd) Suresh Sharma who assisted me proofreading the book.

This book surely portrays Nepal's political profile of those times when I was serving as Military Secretary to the King.

Let it be clear to one and all that the book was not written to please or displease anyone. It is an attempt to record the events and moments of the times.

Vivek Kumar Shah

CHAPTER: ONE

AN UNFINISHED JOURNEY

The Context of My Resignation

One Friday morning in November 2003, I woke up at six, and habitually tuned in on the radio for spiritual refreshening. Fond of the *Kantipur News Diary*, I keenly waited for the news update for the upcoming day.

It was then the telephone rang.

It was a call from the ADC Office of the Royal Palace. The voice said, 'Good Morning, Sir. His Majesty wishes to talk to you, and I'll pass the call.'

Before I could even express my humble greetings, I heard King Gyanendra say, 'Vivek, the Prime Minister yesterday spoke to me about the transfer of the DIGs. Why weren't I informed of it earlier?' [DIG stands for Deputy Inspector General, of Nepal Police.]

Before I could confess my ignorance of the news, the King added, 'Wasn't I supposed to know about it?'

'According to the *Naya Karyasampadan Niyamawali* [New Regulations on Work Procedures], this matter should've been brought to Your Majesty's notice prior to the Cabinet making its decision on it,' I said, 'Your Majesty, I didn't know of it. In fact, neither the Home Ministry nor the Police Headquarters informed me about it.'

'Then, why didn't you tell me?' asked the King, again.

I replied, 'The subject file passes from the Prime Minister to the Secretariat of the Cabinet before it's submitted to the Royal Palace where the Principal Secretary receives it. Therefore, it's the Principal Secretary who must brief Your Majesty on the contents of the file.'

'Don't pass the buck,' the King sounded angry.

'This isn't the first time it has happened, Your Majesty', I said.

Before I could finish my explanation, the King said, 'I've had enough with you. You'd better resign.'

'As you wish, Your Majesty,' I said.

I could hear the phone slam on the other side.

My wife, Geeta, sitting close to me, had heard of my side of the telephone conversation. She said, 'Anything wrong?'

At other times, as and when necessary, King Gyanendra used to call me up straight on my cell phone, and the conversation would take place normally. Unlike other times, today's telephone conversation sounded strange. Geeta had every reason to be concerned.

'I've been asked to resign, and so I'll put in my papers without delay,' I said.

'Why resign?' Geeta was at a total loss. She added, 'Why don't you ask the King, once? It doesn't seem fitting to resign so abruptly for someone who's been so long in service at the Royal Palace. How about making your apology to the King if anything has gone amiss inadvertently?'

'There's no alternative to resignation. Resign I must, to honor the Royal Command. There's no point in offering any apology as I've made no mistakes. Maybe the King has been inconvenienced in some ways. I can sense the undercurrents. Earlier, too, attempts were made to prevent my promotion to Lieutenant General,' I explained.

She said, 'Better consult Ajaya once before you take such an enormous decision.'

Ajaya is our eldest son, sleeping downstairs at that moment. He entered the room like a shot, fright written over his face.

I described everything to him. He looked a bit frantic. 'So that's it?'

'Yes, that's it.'

Having heard everything and with a cool mind, he said, 'There would be no better moment than to resign now. What if situations further deteriorate? Differences have already emerged. Any untoward incident can happen or can be caused to happen. Given the situation, the best option is resignation.'

He also convinced his mother and eased the environment of fright and confusion. We could not consult our youngest son, Dr. Kunda, as he was in the Rukum Barracks. But he knew about it soon enough.

I tried to comfort Geeta. 'It's customary in the Royal Palace for things to proceed according to the King's pleasure. It's for the King himself to decide who is to remain in place and who must leave office. We simply obey his command. There are certain terms of reference under which one works at the Royal Palace. Being a soldier of the Royal Nepal Army, I must by virtue abide by my sense of duty. So it requires of me to resign forthwith.'

I had no inkling of any fresh developments in the situation behind the demand for my resignation. Neither yesterday nor this morning when I spoke to the King did I have an iota of any hint about this situation.

Yesterday was Thursday, Prime Minster Surya Bahadur Thapa's weekly evening of his audience with the King. Normally, the Prime Minister would call on His Majesty once a week to brief him on state affairs. Official formalities required that we, the Secretaries of the concerned Secretariats, be in office during the audience. Just in case there arose any matter of immediate discussion or instruction by the King, the Secretaries were summoned to his office immediately.

I wasn't in my office last evening as His Majesty had granted me permission to attend the dinner at the Chinese Embassy.

The day before, Wednesday, was my weekly schedule of briefing the King. It lasted one and a half hours in which I had referred to the Chinese Ambassador's invitation to me for a dinner. The King had approved of it and instructed me to talk to the Ambassador on certain issues of Nepal's interests and importance. I had not seen any expression of discomfiture or displeasure on his face on the day of the briefing. My appointment with the Chinese Ambassador, with the consent of the King, was the reason I was not in my office on the day of the Prime Minister's weekly briefing, which was last evening.

~~~~~

I called Brigadier General Tika Dhamala, the Principal ADC to the King, and the CoAS [Chief of the Army Staff], General Pyar Jung Thapa to inform them of my imminent resignation, as per the King's command. They sounded surprised at the development of the situation.

As I walked to the office, I was tormented by several thoughts but I tried to avoid them. But I wondered what I should state in my resignation.

I entered the Palace through the usual North Gate to reach the changing room. Walking to the Military Secretariat, I mentioned my resignation to Lt. Col Purna Chandra Thapa, Lt. Col Niranjan Karki, and Maj. Himanshu Khadka who were by my side, and they were staff officers in our office. They were surprised at the fresh turn of events. I told them everything about my intended resignation, asking them to take it easy and keep it to themselves for the time being.
~~~~~

My resignation letter was short, straight, and addressed to His Majesty. I stated my long service at the Royal Palace, thanking him for his trust in me for such an important duty and responsibility. I also expressed my unflinching commitment to remain ever loyal and faithful to the Monarchy, whether in or out of office. By nine o'clock, I got the draft of the resignation letter ready. I showed it to Brig. Gen Tika Dhamala who had arrived in his office.

A call from the ADC Office informed me of my appointment with the King in the garden at half past ten. That was where the King granted audience to the Secretaries in the morning.

I carried with me the envelope containing my resignation and the files of the most confidential papers. Press Secretary Mohan Bahadur Pandey, on the way, jokingly asked me if it was a briefing of immediate necessity. 'It's a briefing on getting fired,' I said in a similar joking manner. On the way, another Secretary, Sudan Prasad Pokharel, also asked me a similar question, and I showed him my resignation letter. He looked uncomfortable reading it and insisted that I rethink before submitting it. He reminded me that the King was a fatherly figure, and it was quite natural sometimes for one to be angry with his children, so 'Please don't submit your resignation letter but do apologize, if necessary.'

I replied that I was too hurt to do that.

The King was already in the garden square at the appointed time. I approached him and said, 'Your Majesty, I've brought my resignation letter as per your wish.' I handed him the envelope. He placed it on the table without opening it. He asked me to sit down.

He asked me, 'Why didn't you inform me about the transfer of the DIGs?'

I replied, 'The works the Military Secretariat attends to regarding the Ministry of Home Affairs concerns only the security situation of the country and the personal wellbeing of Your Majesty, and for that matter, those matters concerning the Royal Palace. The practice since 1990 has it that the office doesn't attend to other administration-related works of the Home Ministry. There's no correspondence between the two, either. The procedure is that the Right Honorable Prime Minister, while wishing to brief, or his office corresponding with His Majesty, ought to move through the Principal Secretariat. Earlier, too, other DIGs had been transferred through the same process. In fact, just about two months ago, two DIGs were transferred by the Home Ministry "against the norm" and the IGP [Inspector General of Police] had complained and talked to me about it, and I had reported the matter to Your Majesty. Your Majesty had told me to keep quiet on that matter. Consecutively, to that effect, I had explicitly remarked that DIGs can't be transferred by the Minister of Home Affairs alone, without a Cabinet decision, while he can at the most send them on deputation. I have no notice and knowledge of who briefed Your Majesty on this matter, and in what manner. Your Majesty had told me that the transfer had been proceeded as per your wish and command,' I reminded the King.

Regarding the new transfers, IGP Shyam Bhakta Thapa had already shared with me, some four months ago, about the pressing need for transferring some DIGs. I had told him that it being a regular process within the system, he could do it in accordance with the norms. I had also told him to keep in mind the special circumstances when operations were progressing as per the concept of a Unified Command against the Maoists. I had also hinted at the less than amicable relations between the Nepal Police and the Army. There was a pressing need to mend relations and to make it happen, he could transfer the best officers for the best possible results. I further recommended the transfer of DIG Ashok Shrestha to the Kathmandu Valley in view of his 16 years of service as ADC in the Royal Palace and his good relations with the officers at the Royal Nepal Army Headquarters. He could be a suitable candidate for Kathmandu Valley. The IGP said, 'We've been thinking along the same line, as Ashok Shrestha also has had a good track record at Dipayal.'

I explained to the King: 'I had briefed Your Majesty previously on the details of my talk with the IGP about the transfer of the DIGs. After that, I've had no knowledge of any further developments.'

When the King heard everything I had said, I saw that he looked unhappy. He remained silent, and so did I.

After a minute or so, His Majesty said, 'Who'll look after the work after your resignation?'

'Whosoever Your Majesty wishes or pleases,' I said. 'I'll formalize my handover to him and brief him as fully as I can. Moreover, he can always contact me any moment if needs arise,' I added.

'Things have gone amiss in the Police and the Ministry of Home Affairs,' His Majesty remarked.

We reviewed the earlier works. Then a long pause followed. I had an urge to walk out because I saw my personal dignity being hurt. After all, I had given everything to my work.

'Given this level of unease and the crisis of confidence that has arisen, it'll be difficult for Your Majesty and me to continue. So it's best that my resignation be accepted with immediate effect', I implored.

The King then said, his eyes meeting mine, 'Let me think about it over the weekend.'

I handed the most confidential files to him and walked out. It turned out to be my last briefing to King Gyanendra.

This could be the beginning of new things to come.

A host of conflicting thoughts surged through in my mind as I walked to my office. I called on Sharada Prasad Pradhan, Chief of the Royal Household, to update him on the fresh developments, and to ask him as to whom I should hand my office over. His face darkened when I explained everything.

An impulse came over and I thought how nice it would be to sing this song for the King:

Lagyo bhanchhan Chaitako mahina
Palayechha pati
Namileko gharbar bhanda
Nabhayekai jati.

April has come, so it's said
New leaves have indeed sprouted
'Tis better not to have a family
That's a dysfunctional one.

Next, I made arrangements of handing over my office in a way to set a precedent. It took me one whole day to set separate files for the documents, and of course, the files of important pending works that would be implemented later on.

The King hadn't demanded my resignation merely casually. Nor had I resigned in any flimsy way. The wording in my resignation letter had put an end to the possibility of any reconsideration to reverse it.

Three days, Friday, Saturday and Sunday, passed without a hint of either acceptance or rejection of my resignation. It came to my knowledge later that the King consulted with Principal Secretary, Pashupati Bhakta

Maharjan, Press Secretary Mohan Bahadur Pandey, and Private Secretary Sagar Prasad Timilsina about it. Allegedly, all the Secretaries advised the King that his move would convey untoward message; it would in fact cause more harm than good. I also came to know that the King even consulted with Sharada Prasad Pradhan, the Chief of the Royal Household, and Sudan Prasad Pokharel, Secretary at the Master of Ceremony's Office. Both reportedly advised the King not to accept my resignation as it would only hurt the Royal Palace. Chief ADC, Brig. Gen Tika Dhamala, offering similar advice, further suggested that the King consult with the CoAS about the matter. Most likely my resignation was accepted subsequently thereafter.

The *Nepal Samacharpatra Daily* on Sunday flashed the news of my resignation. I wondered how it got leaked out, as only two or three in the Royal Palace knew about it. The task of handing over of my responsibilities to my successor was not yet over. In response to my query, the Chief of the Royal Household said there was no hurry about it. He directed me to hand it over to the immediate second man next to me, Joint Military Secretary, Brig. Gen. Gajendra Bahadur Limbu, until the letter to that effect would arrive.

On my part, I had finished all the necessary works for handing over the responsibility. I inquired with the Chief of the Royal Household about the delay of the letter. He said it would arrive soon, and when it did, I handed over my office of the Military Secretary to Brig. Gen Limbu. Next, I collected my belongings from the Royal Guards Brigade and walked out of the North Gate, free.

A stream of memories spun in my mind like a film. It plunged me 31 years in the past. My entry into the Army as an officer cadet; and the subsequent opportunity of training at the Royal Military Academy, at Sandhurst, in Britain; my first entry through the North Gate of the Royal Palace after being appointed as Second Lieutenant in the Gorakh Bahadur Battalion as my mother unit; the joining the Infantry Battalion in the Royal Guards, and today my exit through the same gate as Lt. Gen. All these flipped through my eyes like pages of a bulky book. I had mixed emotions, of both pleasure and annoyance. I also experienced a sense of relief and pride to have resigned.

I reviewed my career in the Army. There was never a moment when I faltered in the performance of my duty.I retained my clean and simple image intact throughout the years.

My rank of Colonel Commandant at the Gorakh Bahadur Battalion was my only link left with the Army, and my tenure would expire in 2009. Meanwhile, I resolved to rid my mind of the hangover of the glamour of the Royal Palace, strenuous and powerful as it had been, and finally shed my military uniform. I hummed a folksong along the way:

Manako dhoko manaima rahyo
Thaha paina ke bhayo, ke bhayo....

My wishlist remained right there, in my heart
I knew not what happened, whatsoever happened.....

Many of my friends and well-wishers still ask me why, all of a sudden, was my resignation sought. What can I tell them when I can't make sense of it myself? The transfer of the DIGs was just an excuse, I think. It still occurs to me that it couldn't have been at all a big enough reason to cause my resignation.

It had only been seven months since I was promoted to Lieutenant General. I had already sensed that some elements wanted to effect my early retirement, and the same force was also against my promotion.

When I muse over the time of my resignation, an incident springs to my mind. It was the day before my resignation. I was at the appointed dinner with the Chinese Ambassador. There had been frequent calls on my cell phone from the Royal Palace ADC Office. One involved Abhishekh Shah, husband of Princess Sitashma, who was asked for his driving license by the traffic police for parking his vehicle illegally on the road. He was angry because he thought it was humiliating that the traffic police did not know who he was [a royal in-law of supra-constitutional entitlement]. He asked the policewoman to come to Jeevan Kunj, his residence and reported the incident to Crown Prince Paras [King Gyanendra's son], and he, in turn, asked the ADC Office to take action. The ADC on duty, Brig. Gen. Tika Dhamala, called me, and I told him that the traffic policewoman was doing her duty, and so it would not be wise to subject her to anything regarding the matter. I said, 'I'll talk to the IGP and you please inform His Majesty.'

Some people at the Palace speculated that the Crown Prince was angry with me about the incident. What I can say is that I was right in my decision, as was the traffic policewoman herself. With regards to Brig. Gen Tika Dhamala's concerns, I had said that summoning the traffic police personnel to the house on such a regulatory case would not result in conveying a good message.

Later, I knew that Her Royal Highness Sitashma had flown into a rage for delaying the action, and instantly called the ADC Office to immediately summon the traffic policewoman. Thus, when the environment turned uncomfortable, Brig. Gen Tika Dhamala called me and explained everything on the telephone. I told him to convey to the Princess that due process would be followed. Just in case there was one false step, it would only harm the image of the Royal family and the Royal Palace. It would not be decent to have the traffic policewoman reprimanded.

The situation softened and the chapter was closed once the King himself commanded that the traffic police 'need not be summoned.'

It occurs to me sometimes whether that very incident had anything at all to do with my forced resignation.

Relieved of the responsibility of Military Secretary and subsequently handing over my official vehicle, weapon and the armed escort of my privilege back to the Royal Palace, I returned home, light and free. There was not a single call on my cellphone the entire day. Presumably being informed of my resignation, Brig. Gen. Tika Dhamala, AIG [Additional Inspector General of Police] Amar Singh Shah, and Col. Dr. Prakash Raj Pandey of Birendra Military Hospital visited me at my place in the evening. We drank late into the night and together had an enjoyable time.

Not many knew of my resignation until Tuesday, November 11, when the *Kantipur Daily* published the news. It really created a stir of sensation.

Former Prime Minister Sher Bahadur Deuba called me in the morning to express his surprise and concerns at the development of the situation.

'The King didn't just single you out. We both suffered the same fate. It's an unfortunate turn of events, not only for the King but also for the entire monarchy,' he tried to soothe me.

Dr. Trailokya Nath Uprety, former Vice Chancellor of Tribhuvan University and Ambassador to France, happened to learn of it on his morning walk while he gleaned the *Kantipur.* He visited me to ask if what the newspaper said was right.

'What's this, General?' he asked.

'But that's it, Sir. The truth,' I said.

'What could be behind asking for your resignation?' he persisted in his curiosity. I explained everything to him and simply said that I did not know anything concrete behind the decision.

'A most unfortunate thing has happened. I still can't believe my ears, as you were so sincere and honest in your duty. This is a bad omen for the Royal Palace and the monarchy. But God has been kind to you and has saved you from the impending evils in the Royal Palace,' he said.

Dr. Uprety is no more, but his remarks will remain as the writing on the wall and in my heart.

A rumor circulated that Prime Minister Surya Bahadur Thapa had proposed the name of Krishna Basnet as the DIG of the Kathmandu Valley while I had suggested the name of Ashok Shrestha. I had no clue whatsoever about this posting except some informal talks I had had with the IGP about it. I had also talked to His Majesty earlier on the topic. That was some five months ago, well before Surya Bahadur Thapa became Prime Minister.

My unwarranted resignation also reminds me of another context. This is about my promotion to Major General on April 14, 1998. Subsequently, I worked as the in-charge of the ADC Office of the King for a couple of months. Next, I moved to the Military Secretariat, in July 1999, where I worked under the Principal Military Secretary, Lt. Gen Shanta Kumar Malla. He was asked to take a leave of one month, effective from March 14, 2000 – an established practice observed in the Royal Palace before the retirement of a top-level officer or official. That lead to his automatic retirement and my appointment as Principal Military Secretary.

The Royal Palace has an annual practice of promoting or transferring its executives around February or March. The decision, as a rule, goes into effect from March 13. I had also heard of the likelihood that Gen Malla's term would be extended for another two years despite the customary leave taken prior to retirement. The speculation about his extension seemed to be true because Gen Malla continued attending his office regularly, and he had not handed over the responsibility to me as yet.

This posed some unease, particularly when it came to signing letters: It would be odd for me to sign the letters while he was still in office.

In the midst all this confusion, the Office of the Royal Household issued a letter to each of the outgoing officials, referring to April 9 as Farewell Day. After this, Gen Malla must have finally sensed that he was indeed retired.

After the farewell party, he called me to the Military Secretariat. Disregarding the common protocol, he handed the cell phone, the bunch of keys, and the confidential files to me without any formal letters of handover and takeover involved in such a transfer of office. Despite that, I took over the charge of Acting Principal Military Secretary – a strange appointment, the first of its kind. By and large, I took over the charge as the Principal Military Secretary from April 13, 2000.

As is customary, a dinner was hosted in honor of the King at the Royal Nepalese Army (RNA) Headquarters on April 13 to mark the occasion of the New Year of the Bikram Sambat (calendar). Many generals and officers congratulated me on my appointment.

As his Military Secretary, my first briefing to His Majesty King Birendra Bir Bikram Shah Dev was on April 19, 2000. I thanked the King for his trust in me and expressed my unflinching dedication, and assured him of my honesty in my work and service to the monarchy and the RNA. The King instructed me to work in a spirit of teamwork in a delicate situation like that.

To make myself clear to the King, I said, 'May I say a few things, Your Majesty?'

'Go ahead,' he said.

'Sarkar, I was born in Rukum and brought up there. I did my SLC from Rukum, and I'm the first ever SLC student from the district. Though I've long been a resident in Kathmandu, I have no roots whatsoever here. Therefore, I'm free to do and say everything in straightforward manners.

'I shall always strive to work impartially, and with a clean conscience. But these very traits of mine are likely to cause unease and inconveniences at times. I know that many groups with their vested interests will try to use me and my office for their advantages. But I shall never sway myself to their selfish ways. My duties shall always be dictated by my uncompromising dedication to the Royal Institution of the Kingdom, my loyalty to the Royal Nepalese Army, and my concerns for the greatest good of the nation. There are bound to be inconvenient circumstances due to my unflinching stands and convictions, and I must hold myself most clearly accountable to Your Majesty for your decision to retain me or have me removed. In such instances, should such delicate circumstances ever materialize, I do a petition for your most just thoughts and actions.'

His Majesty pondered for a long time, and soon, easing the situation, said, 'That's precisely why I wanted you here, Vivek.'

I had similarly explained myself to His Majesty King Gyanendra Bir Bikram Shah Dev when he ascended the throne.

When I walked out of the North Gate of the Royal Palace for the last time, my earlier briefing to King Birendra resonated even more prominently now in my mind than ever. That there most certainly were unhappy elements inside the Royal Palace, and my appointment as the Principal Military Secretary to the King was beginning to make sense to me now.

There were quite a few persons posturing as the favorites of the King and exerting pressures on me to work in line to serve their interests. It was but obvious that they would conspire to have in my place a yesman of their own choice. Now I sense that I must have been a pain in their necks while I was in that particular office. I learned later that they had done everything possible to have my appointment as Principal Military Secretary aborted.

But I always worked in the spirit of King Birendra's abiding stand for 'Fairness and Justice in the Army.' Now I can speculate on similar mounting influences pressurizing the present monarch, King Gyanendra, as well.

A Nepal Police officer, back from Kosovo serving under the UN Peacekeeping Mission there, shared an unbelievable snippet with me:

According to him, 'The commander in charge of our sector, a senior IPS [Indian Police Service] officer, disclosed that things went in Nepal as per India's interests. He claimed, "There are some politicians, bureaucrats, and even some Palace officials, including a few ADCs, who work for Indian interests. But you know, one Military Secretary in the Royal Palace came to know that India is giving clandestine training to the Maoists of Nepal. Before it became public, we [Indians] influenced Prime Minister Surya Bahadur Thapa to ask the King to remove the Military Secretary, and he did just that," he said.'

The Police Officer said he was too stunned to hear what he did. His narration of the story sets me to thinking deeply that it might have been the cause of the demand for my resignation.

Yet another speculation connects itself with the King's October 4 move, regarding which I offered my advice to him to have the CPN (UML) government facilitate to bring the Maoists to mainstream politics to treat them in the avowed 'iron cuts iron' principle. I had forwarded Madhav Kumar Nepal's name for the appointment of Prime Minister. It must have displeased the King, and I have reasons to think so.

Candid political discussions took place between the King and me. Our exchanges covered matters of national and international importance and socio-economic and political issues. Unhesitatingly, I would say that adherence to the 1990 Constitution, in letter and spirit, was the best course of action for the monarchy in Nepal. Consequently, the King began meeting and deliberating with the leaders of different political parties. I would brief the King on many issues of due importance. We would also openly discuss the current security situation of Nepal, as befitted my duty and responsibility as Military Secretary.

The Royal Palace Massacre

The Friday night of June 1, 2001 stands as a witness to a most dreadful and shocking event: the Royal Palace Massacre inside the Narayanhiti Raj Durbar of Kathmandu, a most gruesome and unthinkable episode in the history of Nepal. It sends chills down the spine to recall the bloodletting event. I consider it a black day in the history of the country.

Earlier in the day, Their Majesties had visited the Royal Priest, Nayan Raj Pandey, at his residence at Dilli Bazar for tea. It must have been 5 o'clock in the evening. King Birendra was to visit the Rara Lake in the next five days to inaugurate the Rara Festival. This schedule kept me busy. At around the closing time of the office, the King called me to inform Col. Sundar Pratap Rana, his ADC, to arrange his stay at the No 4 Brigade in Surkhet. Accordingly, I briefed the King that even the smallest details had been attended to for his stay and security. I also spoke to him about the contingent arrangement of emergency stay at the Army Company deployed at Rara Lake in case the weather became unfavorable.

It is sad to recall now that it happened to be my last briefing to King Birendra.

It was the ADC Office that would set the schedules of the King's audience to visitors. When it came to security matters, I would arrange it myself. People of different quarters and prominent walks of life would seek an audience with the King. Normally, there were separate schedules of briefing for the various Secretaries of the Royal Palace, but my briefing could happen at any time of the day when it came to matters of immediate importance or of security.

Today, the King was granting his audience to Madhav Prasad Rimal, the editor of the *Spotlight* magazine, at 6 p.m., and this was arranged by the Press Secretary, Mohan Bahadur Pandey. We did not know that was to be the King's last office function and these were the last persons to see and talk to King Birendra before the dreadful event that took place at 9 p.m.!

We in the service at the Royal Palace don't meet the Crown Prince except when there are works or functions. When I met him just two days earlier, he looked very gloomy and I had wondered the reason behind it. There was infrequent news regarding his marriage. I had noticed a dual personality development in him – one for the Royal Palace, and the other for the outside, in the public. He was slated to visit the National Sports Council Office in Sat Dobato. He looked moody that day, too.

June 1: A Dreadful Night

Throughout that particular Friday, I was busy in my office. I called the ADC Office and had a talk with ADC Rana at around six in the evening. We discussed the King's upcoming visits to Rara. As said before, Mohan Bahadur Pandey, the Press Secretary, had arranged the King's meeting with Madhav Kumar Rimal, the editor of *Spotlight*. As also already said, Their Majesties had been to Dilli Bazar at the Royal Priest's residence where they spent 45 minutes during which the Crown Prince reportedly remained silent. One noticed that he looked very unhappy the entire time he was there.

I called my friend Dr. Duman Thapa to remind him of our get together at a friend's place at Naya Bazaar. It must have been quarter past nine or so when Geeta called to tell me that there was a call for me from the ADC Office of Mahendra Manjil, the residence of Queen Mother Ratna.

I called back to enquire, and a certain Lieutenant reported that he had heard gunshots in the Tribhuvan Sadan. I was alarmed at the mention of firing. So I immediately called the Annapurna ADC's office. There was no answer. It was then when I feared that something had gone wrong there.

We headed to the Royal Palace. On the way, Col. Keshav Mahat called me to tell me to reach the Military Hospital without delay. So I turned my car to the Birendra Military Hospital. Meanwhile, the Master of Ceremonies, Chiran Shamsher Thapa, called me. He also referred to the call from the ADC Office but was ignorant of any developments.

I reached the Military Hospital. It was half-past nine or so. A small crowd had already gathered. The CoAS, Gen Prajwalla SJB Rana was already there. Everyone there turned to me. As I entered the hospital, I met Col. Sundar Pratap Rana and Brig. Gen Tika Dhamala. They said, 'Everything's finished! We're robbed! Nothing remains, Sir!'

The sight inside the trauma center was dreadful. Blood-soaked and mutilated bodies of the members of the Royal Family were all over the place. King Birendra lay unconscious and bleeding. Tremendous efforts were being made to revive him. But there was no sign of life except some warmth in his body. Even his pulse had gone. By that time, Queen Aishwarya had already passed away. The Queen Mother, Ratna, stood there, transfixed, shocked, and speechless.

Col. Sundar described the incident. According to him, finished with his audience to the visitors, King Birendra walked out of his office to join the family private dinner at the Tribhuvan Sadan. The party had already begun.

'We were in the ADC room. Because it was a family get-together, we weren't required to be very vigilant. We heard sounds of gunshots. Crown Prince Dipendra had opened fire at his relatives at point-blank range. We broke open the door and entered to see the heartrending sight of King Birendra and others lying on the floor. He was shot several times and bleeding. There was an immediate need to take the wounded to hospital but confusion followed as to which one the bodies were to be taken. Finally, we rushed them to the Military Hospital. There was no time for discussing the next plan of action. Rising needs would be dealt with along the way,' he said.

Col. Sundar described the events further. The family members had only just assembled for the King's arrival when Crown Prince Dipendra tumbled to the floor. Some thought he had had too much to drink. He was taken to a room in the Tribhuvan Sadan by Prince Nirajan, His Royal Highness Paras, and Kumar Gorakh to make him sleep. No sooner had they returned to the party hall, the Crown Prince appeared there, fully armed. He had the most sophisticated weapons, usually secured in a cabinet, but at that time lying open outside. He entered the billiard room in combat dress. According to the Royal Orderly and private staff, he carried four guns: an MP5 submachine gun, an M16 A2 Commando Rifle, a 12-bore gun and a pistol. First, firing a volley at the ceiling, he aimed and fired a shot at King Birendra who instantly fell to the floor. Before the King could take out his automatic, and as Dhirendra lunged forward to snatch the gun away from the Crown Prince, he fired several shots at his father again and on whomever before him. Having done that, he rushed out to the garden. When Queen Aishwarya followed him, perhaps to grab his gun and shoot him, he opened fire at her with the MP-5, and she fell to the ground. Prince Nirajan rushed to her; he was the next target to be shot.

By that time, the weapons in Crown Prince Dipendra's hands had already emptied of bullets in the clips. Maybe his sixth sense worked at that moment. Normally, a killer at such times shoots himself at the height of remorse. He shot himself into his left temple, his right hand still holding the MP5. The pistol with which he shot himself was found in the garden pond. The other weapons were strewn all around the place when the security personnel took him to hospital. The bodies of the others were picked up in a great hurry and taken to the hospital. The body of Queen Aishwarya was the last one to be found. Crown Prince Dipendra, Her Royal Highness Komal, and Dhirendra Shah were rushed to the operation theater.

Panic had overtaken me. But since it had already happened, we had to work as coolly as possible. Doctors were attending to the bodies with everything possible at their disposal. Dr. Rajeev Shahi was calling Drs. Govinda Sharma, Upendra Devkota, and Bhagwan Koirala to come to the Military Hospital.

In such a critical situation, I consulted with the CoAS, and we decided to keep the Army and other security organizations, such as the Nepal Police, the APF and the National Investigation Department at the ready to deal with any unpleasant eventuality. The CoAS issued his orders to the Army Headquarters accordingly.

Meanwhile, Crown Prince Dipendra's ADC, Raju Karki, contacted His Royal Highness Prince Gyanendra who was in Pokhara at the time of the incident. He briefed him on the shooting. The Army was deployed at Kathmandu and Pokhara airports for security and to airlift Prince Gyanendra to Kathmandu.

Officers and doctors were running in and out frantically. I was shocked and frantic myself. Fright was natural at such a time, but that would not solve the problem. I calmed myself and resolved to work, mustering every bit of courage and intelligence in me.

By that time, the news of the shooting in the Royal Palace had already spread like wildfire, first over the Kathmandu Valley, and then all over the country. I informed the senior officers concerned about the shooting and asked them to remain on standby in their respective offices and circulate the instructions to their units.

On my way to the Royal Palace, I contacted the Chief of the Royal Household, Sharada Prasad Pradhan, the Master of Ceremonies, and all the Secretaries to assemble at the Mangal Sadan, the King's office, as soon as possible. I called the ADC Office in the Royal Palace to tell the Secretaries to open their respective offices and remain in place.

A helicopter flew for Pokhara but bad weather forced it to abandon the flight.

I went to my office and changed into my uniform. By that time, the Principal Secretary, Pashupati Bhakta Maharjan, the Principal Private Secretary, Sudan Prasad Pokharel, Secretary Sagar Prasad Timilsina, Press Secretary Mohan Bahadur Pandey, and Chief of the Royal Household Sharada Prasad Pradhan were already there. We assembled in the Mangal Sadan. I briefed everyone on the unthinkable turn of events. We then discussed the immediate plans of action. The Principal Secretary called Prime Minister Girija Prasad Koirala to inform him that he would come to Baluwatar [the Prime Minister's residence] to discuss the seriousness of the matter. We informed the Chief Justice of the Supreme Court, Keshav Prasad Upadhyaya; the Speaker of the House of Representatives, Taranath Ranabhat, the Chairman of the Upper House, Dr. Mohammad Mohsin, the Chairman of the Standing Committee of the State Council, Dr. Keshar Jung Rayamajhi; and the Leader of the Opposition Party, Madhav Kumar Nepal. Dr. Rayamajhi had already reached the Military Hospital after receiving the information. In no time, the Prime Minister came to the Royal Palace, accompanied by the Principal Secretary.

It was already 11 o'clock. I briefed the Prime Minister with all the details from 9 PM till that time, and the possible course of action to be taken. He gave us instructions to work wisely and cautiously, and also promised the government's help in every way. His cool mind and thought were of great help at that critical time. The Prime Minister wished to go to the Military Hospital, and we accompanied him. He looked most moved at seeing the sight. We were impressed by his presence of mind, restraint, and sense of control. After discussion with the doctors, he met with Queen Mother Ratna. He was anxious about His Royal Highness Gyanendra and gave us instructions to have him back forthwith.

We opened the living room, the meeting hall and other offices of the Military Hospital. The Prime Minister's presence and initiative eased our fright and injected much courage in us to work carefully in the hour of crisis. Other VIPs were also reaching the Hospital, one after the other. The Prime Minister's anxiety about His Royal Highness Gyanendra's arrival from Pokhara could be noticed every two or so minutes in his inquiry. No effort was spared to bring him to Kathmandu. But the bad weather had hindered the operation of every flight. As the helicopter could not fly to Pokhara and had to return each time due to bad weather, I called Brig. Gen Prakash Bahadur Basnyat, the No 3 Brigade Commander, to have Prince Gyanendra travel in an army vehicle with a joint security escort of the Army and Police. Soon I learned that the car had moved from Pokhara at around 2 o'clock in the morning, so we sent a helicopter to Gajuri, midway between Kathmandu and Pokhara, for his airlift to Kathmandu from there.

I came to know that Prince Gyanendra had been to Chitwan a week earlier in connection with the projects undertaken by the King Mahendra Trust for Nature Conservation (KMTNC) in the capacity of its Chairman. I also knew that Prabhu SJB Rana and Birendra Bahadur Shah were with him in the entourage. It also came to my knowledge later that he went straight to Pokhara from Chitwan and was staying at Hima Griha, the bungalow that passed on to him after the passing away of his paternal uncle, His Royal Highness Himalaya Bir Bikram Shah [King Mahendra's second younger brother].

I went to the Royal Palace again at 3 a.m., followed by the King's ADCs and other police officers to inspect the Tribhuvan Sadan, the site of the shooting. The glass on the door to the Billiard Meeting Hall was smashed to splinters. We learned that the ADCs had broken it to intervene. The floor was strewn with bullets, the King's glasses, his hat, a bag and other articles. There was a pool of blood, congealed. The lights, fans, and the AC were on. I instructed the soldiers on duty to strictly guard and watch the site of the shooting to keep it in the 'as is where is' condition.

By the time I reached the hospital again, the hope of the survival of the wounded was growing slender. King Birendra must have already passed away on the way while being carried to the hospital. The norms of the monarchy in Nepal required that the name of the new King be announced before the announcement of the death of the previous ruler. We fully discussed the processes as to who should succeed to the throne and how to announce it. The Prime Minister, the Chairman of the Upper House, the Chairman of the Standing Committee of the State Council/Rajsabha, and the Leader of the Opposition Party took constructive and praiseworthy initiatives to that effect. There also followed several rounds of meeting regarding the very matter. The meeting of the State Council was called on a short notice.

The helicopter landed at the Military Hospital. The VIPs present went out to escort His Royal Highness Gyanendra, whereas we, the Secretaries, and the CoAS were already there. He went straight to the trauma center where the mortal remains of King Birendra, Queen Aishwarya and those of the others lay. Crown Prince Dipendra, Her Royal Highness Komal, Kumar Gorakh, Ketki Chester and Her Royal Highness Shova were undergoing surgery inside the operation theater and dressing room. His Royal Highness Gyanendra looked extremely touched and shocked to see the sight. He talked to the Prime Minister and the other VIPs.

A meeting comprising His Royal Highness Gyanendra and the VIPs followed. It discussed the formality of the announcement of the new King, taking into consideration the critical atmosphere that loomed large in the country. During the meeting, His Royal Highness put his emphasis on the formation of a high-level commission to probe into the mishap and make public the facts objectively and without delay. Madhav Kumar Nepal, the leader of the CPN (UML) and the Opposition Party, underscored a free and fair inquiry into the incident, with himself included in the panel of investigation. He had proposed his own name as a member, and the Prime Minister readily supported his idea. Accordingly, a decision was reached to form a panel of investigation led by the sitting Chief Justice.

The meeting of the State Council concluded, announcing Crown Prince Dipendra as the new king, with King Birendra declared as being expired, and His Royal Highness Prince Gyanendra as Regent. Prior to the funeral procession of the late King and the Queen and for the last rites to be solemnized at the Pashupati Aryaghat, another round of meeting of the Regent and the VIPs followed which discussed the formation of a high-level panel of investigation. Present at the meeting were the Deputy Prime Minister (DPM) and Minister of Home Affairs Ram Chandra Poudel, the Secretaries of the Ministries and high ranking officers of the security organizations. In cognizance of the environment increasingly getting tense outside, the meeting also decided to release the unofficial news about the formation of the high-level panel of enquiry, later to be announced officially by the Regent himself. The proposal was first forwarded by the Prime Minister and readily seconded by all the participants attending the meeting. The Principal Secretary had asked each member personally for their reconfirmation for their participation on the panel of the investigation, and with their consent, the decision to that effect was minted and announced. Later, for reasons unclear, Madhav Kumar Nepal decided to leave the panel.

We lacked foresight and experience to deal with the situation as it concerned with an unprecedented event in the country's history. But we proceeded with cool heads. The Regent expressed his wish to have the postmortem of the deceased performed, and for that purpose, the police officers, lawyers, and doctors were asked to be prepared. But when Queen Mother Ratna objected to the idea of an autopsy, citing the cause and the criminal being already identified, everybody concerned seconded her views.

Regarding the announcement of the new King of the nation, the State Council was the authorized body to do so.

The next morning, on Saturday, a meeting followed, and in accordance with the tradition of the monarchy of Nepal, the eldest son of the late King, Crown Prince Dipendra was proclaimed as the new King of Nepal despite the fact that he was in a coma, and the hope of his survival was slender, but there was no other alternative, either. However, as per the laws of the land and the protocol of the State Council, the meeting unanimously announced Crown Prince Dipendra as the new King of Nepal.

It was our destiny that we got the new King, but his crowning could not be solemnized as he was still unconscious. The tradition of Nepal's monarchy required that the new King succeeded to the throne first, and only then the mortal remains of the late King be carried to the Aryaghat for the last rites. This time, unlike in the past, it could not happen – the first of its kind of happenstance in the history of the monarchy in Nepal.

Next, the mortal remains of the members of the royal family were to be carried in funeral processions for their last rites at the Aryaghat. The arrangements for the last rites of so many dead bodies were in itself a daunting task for the Army, as five funeral pyres had to be prepared, and presenting the last guard of honor had to be planned. The tradition also required that the mortal remains be carried by Brahmins all the way from the start of the funeral procession.

We divided our share of the works: the Office of the Master of Ceremonies and the Office of the Principal Secretary jointly saw to all legal and administrative formalities, and the Army saw to all the funeral processions and last rites. The cooperation and coordination of the Secretary of the Ministry of Home Affairs, Shree Kanta Regmi, and the Chief of the National Intelligence Department (NID), Devi Ram Sharma, and the AIGs, Pradip SJB Rana and Krishna Mohan Shrestha were praiseworthy.

Following the conclusion of the State Council meeting, His Royal Highness Gyanendra was slated to address the countrymen to maintain calm and fortitude in the wake of the unprecedented turn of events. The Royal Palace Secretaries concerned, including me, discussed the contents of the address, word by word. Press Secretary Mohan Bahadur Pandey drafted the address as per the Regent's instruction. As was customary, the draft of the address was first shown to the Prime Minister before the address. The Regent afterward addressed the nation and the citizens.

The Funeral Processions, Followed by Last Rites

The mortal remains of Their Majesties were placed in state at the Military Hospital for visitors to pay their last respects to them. There was a milling crowd outside. Every standing space was occupied from end to end, each with eyes transfixed on the facing side of the hospital. People thronged on the hillock of Swayambhu to have a better view of what would follow. The funeral processions commenced at a very slow pace, as planned.

There was not an inch of space on the road as people from far and wide had gathered to bid their last adieu to their cherished King and Queen. People with their heads shorn, with the portraits of the King and Queen, placards and banners with graffiti, aroma of incense sticks in their hands, stood solemnly, many crying and shouting slogans along the road. Buddhist monks also mingled in the crowds, regardless of race or religion. As the procession inched forward, people from the sides offered flowers and garlands to Their Majesties.

The funeral procession passed very slowly as it was difficult to wade through the crowds. It had just reached Shova Bhagawati when an unsavory incident took place. A bunch of miscreants created a stir there. People shouted slogans against Prime Minister Girija Prasad Koirala. The stones aimed at his vehicle strayed and landed on us. Therefore, we arranged the Prime Minister's car to divert and escort him to the Pashupati Aryaghat via another route. People of all age groups had thronged along the route of the funeral procession. People with moistened eyes were wailing at their irreparable loss. I fathomed the magnitude of the people's reverence and honor showered on their King and Queen. I have no personal knowledge or memory of such a funeral procession and the outpouring of such love and affection on their rulers at their demise. The flowers and the *abir* powder offered to Their Majesties smeared us in red.

The funeral procession reached the Aryaghat at Pashupati where elaborate preparations were made for the last rites and the presentation of the guard of honor, followed by the firing of the canon as the last farewell. The Army had planned the entire funerary details, including the setting of the flame to the royal pyres for the cremation. Five funeral pyres were prepared, decorated, and placed in a row.

Sahebyju Dipak Bikram Shah set the fire to the funeral pyre of the late King Birendra. It was already midnight by the time the last rites were performed and completed.

Busy and buried in work, I had spent two sleepless nights. The whole of June 3 was the busiest day ever, as we had to arrange the observations of death rituals. We had been going on without regular food and rest.

Outside, several rumors were adrift. The situation was growing tenser. Rallies shouting objectionable slogans were taken out from different corners. When I came home, it was already 1 p. m. Uncertainty loomed large. The doctors had given up hope for the survival of the new King,

Dipendra. The news of his expiry could come anytime. An unidentified fear was taking over.

It was June 4; I had only just lain in bed when ADC Shahjan Khand's emergency call came. On reaching the hospital, I discovered that King Dipendra had passed away at around 3.30 o'clock that morning. The meeting of the State Council discussed the fresh development of the situation created by the new King's death. In pursuance of the same process, His Royal Highness Gyanendra was announced the new King of Nepal, and the expiry of King Dipendra. There was no other alternative left than that, as Prince Gyanendra remained as the only successor to the throne. With the death of King Dipendra, the entire succession line of King Birendra came to an end. The heirship thus shifted to the new King, Gyanendra. The meeting of the State Council unanimously decided, likewise, and duly announced His Royal Highness Gyanendra as the new King of Nepal.

Coronation of Gyanendra Bir Bikram Shah Dev

While the preparations for the crowning of the new king were going on, the Maoists issued a press release, declaring the end of monarchy in Nepal. The statement alleged that the new King had resorted to conspiracy to seize the throne. Newspapers carried sensational news of the Maoist intention of disrupting the crowning ceremony. There were also rumors that the Maoists had entered the Kathmandu Valley in large numbers from different points with shaven heads in disguise as royalists. The ugly and unpredictable turn of events was adding more challenges to the maintenance of foolproof security in the capital. But there had been adequate support and help from the government. The Secretary at the Minister of Home Affairs, Shree Kanta Regmi, was doing his best on his part. I was also receiving every support from the NID, the Nepal Police, the APF, and the RNA. Given the threat of disruption of the crowning ceremonies, we, on our part, spared no effort to reinforce the security. As part of our preparations, we planted plainclothes security personnel in every nook and corner of the capital to tip off every bit of information to the intelligence cell. We also deployed the RNA to see to it that no untoward incident happened. There was no question of postponing the crowning ceremony merely on the rumor of Maoist threats. We had done everything possible to maintain unfailing security.

The auspicious time for the coronation of the King was worked out by the astrologers to be between 10:30 to 11 o'clock in the morning of Monday, June 4, 2001. The venue of the ceremony, according to old customs, was set at the Nasal Chowk of the Hanuman Dhoka Durbar. All VIPs and officials concerned had been summoned to the venue in due time. Planning an unassailable security system was my share of duty.

A fear psychosis had taken over the people outside. Terror and gloom also loomed large everywhere. Some Royal Palace people held serious doubts over the successful completion of the crowning ceremony. As is rightly said, rumor travels fast and there were enough of them to spread that the Maoists were all set with their collective strength to disrupt the ceremony. I felt anger, pity, and wonder, or a mixture of all these feelings, at the poverty of thought among people holding positions of power.

The auspicious time was fast approaching. Much time had elapsed in the meeting King Gyanendra was holding with Maj. Gen Tara Bahadur Thapa inside the hospital. I sent an ADC in to brief the King at the auspicious time. I was waiting, holding my breath, outside. The King in an informal dress appeared, followed by Maj. Gen Tara. I assured the King of the security arrangements. Because it was a formal ceremony, Royal tradition required that the new King start his journey to Hanumandhoka from Nirmal Niwas in ceremonial attire.

We reached Nirmal Niwas, the previous residence of the new King, in a hurry. We reviewed the security details of the place. The Secretaries of the Royal Palace were concerned about it. As further precautions, we recommended that the King have his return trip to the Royal Palace in a carcade, and he agreed to it. But later, in deference to Queen Mother Ratna's wish, the newly crowned monarch returned in a horsedrawn carriage, according to an old royal tradition, to the Royal Palace.

We arrived at Hanumandhoka, the venue of the coronation, in time. The Prime Minister, followed by other Cabinet Ministers, the Leader of the Opposition Party, Madhav Kumar Nepal, and former Prime Ministers and other VIPs, had already reached the venue. The coronation commenced and concluded at the appointed time, observing the rituals prescribed in Hinduism and Vedic ceremonies. The VIPs offered their loyal greetings to the King with an offer of coins at his feet in pursuance of the long-observed custom and tradition of the monarchy. This act of loyal greetings to the new King offered by the CPN (UML) leader, Madhav Kumar Nepal, generated much hue and cry among his party members and supporters. Of course, it was just a ceremonial courtesy and so the loud hullabaloo was uncalled for.

Actually, I was ritually mourning the passing away of my mother at the time, and that would ordinarily require me to avoid such functions. But owing to my duty and considering the need of the hour, I had to be on duty, most especially due to the rumor of likely disruptions of the ceremony. There was no room for me to take a leave and be away from my post and responsibilities. As a soldier of the RNA and Military Secretary, I had to be at the forefront of unprecedentedly trying times. Responsibility is not just fulfilling a quota of one's job. In fact, I was proving my worth as Military Secretary to the King. I took little notice if it was recorded somewhere or by somebody in a form of evaluation. As such, with the ceremonies over, and when the time came for the Royal Procession to return to the Royal Palace, the King would be at the forefront of the royal carriage, immediately followed by the Prime Minister and the Field Marshal. That could not happen because firstly, public anger was high against the Prime Minister, and secondly, the Field Marshal was old and feeble. So I was the only person to stand guard of the King in the carriage. The CoAS also sat next to me. Nothing untoward happened as was feared. The rumor, that the Maoists would infiltrate the crowd and foil the crowning ceremony did not take place. We had, on our part, adopted preemptive measures to avert their attempt. The Director General Military Operations (DGMO), Maj. Gen Pradip Pratap Bom Malla and the Director of Military Operations (DMO) had, between the two, strengthened the security. All the security agencies worked unitedly.

Once back inside the Royal Palace, the King entered the Kaski Sadan. The site of the throne was the Gorkha Sadan. But no such program was on the card for the King to ceremonially ascend the throne. The officials of the Royal Palace paid their loyal greetings by offering coins to the King and subsequently they were granted collective audience by him. I briefed the King on the funeral ceremony of the late King Dipendra. He did not stay in the Royal Palace for the night; he went back to his residence, Nirmal Niwas.

That same evening, the King made his royal proclamation, declaring Her Royal Highness Princess Komal as Her Majesty the Queen. He also discussed similar announcement for His Royal Highness Prince Paras and his consort, Her Royal Highness Himani, as Crown Prince and Crown

Princess of the Kingdom of Nepal. The meeting advised that it would be too early to do so, taking into consideration the mood of the populace. It would be solemnized once situations turned to normal.

That same day, on June 4, the King, in accordance with the decision of the earlier meeting, announced the formation of the Panel of Investigations, to be headed by the sitting Chief Justice, Keshav Prasad Upadhyaya, to probe into the shooting in the Royal Palace, and bring the facts found thereof to the light without delay. Accordingly, the new King also addressed the nation for the first time in his new capacity. The broadcast address was reported to have calmed the seething anger of the people to a certain degree.

The Funeral Procession of the Late King Dipendra Amidst a Curfew

With the crowning ceremony over, I visited the Military Hospital where preparations for the late King Dipendra's funerary arrangements were being made, starting from the hospital via the Ring Road to the Aryaghat. His mortal remains were placed the whole day under a tent for mourners and visitors to pay their last respects to him. A special notice was issued to VIPs for this purpose. Given the tense situation, that the general populace would lose their sense of control or take out rallies and unpleasant incidents could take place, we decided on an informal funeral procession and ceremony – unlike past royal practices. The general sentiment of the populace was that the shooting was a sinister conspiracy, and the whole blame was heaped on Dipendra who, now dead, would not otherwise speak up in his defense.

Just when the preparations for the funeral were going on, the King came to the hospital. He expressed his wish to organize a grand funeral ceremony for Dipendra as befitting a departed monarch. But, in view of the briefing by the security officials, he agreed upon the informal one. Later, we agreed that the mortal remains be placed on a flower-bedecked motorcade, with security vehicles in front and behind and on both sides, and driven to the Aryaghat by detouring on the Ring Road. An Army band had already reached the Pashupati to accord the late King the last guard of royal honor and firing of the canon at the time of setting the flame to the funeral pyre. When the funeral procession commenced at around 5 o'clock in the evening, people thronged the route, made noise and threw stones. One stray stone landed on the mortal remains. At Gongabu, the security personnel had to resort to firing in the air to keep the crowd at bay. The CoAS and I were in the car just behind the autocade. It seemed as though the last rites would be impossible to solemnize. The CoAS was giving orders to the Control Room. Despite obstructions, we did reach the Aryaghat. The last rites were performed in accordance with Hindu Vedic traditions due to a monarch.

The last rites for Dhirendra Shah were also performed the same night. No guard of honor was accorded to him as he had long ago renounced his royal title and privileges to live as a commoner.

A funeral procession normally entails complicated formalities in the Hindu religion, even more so if the deceased is the King. We had to keep many religious, traditional and cultural aspects in mind for the funeral procession. And the last rites for the mortal remains, after the legal formalities, were all the more pressing. It involved the Home Ministry, the Army, the Police, and other security agencies for religious, administrative and security aspects. The Adjutant General's Branch of the RNA was on its toe as the entire affair was to be choreographed by them. The Adjutant General, Sadip Bahadur Shah, did a marvelous job. The cooperation of the Secretary of the Ministry of Home Affairs, Shri Kanta Regmi, was no less praiseworthy.

The Hospital as (the King's) Office

King Gyanendra granted me his audience on the 5th day of his assumption of the title of King. On my part, I was prepared with adequate homework for briefing him as Military Secretary. As Queen Komal was undergoing treatment in a VIP cabin at the Military Hospital, the King called me right there for briefing. It was 6 o'clock in the evening when I entered the room.

'Is everything going fine, Vivek?' he asked.

'Yes, Sarkar,' I said.

I added, 'Never before in its history did the monarchy in Nepal pass through a critical period as that of now. The shooting in the Royal Palace has put the monarchy through trying times. We, at such a critical hour, have twin challenges – to bring the situation back to normal, and to strengthen the monarchy. I'm sure it has come to your knowledge that there have been allegations labeled against you. Taking all this into consideration, and for the future of the monarchy, two issues call for immediate action: One, Your Majesty's moves in the interest of the people and the country in a transparent manner; and two, timely changes in the conduct of Prince Paras. The latter is crucial, more so because the future and the perpetuity of the monarchy depend to a great extent on his behavioral patterns.'

True to my nature, I stated everything candidly, without giving a second thought as to whether it would please or displease him. One must be objective when it comes to a briefing on or describing something. There was no fabrication in what I said. It was up to the King to interpret or misinterpret my statements. It was in the interest of and service to the monarchy. I also did not miss to brief the King on the current Maoist statements and their allegations against him. The Maoist leader, Dr. Baburam Bhattarai's piece in *Kantipur* made a frontal attack on him.

Thoughts and counterthoughts as to what must be going on in the King's mind swirled in my mind.

The King observed, 'There should be no need to be frantic about the Maoists' statements. After all, they have made no secret of their ultimate goal, that is, of eliminating the monarchy.'

On the charge of publishing and circulating the previously mentioned article by Baburam Bhattarai in the *Kantipur Daily*, the publishing house's directors, Kailash Shirohiya and Binod Raj Gyawali, and Editor Yubaraj Ghimire were arrested by the police. I sensed a sinister attempt to drag the Royal Palace officials as involved in and behind the arrest of the media. There was a suggestion that, perhaps, the Prime Minister, along with the Deputy Prime Minister (DPM) and the Minister for Home Affairs, Ram Chandra Poudel, had something to do with it. The Prime Minister's move of slamming the media behind bars would amount to inciting the populace against the King. When I briefed the King with everything on the issue, he sounded ignorant of it all. Instead, he asked me to find clues to the elements, if any, behind such a sinister plan, and also hinting that I convey to the media house the assurance that the Royal Palace had had no part in such a lowly act.

To buttress the message of the non-involvement of the Royal Palace in the arrest of the said media executives, the CoAS, in the same newspaper, published his loyal felicitations, congratulating His Majesty the King on his ascending the throne of the Kingdom of Nepal.

Because Queen Komal was still undergoing treatment at the Military Hospital and the King also remained there till late in the evening until her discharge, I had to be there most of the time. Most of the Royal Palace functions and official works, including Royal audiences to visitors, and issuance of instructions to us were all being done from there. The Prime Minister, the leader of the opposition party and other VIPs also visited Their Majesties inside the hospital. In many ways, thus, the hospital had turned into a kind of an office where the King showed his keen concerns and interests in the fresh developments.

The Secretary of the Ministry of Home Affairs assured me of all possible help and support for the smooth running of the nation's affairs.

In the observance of the passing away of Kings Birendra and Dipendra, the government declared a national mourning period of 11 days.

Dipendra and Royal Palace Massacre: The Undercurrents and the Enigmas

The fateful night of June 1, 2001 put an end to the dynastic line of King Birendra, and ultimately that of the monarchy itself, too, in Nepal.

There have been many speculations on the Royal Palace Massacre. Questions, such as who did it, how it happened, and in whose interest was it caused to happen still remain unanswered. However hard to believe and digest it may be for the common populace, the blemish of the carnage, based on received facts and figures and the information obtained thereof, all squarely rests on the then Crown Prince. I believe it is hard to accept easily that such a popular young man with such a good public image could have ever done such harrowing misdeeds.

But those who knew him to be emotional, moody and bad-tempered didn't find it hard accepting the seemingly unacceptable.

However, having been in the service of the Royal Palace for well over 30 years, I also contemplate another hand or a sinister conspiracy behind the massacre. That the Nepali monarchy was a hindrance to certain power centers outside Nepal in securing their interests here, and also that attempts to eliminate the institution were going on from some quarters were apparent.

To make matters even worse, and along the way, the contentions brewing up in the First Family over the question of the Crown Prince's marriage was vitiating the atmosphere to the extremes in the Royal Palace.

It is quite natural that, like any other person, Dipendra, too, had some human flaws. It was those flaws that the outside elements played upon and caused the carnage to happen.

I think so.

I knew Crown Prince Dipendra from April 1974 when he was a child. He was born on June 27, 1971. Naturally, he was destined to be King. But sadly, as it looked to me, he grew up without parental care and mentoring to assume the position of King in the future. I had heard how he passed through his infant years without proper upbringing and training which are accorded to young royals where there is a monarchy. I was amazed to see a sheer lack of these essentials in the formative stages of his life.

His childhood coincided with the apex of the Panchayat, the partyless political system in Nepal, introduced by King Mahendra, the present monarch's father. This position called for active role and involvement of King Birendra in the political, economic and administrative affairs of his kingdom. In fact, both King Birendra and Queen Aishwarya remained extremely busy in the political firmament of the Kingdom. Therefore, Crown Prince Dipendra grew up without direct parental care and stewardship. He was actually attended to by the Palace's orderlies or runners and housemaids, leaving the heir untutored and uncultured from his very infancy. Most obviously, therefore, he was largely denied affection and tutelage in royal ways. Breaking himself free of his minders, he would bounce and leap out of the house and dash for the ADC Office.

I wondered at the difficult and stubborn nature of the Crown Prince. He grew fond of sadistic pleasures at the pains of his attendants. He would land a powerful punch on a manservant and fling him down to the floor and laugh, while the man cried. He would also similarly slap or knock a classmate down. I had noticed difficult and sulky temperaments develop in him. As Crown Prince, and in order that he be groomed for the future role of King, he was to be taught and trained in the basics by a host of tutors and scholars.

Sadly, it did not happen. He did not even receive the sensitive training to be a sensible commoner to be able to live a normal life, let alone assume the guiding role of a future King.

The temperament, understanding, fortitude and the capacity for affection in a child depend to a great extent on his receipt of the same attributes from his parents and immediate caregivers. His character accordingly takes shape in him.

To us, the ADCs at the Royal Palace, however, it really seemed strange the way he was deprived of parental care and concerns. We knew from his private staff that he hardly enjoyed the tender loving care of his mother beyond the first two years of his life. Having spent much of his early years with the Palace caretakers, he was used to them. Thus he grew up in the care and casual attendance of the Royal Palace orderlies and attendants.

One day, years later, when he was old enough, he shared with me his hurt feelings and expressed his grievances at being deprived of love and affection of his parents. That lack of affection drove him into a sadist. I had taken care to convey my observances to the King during my regular audiences with him. The Crown Prince had complaints against the father, too. I noticed that he was estranged from and aloof of his parents, particularly his mother, Queen Aishwarya.

Both Queen Aishwarya and Crown Prince Dipendra were hotheaded, and so would not budge an inch from their desire and wish. Thus, although a royal child growing up in a royal household, the parent-child relations remained purely formal and mechanical.

Being Military Secretary with a long tenure at the Royal Palace, I saw up close and from the very beginning of his life the character and temperamental buildup of the Crown Prince. His ego, whim, emotion, and mischiefs grew worse as he came of age. While a kindergarten pupil at the Kanti Ishwori Vidhyalaya, he was an adamant child with too much mischiefmaking. He would create an excuse not to go to school. Nothing we said or would calm and convince him. Queen Aishwarya, who was an adamant person herself, would often slap him hard. We ADCs had a tough time taking him to and from school. One day, he would wish to get out of the car transporting him to school, at other times, he would grow terribly restless. Some ADCs would rather avoid him. But he would sneak into the ADC Office and do one mischief after another. He would ask the ADCs to do the impossible. Those who would not comply would receive a slap, or a punch would land on his back or sides. I have had such experiences myself. His mischievousness included breaking glasses, pens or whatever came to hand.

From the fourth grade, Crown Prince Dipendra was moved to the Budhanilkantha School as a boarder. It was a school for the elites. It was hoped that the change of school and environment would bring positive growth and development in his behavior. But his unruly habits did not diminish one bit. As he was a blue blood, he grew up muscular, healthy and handsome. Later it came to our knowledge that he once incited gang fights between the students of the Budhanilkantha and the Siddhartha Banasthali School. He was reported to have run wild all the way from Budhanilkantha through the rice fields to fight the students of the latter school. He would flex his muscles and prove his strength to others, including us at the ADC Office.

His sadistic shenanigans would never diminish; rather his rascalities increased all the more with the passing years. He would find delight in shooting down flying pigeons or crows in the royal complex, bringing them down to the ground. He would catch a live mouse, set its tail on fire and let it go.

But he was also grave and serene at times. He was fond of travel, too, and was known to write poems with clear marks of mature sensitivities. His poems in Nepali, written at Budhanilkantha, are excellent, and tutors were hired to further improve his excellence in the language.

Once, he openly told me that his mother controlled him to the extremes, and even asked me to appeal to her not to do so. Sadly, there was nothing I could say or do in this matter.

It so transpired that he had developed a fondness for Supriya Shah, his relative cousin –both being grandchildren of two biological sisters. I found out that Queen Aishwarya, however, got wind of his having written love letters to her since he was a seventh-grade student. But it was obvious that the Queen would have known of it and would not like to see the romance turn into matrimonial relations between the two families in any respects. Her stand was, 'Supriya [isn't] competent enough to be queen.' But the son would not listen to her. He would rather retort, 'Who does she think she is?' or 'Where and what was she before her marriage?'

On the other hand, Queen Mother Ratna was in a strong favor of the romantic relations, as Supriya was the granddaughter of her own older sister. It was also heard that the Queen Mother was promoting her choice through King Birendra himself. The King, thus, was under twin pressures from Queen Aishwarya and Queen Mother Ratna. Well aware of Queen Aishwarya's obstinate nature, no one in the family would dare oppose her thoughts and actions; it would be best to toe her line. On the other hand, due to Supriya, the relations between the Crown Prince and the Queen Mother were excellent.

Meanwhile, the photographs of other suitable girls, for marriage, were sent to the Royal Palace from different quarters and by different individuals, particularly from the residence of Gen Kendra SJB Rana, the parental house of Queen Aishwarya.

The Queen actually wanted Dipendra to marry a fellow Rana maiden, particularly Garima Rana, the daughter of Lokbhakta SJB Rana.

In the same connection of marriage, the Maharaja of Jaipur [in India] had invited Their Majesties to consider his only daughter for Crown Prince Dipendra. In those days, Crown Princess Shruti was in the Mayo College of Ajmer, in Rajasthan, India, for her studies, and Their Majesties had been there to see her.

While talks of marriage were in progress at home, the expected event took place. After the Crown Prince's SLC (School Leaving Certificate) Examinations and graduation, he joined the Eton College of Britain for his study at the A Level.

Coincidently, Devyani, daughter of Pashupati SJB Rana, was also at Eton for her studies. They met and befriended each other though they were not classmates. Not surprisingly, he developed a crush on her. The public also came to know later about the romance.

Lord Camoys, an acquaintance of Pashupati SJB Rana, was the local guardian to both Dipendra and Devyani, as he was to King Birendra, too, when he also went to Eton when he was Crown Prince. Lord Camoys himself had three daughters, and Dipendra had most cordial relationships with them. Because of the peer of the realm's stewardship, no ADC was appointed for Dipendra at Eton.

Back in Nepal, the Crown Prince maintained his romance with both Devyani and Supriya. It was alleged, however, that once Supriya came to know of his affair with Devyani, she had a serious row with him and severed the relationship. From Devyani's side, nothing untoward was heard. She was reputedly of higher learning and an upper social standing.

Queen Aishwarya, however, would have nothing to do with Devyani. Queen Mother Ratna also opposed any matrimonial relations with the family for the reason that her maternal home being in India. The Royal Palace, since the time of the late King Mahendra, adopted a policy not to have any matrimonial relations with India.He himself had arranged all his children's marital ties with Nepalis at home. But the Crown Prince's dogged wish to marry Devyani and the rejection of his design by many other family members was the real seed of discord in the royal family.

It also came to my knowledge that while at Eton, Crown Prince Dipendra loved a Japanese girl whom he had allegedly promised to marry one day. That love affair must have been special because the girl actually came to Nepal a couple of times to meet him. Perhaps she asked him to honor his promise, and when the pledge was broken, they had a row. I only heard about it from other ADCs and did not know much of it myself because I was the ADC to King Birendra then. Two ADCs were accredited to the Crown Prince. He once instructed his ADC, Major Gajendra Bohara, to escort the girl to the airport. Furious, she allegedly said, 'Your Crown Prince is mad. He'll die a dog's death.'

It turned out to be quite true for the Tragic Prince !

The Crown Prince and I were very close in later years. We were so open to each other that he would candidly speak his mind and share his romantic experiences with me. We would at times discuss politics and the current problems of the country. He would also ask me to join him in drinking. He displayed an adamant, insensitive, mischievous and a sadistic nature within the Royal Palace complex whereas he was a placid, easygoing and considerate young man in public.

I take readers to 1989 when King Birendra was scheduled to visit Belgrade of the then Yugoslavia to participate in the Non-Aligned Nations' Conference. Crown Prince Dipendra was back in Nepal on a vacation from Eton. He was to accompany King Birendra on his return to London. Being ADC to King Birendra, I accompanied him on the trip.

While we were on board the plane, we discussed the current politics, and the problems thereof, particularly that of the ways the Panchayat Policy and Evaluation Committee (PPEC, Panchayat Niti tatha Janchbujh Samiti) was overstepping its limits, causing, in the long run, the end of the Panchayat System itself, particularly when it came to the selection of candidates for the general elections.

We talked about non-political matters, too.

In a drunken state, all of a sudden, he said, 'Vivek, I don't know what love is. I've never received it.' I presumed that he was referring to his being deprived of parental affections.

The PPEC was formed by King Birendra in 1975. Given the nature and manner of its functions from the beginning, I was convinced that it would eventually paralyze and cause the demise of the Panchayat System itself.

Due to the prolonged and violent agitations for democracy in Nepal in the mid-'70s, the King called for a historic Referendum on the system of government in 1979. The Panchayat won, with 55% in its favor, as against 45% who favored multiparty democracy.

I had recommended timely reforms in Nepali politics. It fell on deaf ears. I had shared my views with my superiors at the Palace, Generals Shanta Kumar Malla and Tara Bahadur Thapa. I was not in a position to express my views to the King on this as I was just a Captain then. True enough, another mass political movement gathered momentum, and the Panchayat System became history in 1990. In greater truth, the very downfall of the system can actually be attributed more to its own venal leaders rather than the force of the mass movements.

On our trip to Belgrade, we spent three days in Bonn where Singha Pratap Shah, my father-in-law's younger brother, was the Royal Nepalese Ambassador to West Germany. He and Crown Prince Dipendra adored each other. The Crown Prince instructed me to call DSP (Deputy Superintendent of Police) Rabi Raj Thapa forthwith, who was currently a student in Frankfurt. He was a confidant of the Crown Prince, as Thapa had served as ADC at the Royal Palace and also taught at the Budhanilkantha School when the Crown Prince was a student. Thapa was equally close to me, too. We got together, drank, sang and shared our experiences. We did a little jousting, too, for the Crown Prince was also a boxer. We enjoyed our jaunts to our hearts' content before Their Majesties would arrive on the third day. The royal visitors had been provided with a villa, but the Crown Prince preferred to stay in the hotel with us. I made every arrangement for King Birendra's preparations for his participation at the conference.

From the trip to Belgrade, I learned that Crown Prince Dipendra could go to any extreme when it came to girls. One night, to our fright, he drank excessively with a girl at his side and spent a long time with her in a lift of the hotel. We were scared and dreaded for what hell would break loose on us in that foreign land.

The Conference was over. Crown Prince Dipendra was to return to Eton the next morning. That evening, he forced me into drinking. Excessively drunk, we overslept and woke up late the next morning. Hurriedly, we reached the Villa to see Their Majesties. The Crown Prince had a hangover. Queen Aishwarya was quite displeased when she saw it. She reproached me for our recklessness. We managed to reach the airport, and luckily, the flight was delayed for an hour. We reached Eton in the car sent by the Royal Nepalese Embassy. He wished me to be with him at Eton for the entire term of his study there. But I had to leave him under the care of Badri Prasad Khanal, my previous college classmate and the acting Ambassador of Nepal in Britain. Coincidently, ADC Gen. Tika Dhamala was also in the UK for his Staff College course. He hosted a dinner for us at the Nepali restaurant owned by one Hari Bibhor Karki at Aldershot. At 5 o'clock the next morning, I hurriedly collected my bags from the Royal Nepalese Embassy and flew to join Their Majesties in Prague on the eve of their official visits to Finland and other European countries. Only then I had some respite from the previous hectic week.

Let me relate here one particular incident on Crown Prince Dipendra's love affairs which caused such dissensions in the family. When his insistence to marry Supriya grew more and more, Their Majesties discovered a good excuse to send him to Britain for studies and to distance him from the girl. They wanted to prolong his stay and studies in the UK, but after his completion of the A-Level, he returned to Nepal, started his Bachelor's Studies at the Tri-Chandra Campus, took the Board exams as a regular student and secured the position of first class first. He also secured first class first in Masters' Degree, majoring in Geography, from the University Campus of Kritipur. His academic achievements created a good image for him in public.

To walk in his shoes, one would find it really unfortunate and frustrating because more than half of his adult life passed in marriage wrangles. The rest of the family would not let him marry the girl of his choice and he would not compromise his stance with their preferences. Princess Shruti was also interested in Garima Rana because she was her sister-in-law by marriage

It was June 27, 2000. The entire family was at a Royal Banquet to celebrate his birthday. Recently appointed as Military Secretary to the King, I also attended the party. As usual, the topic of marriage emerged there, too. The Crown Prince sounded angry with me and asked what role I could play to create an environment to convince Their Majesties to accept his choice. I said that it being a purely family matter, there was very little I could do. But I added that if he so wished, I would brief the King on this issue the next day. He said he had already asked the likes of 'Bharat, Tara Bahadur, and Dharmapal' (Honorary ADC Bharat Keshar

Simha, former ADC Tara Bahadur Thapa, and former CoAS Dharmapal Bar Singh Thapa) to sway Their Majesties' hearts to his choice. He wanted me to also join the team. I heard that the celebrations ended at midnight. The Crown Prince, with the closest relatives, entered the Tribhuvan Sadan. He acted in shocking manners, even slapping Princess Shruti, his own younger sister when the latter suggested that he would do well to marry Garima Rana. I learned of it only the following day.

He summoned me to his office, the Tribhuvan Sadan, a couple of days later, to say I was dropped from the team, taking into consideration my post and duty. I thought it was wise of him to understand my inconvenience. He disclosed his plan to marry Devyani on the quiet, with only the closest friends invited to the Nagarjun Bungalow. He also added he had already sent someone to the Civil Marriage Registry Office to enquire about the process of legal marriage.

What I could not understand was why Queen Aishwarya objected to her son's wish to marry Devyani. Born to a noble family, she was beautiful, educated, and intelligent. She was, therefore, ideal to be queen in every respect. It was, however, beyond us, the ADCs, to make peace in the family. Differences developed sharply and the chasms widened with every passing day. Solutions seemed nowhere in sight.

Queen Aishwarya, Queen Mother Ratna, Crown Prince Dipendra, Prince Nirajan, and Princess Shruti all held their own respective positions on Devyani with other eligible prospects, thus turning in their diametrically opposite directions. King Birendra would just bypass it as 'the ladies' department. They are working on it.' Over the overall question of marriage, the relations between Queen Mother Ratna and Queen Aishwarya had turned so sour that it was reduced to a mere formality, each trying to avoid meeting the other.

The prolonged debate on marriage seemed to precipitate and lead to very portentous omens. An unpleasantness on a certain Friday in 1997 suggested as much when the King and Queen attended the dinner at Princess Shruti's residence at Keshar Mahal. Their Majesties did not speak a word during their drive to the Nagarjun Bungalow. According to the ADC on duty, Brig. Gen Tika Dhamala, the reference to marriage surfaced at the dinner, too, and spoilt the entire atmosphere of joy and fun. At the unusual hour of 3 o'clock in the morning, the King summoned Tika to come with a car to go on a drive to Trishuli in Nuwakot. In our office, Maj. Gen Tara and I discussed the return trip of the King to Nagarjun.

Not long after that incident, the King had a heart attack, requiring medical treatment in London.

King Birendra's most adorable child was the daughter, Princess Shruti, while Queen Aishwarya's most favorite son was Prince Nirajan, the youngest of three children. Reading between the lines, one can easily figure out the reasons for Crown Prince Dipendra's displeasure. He was also opposed to Shruti's marriage taking place before his own, and so had remained disconsolate through the years.

Over lunch one day, a month prior to the Friday, June 1 Royal Palace Massacre, there had been a very unpleasant and hurtful exchange between Queen Aishwarya and Crown Prince Dipendra over the same chronic question of marriage. The Queen was reported to have expressed her wish to have Prince Nirajan married and declared Crown Prince if Dipendra did not marry the girl of her choice. And Crown Prince Dipendra obviously felt hurt at the nonintervention by his father, King Birendra, over such an intolerable remarks made by the Queen.

I had heard about a certain reference to Crown Prince Dipendra's birth: It had taken place under inauspicious stars. This meant a clear indication of bad omens to beset the monarchy. The astrologist, Dibya Mangal Joshi, who had prepared the horoscope, was asked to make the necessary and convenient changes in the birth chart. When he reportedly refused to oblige, another astrologer, Dr Mangal Raj Joshi, was commissioned to make the suitable amendments. On his part, Dibya Mangal Joshi would never visit the Royal Palace again.

~~~~~

There took place drastic political changes in Nepal in 1990. The umpteenth 'democratic revolution' caused Nepal's absolute monarchy to give way to a constitutional monarchy. The old partyless Panchayat political system was replaced by multiparty democracy.

However, despite all the vaunted transformations, there appeared to be no substantial changes in the lives of the people. So, pressures to take bold steps to reassert the traditional powers of the crown were mounting on the King from, among other quarters, the members of the Royal family itself. Such views were attributed particularly to Queen Aishwarya, Crown Prince Dipendra, His Royal Highness Gyanendra, and Queen Mother
~~~~~

Ratna, all of whom considered the late King Mahendra their role model.

However, while diehard royal members resumed their pressures on the King to take stringent measures, the King demonstrated his determination to remain committed to constitutional monarchy, with his firm belief in the future of democracy, peace, and prosperity in Nepal through its multiparty polity. He would not flinch from his ideals, regardless of the mounting influences.

Thus, Crown Prince Dipendra would contemplate on becoming a 'dynamic' leader. Those elements at home and from abroad propelling him to do so would obviously go to any extreme, including causing to unleash what would turn out to be the Royal Palace Massacre on the night of Friday, June 1, 2001.

Some Unanswered Questions

World history is replete with political conspiracies and strategic game plans played against a country's power centers by such elements contending for takeover and domination. Such provocations emerge organically at home or are sourced from abroad. Such agencies wait for chinks to appear in the targeted power bases to mount their coups and missions.

It was obvious that Nepal's monarchy had been an eyesore to and the envy of certain elements and power centers at home and abroad. It occurs to me that these very bureaux played upon the grave weaknesses of the Crown Prince – a particular shortcoming being his licentious liaisons with women – and caused the Royal Palace Massacre to happen, which was the result of a political conspiracy.

As already referred to in this book, the high-level Commission to look into the massacre merely studied what had happened at the site of the mass shooting that night. Frankly, as it was, the Commission was provided with a limited scope: merely to look into what had taken place. Further, the investigation team did not include experts and specialists in crime. The Commission, therefore, did not bother to ascertain the background/s, reason/s and the accomplices involved in the massacre. The Commission's report stated that Crown Prince Dipendra had consumed a narcotic substance but it did not point out as to who entrapped him into taking it.

Therefore, the Royal Palace Massacre calls for another high-level and scientific inquiry into the shooting, such an investigation commission comprising experts and specialists from home and abroad.

As Military Secretary, I had taken the initiative to form a committee of investigation, led by Lt. Gen Shanta Kumar Malla, in accordance with the King's consent. The committee looked into the role of the Royal Palace ADCs and the security situation of that night.

Given the fact that the Royal Palace Massacre took place on my watch and in my tenure as Military Secretary, I forthwith tendered my resignation from the post, on moral and ethical grounds. But King Gyanendra would not accept it.

Following the Royal Palace Massacre, speculations of various shades and types surfaced, with accusing fingers pointing at many individuals. People pointed at the laxity and incompetence of the security system, which fell within the scope of my duties and responsibilities. There is no flinching from this fact.

What is also equally true is that the Crown Prince lived in the Tribhuvan Sadan where foolproof protection was in place round the clock, with security details mounted in and around the perimeters. The type of security arrangement in place at the Tribhuvan Sadan, therefore, rules out any possibility of anyone's entry, even one wearing a Dipendra-like mask, as imaginatively purported by some, to open fire there.

The high-level panel of investigation has already made public the actual ground realities of the area in question. As far as Dipendra's consumption of narcotic substances is concerned, some young members and relatives of the royal family had that addiction. We ADCs had seen the Crown Prince taking drugs on the pretext of going to the restroom. And we, as ADCs, were not in any position to prevent it or intervene. The high-level panel of investigation should have looked into the sources of drugs and its entry into the Royal Palace.

Prayers and religious rituals, including some Tantric rites, were routinely performed and held at the Royal Palace for the desired results. For one to obtain Tantric knowledge and skills – loosely, to have the power of casting spells – the initiate must strictly observe certain ways and methods. For instance, it is said that it requires thorough performances of puja/worship and rituals from the novitiate. Tantric masters can allegedly summon dead spirits to reappear and speak like normal people and have them work in one's interest. Incomplete or imperfect learning and practice of Tantrism harm the practitioner himself/herself.

King Birendra had begun learning Tantrism from 1968 under Dhana SJB Rana. But as the royal tutor grew too old to teach it, King Birendra stopped practicing it in 1978. I know about it, but I still don't know to what extent the King had progressed.

I also remember Queen Aishwarya performing a ritual worship to fend off the spells of a girl on Dipendra in 1988.

There was no secret in the Royal Palace and its circles that various occult rituals were held and sacrifices made with the active presence of Their Majesties and other Royal Family members.

King Birendra's Plans for the Immediate Future

During the preceding year – the last year of his reign, as it would turn out to be – the modernization plan for the Royal Nepalese Army (RNA) was progressing. It included the replacement of obsolete, worn and inferior weapons with modern equipment. King Birendra was keenly supervising it. Given the Maoist problems increasingly getting from bad to worse, and the likely mobilization of the army any time, such a timely priority was gaining its urgency. Nepal had forwarded the proposal of the purchase and assemblage in Nepal and the sale of the German-made HK G36 ultramodern rifles in the South Asian countries. Agreement to that effect had been reached with the manufacturers, Heckler & Koch GmbH. It was not difficult to perceive, however, that the immediate neighbor, India, was not happy about this deal, as it was seeking to sell its own but relatively inferior Insas Rifles to Nepal.

It is apparent from the proven series of instances that India wants to impose its exclusive influence over Nepal, and is displeased as and when things go against its hegemonic interests.

Hot on the heels on this very issue, the Citizenship Bill was passed by Nepal's Parliament and forwarded to the King for the Royal Seal of Approval. Because it was deemed controversial and incompatible with Nepal's own national interests, the King forwarded the Bill to the Supreme Court to seek its judgment on it in case it was deemed to be against the spirit of the Constitution. The Supreme Court declared the provision unconstitutional.

Unsurprisingly, India, with its opinions broadcast through its officials, was openly concerned about the future of the Citizenship Bill.

India had continued criticizing and opposing Nepal's moves and actions all through the years, since 1950 when Nepal entered the democratic age.

When Nepal purchased antiaircraft guns from China in 1986/87 to address the Nepal Army's defiency in air defense capabality, India countered the decision by a long economic blockade on Nepal all along the Nepal-India borderline, causing extreme sufferings to Nepalis.

Nepal's proposal to have the country declared a 'Zone of Peace' by King Birendra on the occasion of his coronation in 1975 was endorsed by as many as 116 countries, but the very national sentiment was vetoed by India with one excuse or another argument.

In the beginning of 2001, the King went on an official visit to China with a view to resolving the Maoist problem and putting an end to the ongoing political deadlock in Nepal. This particular situation, as it was developing in certain manners, called for a timely step to be taken by the King. He was consulting with the nationalist leaders of the political parties. He had given instructions to form a task force comprising some senior retired administrators and officers of the security agencies to develop a strategy in order to initiate dialogue with the Maoists to find a mutual way out of the political standoff.

As part of the strategy, he had deputed Dhirendra Shah, his youngest brother, to quietly initiate informal talks with the Maoists. The RNA had also deputized Col. Dilip Rayamajhi to establish a channel of communication with them. Once an understanding would be reached, the King was perhaps contemplating the dissolution of the Parliament and forming a national government, thereupon taking the political parties and the Maoists on board via a roundtable conference, a similar demand put forward by the Maoists, that of holding of general elections after two years, and amending the Constitution while retaining in its entirety the Preamble to the Charter.

The King was even contemplating of having the Nepali Monarchy redefined in the image of, say, the royal and imperial institutions of Britain and Japan. Simultaneously, he was keen to have the Maoists enter the country's mainstream and competitive politics. For this purpose, and if needs arose, he was determined even to solicit the cooperation of friendly powers.

The leaders of the immediate neighboring nations, in their meetings with the King, would always note Nepal's poor governance, its extreme corruption, and the air of insecurity, and strongly suggest that he take decisive steps to cure the national ills.

The King always expected friendly support from China and India as Nepal's good neighbors. The King even read Nepal's Maoist problems as a lurking threat to the vast regions' own peace and security.

In this connection, in 2000, King Birendra handed me a confidential file to study. This was on the strategy to cope with the Maoist problems in any eventuality. The name of the strategy paper was 'Vaaj' (hawk).

He was also seriously considering raising the former CoAS, General Dharmapal Bar Singh Thapa to Field Marshal, thereby empowering him as the coordinator of all the security agencies.

Such a comprehensive national strategy would also monitor the movements and activities of the political parties and their associates.

Above all, if the envisaged plan went into effect too, among other aims and objectives, solve the Maoist problems, it would also doubtlessly create a good image for the monarchy itself.

The Royal Palace Massacre would demolish the wide-ranging concepts of King Birendra.

The Maoist Rebellion: Its Genesis and Growth

The Panchayat Political System prevailed in Nepal for thirty years – from 1960 to 1990 – during which everything was controlled at the center by the state. Political parties were banned, and the press censored.

Nevertheless, deep yearnings for freedom remained in the general populace. Consequently, all Communist forces inched closer for a common cause and joined hands with the Nepali Congress for a mass movement, most specifically in 1989, to liberate and emancipate the people from all the decades-long restrictive politics and controls imposed on them. The CPN Mashal group, led by Mohan Bikram Singh, and the CPN Maoist faction under 'Prachanda' (later to appear as Pushpa Kamal Dahal), Mohan Vaidhya and Nirmal Lama formed their joint mass movement committee to wage their political struggles against the Panchayat System under the King.

On April 8, 1990, as a result, an agreement was reached between the King and the joint forces led by the Nepali Congress and the Communist

Party of Nepal (Left Front). The accord put an end to the Panchayat System for once and for all.

People felt a measure of freedom, with soaring expectations for changes for the better in their lives. A consensually agreed-upon Interim Government, under Krishna Prasad (KP) Bhattarai as caretaker Prime Minister to administer the country, was formed in 1990, the penning of the subsequent Constitution was put on the agenda, and general elections were earmarked for holding in 1991. All of these unprecedented political developments created conducive environments for democracy in the country.

The Communist forces outside the Left Front, however, called the comprehensive agreement a betrayal, and wanted to perpetuate their avowed mass movement. But they had no success to steer it ahead. The CPN Mashal of the Mohan Bikram Singh faction boycotted the elections. The People's Front contested the elections, won nine seats, and embraced the tactic of influencing the Parliament to unmask the anti-poor policies of the multiparty system. The Nepali Congress won a simple majority of votes in the Parliament and formed the government led by Girija Prasad (GP) Koirala, defeating his fellow Nepali Congress contender, Krishna Prasad Bhattarai.

But mired in the infighting and internal feuds within the Nepali Congress Party itself, and compounded by the mounting opposition from the CPN (UML), democratic norms and exercises suffered in Nepal.

Against this backdrop, PM Girija Prasad Koirala played negative politics. He had every opportunity to have the common people on his side by giving them a feel of positive political change and a sense of democracy.

But the reverse happened. He displayed his narrowmindedness, and excessively politicized the institutions of the state structure, such as the Nepal Police, the bureaucracy of His Majesty's Government (HMG), the National Investigation Department (NID), but altogether ignored the Army. He propagated an unsavory form of democratic propaganda of [the Nepali] 'Congress is democracy and democracy is the Congress.'

As if that were not enough, it was during his premiership – to illustrate but one instance – that as many as 334 cases of public offences or criminal charges were slapped on the leaders of the other political parties in the Rukum District alone. This figure stands out in sharp contrast to just three such offences registered against the opposition in Rukum – which incidentally happens to be my home district – during the entire thirty years of the Panchayat Raj in Nepal.

Amidst, or due to, the growing infightings and acrimonies within his own party, PM Koirala opted for the dissolution of the Parliament for a fresh mandate in 1994. But neither the Congress nor the UML got an absolute majority in the elections.

The Jana Morcha (The People's Front), which had secured nine seats in the earlier elections, decided to wage their 'People's War.' The largest party in the Parliament, the CPN (UML), formed a minority government, led by Manmohan Adhikari. From the first day of the formation of the first Communist government, this too while the country had a monarchy, the Nepali Congress Party stepped up its noncooperation, forcing Prime Minister Adhikari to finally dissolve the Parliament and call for snap elections. The move was challenged in the Supreme Court which, by its verdict of August 28, 1995, reinstated the Parliament.

The judgment of the Supreme Court, I do think, gave rise to newer political instability and undermined parliamentary democracy. Thus, the Constitution of 1990 was rendered null and void, as no party or party leaders adhered to it in its letter and spirit. This situation caused a rise in the total disillusionment of the people with the entire democratic system. It was akin to lending credence to the Maoists to wage their so-called 'People's War' (Jana Yuddha) which officially began from February 13, 1996.

The Royal Palace's Perceptions of the Maoist Problem

The Narayanhiti Raj Durbar had its own reading/perception of the problem. It believed that the Maoist problem did not merely emerge out of the blue. It was viewed as a homegrown byproduct of poverty, lack of good governance, and unemployment. With the submission of the much publicized 40-Point Memorandum to the then Prime Minister, Sher Bahadur Deuba, and calling for its immediate implementation, the Maoists went underground even before the expiry of the ultimatum and its timeline given in the said Memo.

In its early stage, however, few political actors took any notice of what would eventually turn out to be a total civil war in Nepal.

Once, in 1996, in a talk with King Birendra, I asked, 'How would Your Majesty observe the Maoist Problem?'

'What do you think our observation of it should be like?' he put a counter question, in jest. 'Aren't they also Nepalis? They have explicit objectives of massive changes in the polity of the country. Yes, a few of their demands don't seem feasible for implementation. But the country can't afford to go into a conflict nor can it sustain it. The government would do well to address the problem through social development programs, rather than taking recourse to its suppression,' he elaborated. 'When an infighting [of this kind] takes place, whoever dies is, after all, a Nepali. So bloodshed must end without delay.'

I was pretty clear that he abhorred conflict which would only harm the country and the people.

After the unmaking of the Deuba Government, and with Lokendra Bahadur Chand as the next Prime Minister, whose candidacy was backed by the CPN UML [Communist Party of Nepal, United Marxist Leninist], the new Deputy Prime Minister (DPM), Bam Dev Gautam, took an initiative to address the problem with utmost official flexibility to bring the Maoists into mainstream politics. Should it fail, he was for resorting to the use of draconian laws to suppress their rebellion.

The situation, meanwhile, was rapidly deteriorating, even warranting the mobilization of the Army. The government formed a task force led by CPN UML member, Prem Singh Dhami. The team included another leader of the same party, Yogesh Bhattarai, and the NID Deputy Chief Devi Ram Sharma, AIG [Additional Inspector General of Nepal Police) Pradip SJB Rana, and the Joint Secretary at the Ministry of Home Affairs, Krishna Murari Sharma. The official group was formed to study the Maoist 'Problem.'

'The Police alone can't effectively address the Maoist problem' was the perception of the government while the situation was getting increasingly complex, and getting out of the government's grip. So it contemplated framing the 'Anti-Terrorism & Subversion Act'.

Meanwhile, the CPN UML, on board the government's bandwagon, and other political parties objected to the government's moot idea to promulgate such an act, citing the likely curtailment of civil rights and democracy.

At the same time, attempts to build good relations between the police and the army were also progressing. The Army was, in fact, training Nepal

Police personnel to fight the Maoists, and the two forces also had set up a joint communications center for the monitoring of the rebels' activities.

The government, however, took too much time in objectively fathoming the intensity of the Maoist penetration across the countrysides, and by the time it did, it was already too late. The Royal Palace was averse to army mobilization against the Maoists because that would be tantamount to setting one Nepali to kill another Nepali. In its reasoning, the wise course of action would rather be to take steps to alienate the Maoists from the people for whom they had drawn up their own social, political and economic goals at the grassroots level. For this countermeasure to be effective, development programs must be taken to the far-flung villages where the Maoists had established their solid bases.

But by 1997, the problems had further worsened, rendering the police actions fickle and ineffective. The common people faced untold sufferings in the civil war. The Nepal Police operations, codenamed 'Kilo Sierra-2' and 'Romeo' rather aggravated the situation instead of ameliorating the vicissitudes of the innocent and powerless people wedged between the state and the Maoists. Police atrocities actually tilted the people's sympathies to the Maoists, contributing to the growth of their guerilla cadres, rather than eliminating them.

The Royal Palace was naturally worried and was closely watching the devolving situations. Taking heed of the King's remarks made in 1997, the CoAS [Chief of Army Staff], General Dharmapal Bar Singh Thapa formed a team to carry out a detailed study of the Maoist problem. Both the Royal Palace and the Royal Nepalese Army (RNA) Headquarters reached an identical conclusion to create an effective mechanism to mobilize the Army as the last resort. But it would follow only after the implementation of social and economic development programs in the Maoist-affected areas.

Once prepared, the detailed plan was forwarded to the government. Since the beginning of the Maoist rebellion, the King had advised each successive government to activate the National Security Council (NSC) and set up and operate a Secretariat to analyze all security dynamics and likewise brief the Prime Minister on the developments. But smelling a conspiracy behind the concept, the Prime Ministers in the successive governments ignored the advice, pushing it under the carpet. The King was well aware of the immediate need for weapons, training, and logistics for the Nepal Army, as well as the occasional views made by the officers concerned. He was equally worried about the government's mistaken perception of the Army with its discouraging comments. He was also aware that the Army lacked arms and ammunition and logistics. Whenever the CoAS boasted about the strength of the Army, the King declared, 'The Army is only a paper tiger, and for it to learn its lessons, it needs to face a calamity.' He had also seen the need to strengthen the intelligence system, and special attention to be given to counterinsurgency training.

The King was convinced that the Maoist problem was solvable through dialogues. On the pragmatic front, development activities should be begun in the very areas controlled by the Maoists with local leaders at the helms.

The Maoists, all along and on their part, were fortifying themselves. It was seen in their attack on Dunai of Dolpa District in September 2000.

The assault effectively caused the immediate resignation of the then Home Minister, Govinda Raj Joshi, and the IGP, Achyut Krishna Kharel, and resulted in the transfer of the Home Ministry Secretary, Padam Prasad Pokharel.

However, the diagnosis of and the remedies prescribed for the Maoist problem widely differed between the King and the government; and these diametrically different opinions further stymied any possible peace process to begin.

I got my opportunities to learn about the Maoist problem from a close perspective from the time I joined the Military Secretariat of the Royal Palace. So far, not a single Secretariat had been considered for a look into the Maoist question. I think that the Principal Military Secretariat should have been assigned to study it, but it had poor coordination with other security agencies of the government.

As it was, and in any case, the then Principal Military Secretary, Lt. Gen Shanta Kumar Malla, would prepare his regular briefings to the King in an ad hoc manner. Until then, there was neither a system of regular briefings to the King by other security officers, except the CoAS, nor was there a proviso for the King to grant his audience to, or issuing instructions, or holding consultations with the other security officers in the Secretariat.

Therefore, with the realization of the need for the King to avail himself of direct consultations with the security officers of the country on a regular basis, I took the initiative to commence scheduled royal audiences and briefings. Thus, the process of briefing started on a regular basis, beginning with the fresh developments on the Maoist rebellion to be reported to the King by the IGPs of the Nepal Police and the Armed Police Force, and also by the Chief of the NID.

The unthinkable and unfortunate Royal Palace Massacre on Friday, June 1, 2001, and the Prime Minister caught up in party infighting and the resultant climate of noncooperation from the other political parties further complicated the national panorama, putting an end to the progress made on peace initiative till that point in time.

Subsequently, too, the Maoists launched a daring attack on the Rolpa Police Base Camp in Holeri. This confrontation proved to be a big blow to the peace initiative. Prime Minister Girija Prasad Koirala customarily accused the Army of noncooperation and noted the King's refusal to have the Army mobilized to isolate the Maoists at Holeri.

Koirala, and for that matter, all successive governments of the day, wanted to have the Army deployed against the Maoists, in the same manner in which the Police had previously been ordered. But the King cited the Constitution for the necessary ways and means, which would effectively be materialised by activating the NSC.

The Sher Bahadur Deuba Government took an initiative to start a dialogue with the Maoists. The next monarch, King Gyanendra, on his part, began consulting with several political leaders, intellectuals, and journalists, civil society activists, and diplomats, all in order to find ways and means to break the national stalemate and thereby resolve the Maoist problem and other serious issues besetting the country.

It seemed that the King was developing his strategic game plan to end the political deadlock first, and then to resolve the Maoist problem.

It occurred to me that King Gyanendra was deftly playing with the prevailing situations to get things done to suit his agenda, and he would make his moves through PM Deuba. After all, it so happened that the political parties were thoroughly weakening, and circumstances increasingly moved the Maoist problem over to the government's table. Therefore, it would be the government in place, and not the political parties, who would be obliged to resolve the stalemate by inviting the Maoists to the mainstream politics of the nation.

However, the first round of dialogue between the government and the Maoists failed.

But when the rebels attacked the Army's Dang Barracks and made off with a large cache of arms and ammunition, the politics of the country took a different turn and course. With the consequent declaration of the State of Emergency, the Army was finally mobilized against the Maoists. The operations were conducted through the NSC when the NSC Secretariat was at last established.

The country hoped that the political parties and the general public would support the Army. But in reality, the RNA sensed noncooperation from the populace, the media, intellectuals, and the government itself. It was, in fact, long taken as the King's Army. As if that were not enough, the international community raised the issue of human rights violations. They alleged that the Army, with admirable track records in the UN Peacekeeping Missions, could not display the same virtues at home.

No section of the populace, in addition, bothered to realize the extents to which the Army lacked adequate logistics to fight the Maoists. But, in retrospect, it was the Army which did prevent the Maoists from capturing state power.

Nepal's Political Parties and the Maoist Problem

On February 13, 1996, when the Maoists declared their People's War, or on November 23, 2001, when they attacked the Dang Army Barracks and made off with the company's considerable arms and ammunition, Sher Bahadur Deuba was coincidentally the Prime Minister on both occasions.

In these five years, the political instability reached its nadir: Nepal saw seven successive governments and five Prime Ministers during this particular period. Though each government that succeeded in ousting the previous one professed to resolve the very civil war situation, however, even in all of these political dire straits, invariably, the political parties displayed their typical lethargy and maintained their indifference to the Maoist problem. Even when the situation deteriorated to the worst imaginable, and the government felt fire set right on its tail, they still resorted to their textbook alibi of noncooperation from the Royal Palace and the Army. In fact, it was clear that no party in power ever took the problem seriously. Nor did they bother to accept the problem as the byproduct of poor socio-economic conditions as a result of chronic bad governance. It was rather transparent that the political parties actually schemed to cash in on the problem to use it as a ladder to ascend to power. History is a testimony to the fact that the Maoists caused the making and unmaking of the governments in the latter half of the 1990s. It was because not a single government down the line in the 1990s woke up to the need of passing an Act pertaining to the Army and its deployment in the changed context; instead, their politicians harbored apprehensions and doubts. The culture of preferring to remain incommunicado and lack of warmth between the successive governments and the Army rather added to the old doubts and fears. In short, the already bleak situation was further fueled by a mounting crisis of confidence, in which the political parties always beheld the Army with suspicion.

A formal attempt to address the Maoist Problem was somewhat made only in 2001. By that very year, however, the Maoists had become so strong that its cadres were able to demoralize the Nepal Police even to desertion because of their capacity to surprise them anywhere and at any time.

In the general elections of 2000, the Nepali Congress received the majority votes, and Krishna Prasad Bhattarai headed the government as Prime Minister. He duly appointed Sher Bahadur Deuba to head the task force to thoroughly study the Maoist issue and to negotiate with the Maoists for resolving the problem. Accordingly, it started laying the groundwork for its concept, concurrently trying to create a congenial environment to initiate talks with the Maoists. But the internal squabble and rift within the governing party caused Bhattarai to resign, and Girija Prasad Koirala reappeared as Prime Minister, to be in turn ousted by Deuba.

When Deuba became Prime Minister for the second time, in July 2001, he tried to mend the government's relations with the Army by asking the Cabinet to allocate enough budgets for the purchase of weapons and equipment for the Army. By the same year, 2001, the preponderance of the Maoists across Nepal had shrunk the influence as well as the presence of the political parties merely to district headquarters and the capital city of Kathmandu. Not only this, but each political party's ploy to attract leaders from rival political groups to their side reduced the parties to a defensive position. The Maoists termed the elected government the 'servant' and the King the 'master'. Having disarrayed the political parties, they wove a strategy to go into direct confrontation with the King. The Royal Palace, apprehensive of an unpleasant eventuality, wanted the government to start a dialogue with the Maoists.

Against this backdrop, and with the due realization of a propitious time to address the problem, King Gyanendra took a bold step on October 4, 2003. The political parties happened to misread the move as a ploy to sabotage democracy by teaming up with the Maoists. It was a misinterpretation because the Maoists in their 40-Point Memorandum had explicitly stated the elimination of the monarchy. Some political party leaders passed very lighthearted comments on the 'Maoists being operated by the Royal Palace itself.' Even while the leaders and cadres of the Nepali Congress and the CPN (UML) were being targeted for assassination by the Maoists, the UML leader, Madhav Kumar Nepal, and other high-level leaders were conducting several rounds of clandestine talks with the Maoist leadership in Siliguri, Gorakhpur and Lucknow of India. However, despite such endeavors, it was discernible that no political party bothered to discuss the political agenda of the Maoists which had become so pervasive in the country. They did not give any serious thought to solving the problem by negotiating; they instead turned their eyes to the King for deliverance. For themselves, they would rather be more concerned to grab and cling to power than resolving the problem.

The Maoist Problem and the Royal Nepalese Army (RNA)

By its very scope, nature, and structure of the organization, the RNA is not normally deployed to frontline actions and everyday security of the people until a national crisis and a state of emergency are formally declared by the government. The RNA was not ignorant of the fact that the Maoists attacked government structures, police posts, and seized weapons. It made its view clear, that the right procedures were to be followed for the Army to move out of the barracks. It perceived and interpreted the Maoist rebellion as a reaction to the socio-economic discriminations and poor delivery of services and the absence of answerable governance. The conviction of the RNA was that since the Maoist rebellion was a homegrown problem, its solution should be sought politically, with a focus on the vaunted integrated Internal Security and Development Plan (ISDP). The Maoists could also be ideologically weakened in their own base areas by the leaders of the mainstream political parties and could be disarmed by the police, backed by the army, to ultimately bring them to the mainstream and competitive politics. That would prevent the eventuality of loss of lives and property. It would not be any heroic on the part of the country to enter into a situation of one Nepali set to kill another Nepali.

Deputy Prime Minister (DPM) Bam Dev Gautam, in 1996, had sought to have a joint Anti-Terrorist Cell with a view to fighting the Maoists. The army provided essential training to the police on the know-how of fighting terrorism. Later, in 1997, Prime Minister Surya Bahadur Thapa realized the need to have adequate budget allocated for the Army to purchase logistics and communication equipment. As part of the preparation to disarm the Maoists, the RNA was to establish a Divisional Headquarters in Surkhet, and a Brigade each in the districts of Rukum and Rolpa, the very base areas and the origin of the rebels. The policy of disarming the Maoists would be a viable way to solve the problem with the least loss of life and property. The Royal Palace and the RNA had identical views regarding the question of addressing the problem. The political leaders, on their part, did not bother to pursue the correct constitutional procedures to mobilize the army. They rather resorted to their customary unserious comments. They feared that once the army moved out, it would not easily go back to the barracks. In such a case, democracy itself would be imperiled.

Unlike the previous PMs, Lokendra Bahadur Chand and Surya Bahadur Thapa, Girija Prasad Koirala sought to mobilize the army by direct commands, quite like the way he was accustomed to assigning and reassigning his party members and workers. He seemed to be haunted by the ghost of noncooperation and imagined ulterior motives of the Royal Palace regarding the question of the mobilization of the army against the Maoists. Apparently, there appeared a serious lack of rapport between him and the two monarchs – first, with King Birendra, and with King Gyanendra later.

The Maoist attack on the Dang Army Barracks with the capture of much of the unit's weapons finally shook up the government to realize the necessity of army mobilization as soon as possible. In the days that followed, other army barracks also faced the attacks of the rebel. It seemed that the army could not even protect its weapons from being borne off by the enemy. The recurrent incidents lent space for much adverse comments and criticism.

It was acknowledged, but only in retrospection, and years later, that it was only the RNA that foiled the Maoists' mission to capture the entire state, its powers, and politics.

For three and a half years, the RNA, in pursuance of the concept of the Unified Command, with the National Security Committee headed by the CoAS, went into action against the Maoists following the declaration of the State of Emergency. The Committee comprised all the security agency chiefs, CGS, DGMO, operation heads of the security agencies as members, and similar security coordination committees active at the zonal and district levels. Simultaneously, the high-level national security meetings, headed by the Minister of Home Affairs, were also operative. The CoAS and the DGMO (the Director General of the Military Operations) of the RNA attended these regular meetings. Before November 2001, no such provisions were there.

Taking into consideration the sensitivity of security, and in agreement with the related committees, a regional-level security coordination committee under the Ministry of Home Affairs was also set up.

The RNA believes that its actions against the Maoists compelled the rebels to accept the peace deal, because, among many other factors, they also obviously suffered colossal losses of their guerillas and resources.

Personally, it is my opinion that the political parties, the civil society, the media, and human rights activists only halfheartedly cooperated with the RNA during this critical period.

The Maoist Problem and the Nepal Police

Ever since the declaration of the Maoists' 'People's War' (Jana Yuddha), the Nepal Police singly battled them and suffered both physical and psychological loss in the confrontation and operations. In place generally for policing to maintain law and order in the society, it was not pitched for actions against the kinds of rebels that the Maoists were. Therefore, the Nepal Police were given an altogether different scope of duty, that too at a time when it was not equipped well and trained with adequate know-how. Despite these odds, it put up great fights against the rebels and held them up for ten years, but not without great losses and casualties.

The government in 1996 and 1997 deployed the police which, in the early years with its strategies called 'Romeo' and Kilo Sierra-2, which succeeded in causing damage to the Maoists. During these operations, the non-combatant public also suffered casualties. These incidents happened more accidentally than deliberately. However, these instances cost the police a gradual loss of its image in the years that followed. The police leadership and the entire organization were charged with extreme brutality, causing collateral damages.

On their own, the Maoists, too, had grown bold and belligerent.

Of the two security organs in the country, only the police being singled out for deployment to launch its actions against the Maoists, with the army still in barracks, speaks of sheer neglect and indifference of the political leadership of the times. The infrequent and occasional and inadequate allocation of budget and purchase of weapons and logistics for the police came under criticism from within and outside.

To those involved in it, the purchase deals proved to be blessed occasions to make money. Those dealmakers and arms dealers wished the insurgency to prolong for them to fish big in the troubled waters.

It was also apparent that the political leadership and the leadership of the Nepal Police and the RNA were unable to go ahead with the Integrated National Security concept to cope with the Maoist problem. They, therefore, understandably lacked a uniform policy and approach; consequently, dealing with it in piecemeal and ad hoc manner was the only choice. It was so because there seemed to be a perceptible absence of harmony between the indispensable security agencies in their uniformity of concept and strategies of actions against the rebels.

Not amusingly, too, the Nepal Police and the RNA hierarchies traded accusations and rebuttals of noncooperation against each other. They rather seemed to rejoice in the blame game. The police accused the RNA of disallowing the purchase of sophisticated weapons from abroad. The RNA counterclaimed that, on its part, they did supply enough weapons and offered training to the police as and when so required.

It was, nevertheless, quite clear that the RNA did not wish the Nepal Police to possess sophisticated arms and ammunition.

The Maoist Problem and the Armed Police Force (AFP)

The APF was formed in 2001 by the government through an

Ordinance during the critical and pressing times of the Maoist insurgency at its extremes. Most leaders, the civil society, the intellectuals, and even most security officers were found to support the formation of such a force because of the delay experienced at the Army's mobilization to deal with the Maoist menaces. Understandably, the novel setup would take time to adjust to the ground realities and to strengthen and equip itself in every respect to counter the rebels.

Although the APF was formed at this late date, the background for the same was prepared much earlier, around the end of the Panchayat era itself. There was an ever-increasing realization of the need to expand the police force from the grassroots level of the villages of the districts, with weapons and paramilitary structure.

The formation of the APF also solved the dual dilemmas facing Nepal for quite some time as the RNA could not be manifestly mobilized for internal actions, and the Nepal Police being traditionally in place for the maintenance of civil law and order in the land and also being insufficiently equipped in manpower and firepower to address something like the Maoist menace.

The 1990 political changes in Nepal guaranteed the people their rights of expression, assembly, and organization as enshrined in the Interim Constitution. But no concrete changes came in place for the concept and scope of the police in the changed contexts that followed. Oddly, the volume of the budget, recruitment, and size of the organization swelled up.

The Nepal Police, on its own, suffered from extreme politicization right from the point of recruitment to promotions and transfers. Its leadership could not insulate the organization from political interventions. It instead enjoyed political patronage. The Nepal Police leadership – for instance, the IGP, Achyut Krishna Kharel – reportedly helped certain political party candidates and hindered others in the general elections.

The idea of the formation of the APF also dates back to the time of the realization to downsize the Nepal Police, decentralize it down to the local level, and to establish an altogether integral force with a provision of a Major General of the Army appointed as the first Chief of the APF, as per the King's wishes. Interestingly, while all the Maj. Gens declined to accept the offer on the one hand, on the other, the Nepal Police AIGs, Krishna Mohan Shrestha, and Ram Kaji Banatawa, raced against each other for the coveted post.

When asked for my opinion, I forwarded to the King the name of Krishna Mohan Shrestha on the grounds of seniority and his good rapport with the Army officers. Later on, I knew that PM Koirala was seeking to place AIG Ram Kaji Bantawa in that post. Only after much consultations, he agreed to the candidacy of Krishna Mohan Shrestha. And once the APF was formed Krishna Mohan Shrestha became its IGP.

To insulate the new organization from politicization, special care was taken with the formation of the APF Service Commission.

India expressed its willingness to aid the AFP with weapon and training. But King Birendra was not in favor of equipping the force with sophisticated arms and ammunition. As it was, the APF chief's clandestine visits to India in connection with arms deals drew criticism, as it violated the official and Royal Palace norms.

The then Minister of the Home Affairs, Khum Bahadur Khadka, also came under criticism for similar violations of the standard norms.

The Maoist Problem and the National Investigation Department (NID)

Espionage, in South Asia at least, goes back to the time of Chanakya, or even earlier. Its scope and importance have further increased in modern times. There is no doubt that intelligence gathering is essential for successful statecraft.

The NID was comparatively more efficient during the Panchayat Period. Excessive politicization, however, paralyzed the institution after the political changes in 1990.

In the wake of the Maoist insurgency, the political leadership did realize the need to equip the NID with resources and training to enhance its capacity to obtain and deliver information to the government. The King even advised the government to retain the Directorate under the Prime Minister's Office (PMO) to safeguard it from external elements as well as leaks from within. Sadly, the myopia of the political leadership hindered the process of capacity-building of the NID to the required and desired degrees.

Despite these chronic ills, the NID Chief, Devi Ram Sharma, and I managed to gather information on the size and structure of the Maoist combatant force and their immediate plan of likely attacks.

It is true that the NID suffered from the lack of adequate resources and equipment. But it spared no efforts to extract confidential information out of the closed-door and clandestine meetings and deliberations of the political leaders and their hotlinks to the Maoists.

One such scoop was on the Nepali Congress (NC) Party President and former Prime Minister Girija Prasad Koirala's dash to New Delhi for a meeting with the Maoist leader 'Prachanda' while still on a visit to China. The NID obtained the confidential details of the closed-door meetings between the two leaders. It is another matter that the information the NID disseminated was not much followed, or taken notice of and attended to.

The NID set a good track record of strategic intelligence and Maoist-related political analyses.

The Maoist Problem and India

In 1994, SD Muni, the Nepal 'specialist' and Professor at the Jawaharlal Nehru University in Delhi, published an analytical paper with a subtle analysis of the dimensions of Nepali politics of the past, the present, and the predictable future. The piece, also regarded as a policy paper of the South Block [the foreign policy nerve center of the Indian Government in New Delhi], discussed the preferred option for a republican Nepal, doing away with its monarchy for democracy for her better future, particularly for securing the vested interests of India in Nepal.

It is said that India saw Nepal through the very strategic policy paper. Nepal's Leftist leaders are said to have good liaisons with Prof. Muni who often visits Nepal, or Nepali leaders visit India to meet him. So much so that he even came to Nepal and met Dr. Baburam Bhattarai [the second most powerful Comrade, after 'Prachanda', in the Maoist hierarchy] at The Everest Hotel at a time when the government-Maoist peace talk was in progress.

It was also heard that after he left the government, Prachanda, in November 2009, headed to Singapore to hold confidential talks with AV Mathur, the senior official of the Indian Intelligence Agency, RAW [Research & Analysis Wing], and SD Muni was also present at the rendezvous.

It also cannot be a mere coincidence that Girija Prasad Koirala himself was already there, ostensibly in connection with his medical treatment.

The speculations that India was inciting the Maoists in Nepal to have the monarchy removed and thereby serve its interests were ringing louder all through the years.

It was not a matter to be proven or disproved by anyone. However, it is no secret that ever since its independence in 1947, India has constantly faced threats from its Leftist renegades. In recent times, there has been the merger of the People's War Group (PWG) and the Maoist Communist Center (MCC) and is known as the Indian Communist Party (ICP) Maoist.

Successive Indian Prime Ministers in their key speeches have expressed fears and have accused several terrorist groups of receiving training in Pakistan [India's avowed enemy], then infiltrating Indian territories and perpetrating terror in the country. On its part, Pakistan constantly accuses India [its professed foe] of having its hand in terrorist and violent activities in that country.

All the while, for India, the Kashmir problem is chronically aching like a loose molar.

If history is anything to go by, it is still fresh in one's memory that India did indeed play a proactive role in splintering Pakistan, by clandestinely giving every help, moral and material, to the freedom fighters of the then East Pakistan, lending its members safe havens in India. The grand design helped create an independent nation, now called Bangladesh, wrested from Pakistan. India and Pakistan have fought three large-scale wars since their independence from Britain in 1947.

Fingers also point at India for actively backing the Tamil separatists, known as the LTTE (Liberation Tigers of Tamil Elam), in Sri Lanka. It is also known that India tacitly extends its support to the 'Free Tibet' Tibetans, too. Amidst all these volatilities in South Asia at large, the Maoist rebellion exploded in Nepal and soon spread its tentacles to many districts.

However, regarding the Maoists, Nepalis have continually remained unable to ascertain how India still takes them. It is apparent that those CPN (UML) leaders and even Girija Prasad Koirala of the Nepali Congrees could not have been able to meet the Maoist leaders in Siliguri, Lucknow and New Delhi without the prior knowledge of the Indian security apparatus in these far-flung places at a time when the Maoists were on the terrorist watchlist in Nepal. Given Nepal's geopolitical location in South Asia and the porous borders between Nepal and India, the Maoists could never have been able to grow in number and strength, as it did, without the nod from India.

This riddle hovers over Nepal all the time.

The emergence of the Maoists as a strong political force brought about the restructuring of the power equations: Nepal's monarchy, backed by the Army, with the mainstream political parties in the middle, and the Maoists waging their armed rebellion from the peripheries. An alliance or propinquity between any two players could obviously weaken the third.

There were two perceptions about this fluid situation. The Royal Palace always believed that the Maoists and the mainstream political parties would never make alliances; rather the rebels would inch closer to the Royal Palace, given the nationalistic position maintained by both camps. On their own, the mainstream political parties flirted with their reading that the interests of the Royal Palace and the Maoists would never converge, as the Royal Palace would wish to perpetuate monarchy whereas the Maoists' goal was to eradicate the institution.

The Royal Palace and the Maoists, separately and on their own, alienated the mainstream political parties. Whereas India wanted the Royal Palace and the mainstream political parties to team up to blunt the American and Chinese influences in Nepal. It seems that India also wanted to play the Maoist trump card to secure its interests in Nepal.

Following the February 1, 2005 royal takeover by King Gyanendra, India dispatched the chief of the Intelligence Agency, RAW, to Kathmandu to effect an alliance of the Royal Palace and the mainstream political parties. India also sensed the King of playing his China Card when the latter openly proposed that China be invited as an observer at the imminent SAARC Summit to be held in Bangladesh.

That the Maoists were given safe shelters in India was already felt in Nepal. An instance may prove this in real terms:

A company of Nepal's APF personnel had been sent to Chakrata in Dehradun, India, for commando training.

A Police Inspector, back from the said training there, shared with me his knowledge of some Maoist cadres also having received similar training there a month earlier !

The Maoists, while launching their attacks on Satbaria of Dang and Mangalsen of Achham and Chainpur of Sankhuwasabha districts had employed the very same Sand Model as had been used for training the Nepal Armed Police Force group at the aforementioned training center in India.

As reported by the said Inspector, a local tea stall operator outside the training center jokingly expressed his wonder that the earlier training group had also included females whereas there were only men in the second one.

One camp instructor had also allegedly expressed his surprise at the double standards of India by conducting training to rebels in one round and the police force in another. Needless to say, both training contingents were from the same country: Nepal.

I was taken by surprise to hear the Inspector share his firsthand knowledge and experience. I could not help sharing my surprise with the IGP, Sahabir Thapa, of the APF.

The said camp was the Training Complex of Establishment 22 and was constructed in Chakrata, with American assistance in a particular Indo-US joint strategy to train Tibetan Khampa guerillas to be deployed in Tibet against China.

Chakrata is some 70 kilometers away from Dehradun. It was also there that India provided training to the Tamil Separatist guerillas of Sri Lanka, the Liberation Front of the Freedom Fighters of Bangladesh, and certainly to the Maoists of Nepal, too.

It made sense to me that the Maoists would never have been able to grow to this scale of strength without training in or the backing of India.

I updated the King on these equivalences and pointed to the need for further ascertaining of the facts and details thereof. He sounded serious after hearing this, and instructed me to start consultations with the security officials, and proceed with the needful preparations to that effect. Accordingly, I consulted with the IGP of the APF Sahabir Thapa to discuss the matter with other security officers.

Lal Bahadur Thapa, the retired Deputy Chief of the NID, had visited me one day and had expressed his strong desire to serve the country in whatever capacity the King considered him fit. In my briefing regarding the matter, the King instructed me to send him quietly to Chakrata and find the facts behind our findings. In August 2003, having briefed him on everything about the Maoist training, I assigned him to go there and clandestinely collect facts and thereafter report the finding to me. He seemed to be the right person for the job because, first, his service in the intelligence department would help him do his sleuthing dextrously, and secondly, because he was not new to Dehradun, it was a familiar terrain for him.

It was a confidential three-man operation. Accordingly, he was assigned to go to Chakrata during the coming Dashain and Tihar, the two great Hindu festivals in Nepal, and report to me about the facts gathered there.

Amidst this engagement, as it happened, I resigned, in November 2003. He neither met me nor did he report his finding to Brig. Gen Gajendra Bahadur Limbu who replaced me in my previous office.

In the six decades that have followed its independence, in 1947, India has been constantly causing troubles to keep Nepal under its umbrella of influence. As has already been referred to, India never supported King Birendra's proposal for Nepal as a 'Zone of Peace' even when as many as 116 countries had endorsed it. India seldom acted in good faith toward Nepal, and at times did not even flinch to interfere in Nepal's own internal matters. As and when Nepal sought to purchase weapons to strengthen and equip its army and the police, India stalled the process under one pretext or another, thus disregarding the 1950 accords between the two nations.

Other Indian pressures were also applied from time to time: India wished to have a consular office in Birgunj and wanted to set up Indian pension camps for ex-Indian Army retirees in places like Butwal. Another insistence was on the immediate Cabinet decision on the Citizenship Act in Nepal.

These are but just a few cases in point of how India intervenes in Nepal.

After the February 1 Royal takeover, India brokered the 12-Point Peace Accord between the Maoists and the mainstream political parties, on November 21, 2005, in New Delhi, costing Nepal its monarchy.

The West also developed its strategy to see Nepal through India's eyes.

There is no end to India's power play and interference in Nepal. It played the Maoist card to displace the monarchy. Later, it did not wish to see the Maoists emerge as a big and strong force; so it played the Madhesh Card to cut it down to size.

With a couple of friends, I was at a dinner in the house of Dhiraj Mukhiya. He was a Second Secretary at the Indian Embassy in Kathmandu and was from Darjeeling. At the said dinner, held on March

21, 2007, some officials of the Indians Embassy also joined us. One of them was repeatedly talking on his cell phone and taking every update on the Rautahat incident of the mass murder of over 27 Maoist cadres. He seemed to be giving information and instructions to someone over the phone. It did not take me long to learn that the person belonged to the Indian Intelligence Agency, the RAW.

Coincidentally, on the following day, I happened to meet the CPN Maoist President 'Prachanda' and Dr. Baburam Bhattarai, the second Comrade in the Maoist hierarchy. They sounded upset and spoke their mind that India wanted to use them as stepping stones to serve its interests. I told them they would not be able to make any drastic difference alone without taking others into confidence. They would do well to join hands with the pro-Palace political forces to achieve their goals because, I added, their political aims and objectives were similar to the nationalistic sentiments nurtured by the Royal Palace. True, their ways and means were different, but there was a doubt that they would get help from India, as expected, in the changed context of their joining Nepal's mainstream politics. It is likely that they never knew that India merely wished them to shun violence and join Nepal's mainstream competitive politics but not to emerge as the largest party, as had happened in the previous elections to the Constituent Assembly (CA). India might well turn hostile to them the moment it sensed obstructions posed to its interests by the Maoists.

India is also not happy about the growing influences of the West in Nepal. It was clear that India was not approving of the presence of the United Nations Mission in Nepal (UNMIN) monitoring the rehabilitation of the Maoist combatants and inventorying their surrendered weapons.

Matters just don't end there. India also declined the Carter Center's proposal of brokering peace in Nepal, as extended by the Carter Center's peace initiative. Three developments – the emergence of the Maoists as the largest political party in the Constituent Assembly elections, its nationalistic agenda, and Prachanda's choice of China for his first official foreign visit as Prime Minister – alarmed India in the pursuit of its vested interests. India playing the Madhesh Card, therefore, appeared to be directed to downsize the Maoists.

In the interregnum, China termed unnatural the coalition of the Madhesh Movement and the Madhesh-based political parties.

Maoist Problem and China

Nepal's immediate northern neighbor, China, before and even after the political changes of 1990, continued to recognize the monarchy in Nepal as its most trusted and reliable friend, without bothering to cultivate other alternative allies.

There was a reason behind this. Clearly enough, China read the minds of the major political parties and sensed their extra tilt to India to please it as though their ascent to power was impossible without its blessings. The Chinese correctly reasoned that the political parties feared that any closeness with China would displease India.

China always accorded its high regards and goodwill to King Mahendra and King Birendra alike, and consistently promoted cordial relations between the two countries. It highly appreciated King Birendra's move to quash the Khampa rebellion against China. It displayed its good neighborly behavior and largeness of heart even as the King cancelled its tender to construct the Kohalpur-Banbasa sector of the Mahendra Highway and awarded the contract instead to India.

However, India, Nepal's immediate southern neighbor, strongly objected to Nepal's purchase of antiaircraft weapons from China, going even to the extent of clamping economic blockade against Nepal.

China always viewed the Maoists' 'People's War' as an internal problem of Nepal. It consistently expressed the belief that Nepal alone was capable of settling it on its own, and continued its support to the country. Instead, China objected to the way the Maoists abused the great name of Mao Zedong while they themselves received resources from elsewhere.

Chinese officials wished to have King Gyanendra make his first official foreign visit to China as an attempt to remove their skepticism regarding the Royal Palace Massacre, because the Chinese people still would not accept the claim that it was the doing of Crown Prince Dipendra. But that did not happen, as the King went to India first. Even so, regarding the King's moves on October 4, 2002 and February 1, 2005, China did not issue any negative or positive comments; instead, it said it was the matter for the Nepalis themselves to settle. It advised the King to go along with the political parties for the stability and development of the country.

China did not recognize the Maoists as a credible political force until they came out in the open, in public. It continued to remain aloof even as they tried to contact the officials at the Chinese Embassy during the first and second rounds of the peace talks. I have no knowledge of China ever extending its moral or material support to the Maoists. It had the knowledge of the Maoists taking safe havens in India and receiving every help from there. China rather continued its support of the Nepal Government with logistics and communication equipment, if not weapons.

In the changed context following the dissolution of the monarchy and the emergence of the Maoists as the largest party in the Constituent Assembly elections, China appeared to be building rapport with them. It invariably sees to it that there are no anti-China activities from the soil of Nepal. However, China still does not appear to be fully convinced of the political parties' 'One China' policy.

The month-long anti-China activities in Nepal on the eve of the 2008 Beijing Olympics heightened its fears. Taking into consideration the fiasco of its Sikkim and Bhutan policy, China seems to be observing things cautiously in Nepal. It seems to be moving out from a 'reactive' to 'proactive' policy.

When it comes to Nepal's own national interests, these are interpreted otherwise. If China further fears about India's overt meddling in Nepal as its security concern, there is every danger of Nepal being trapped in the interests of its two immediate neighbors.

The Politics of Conspiracies: The Kismat of Nepal

Though many may say, 'Politics is a dirty game,' quite a few other common sayings go:'No one joins politics without any prospect of reaching to power.' Or, 'No one enters politics merely to be a saint.' And, 'There's no permanent foe or friend in politics.' And so on it goes.

These adages apply perfectly truly to Nepal, like anywhere else. The political course and history of Nepal for the last two and a half centuries at least have gone round and round with conspiracies to gain and hold on to power, disregarding the wellbeing of the people and the country. There has been a series of massacres or coups, be it the Kot Parba (Massacre) of 1846, or the Bhandarkhal Parba thereafter, or the royal coup d'état mounted by King Mahendra in 1960, as political parties call it, or the Maoists' declaring their 'People's War' (Jana Yuddha) in 1996.

But in essence, all such acts are committed to grabbing power in one way or the other. While each move or step is dubbed 'necessary for the greater good and in the interests of people,' the mantra is meaningful only to the perpetrators and their acolytes.

The Downfall of the Panchayat Polity and the Restoration of Democracy

The end of the Panchayat System, a partyless polity, is often attributed to the strategy of denial and non-transparency of the managers of the political system. In one sense, the absence of decentralization was the cause of the fall of the Panchayat Polity.

I personally felt for a long time that the very concept of having in place the PPEC digging the very grave of this system of politics. In my briefing to King Birendra, I had tacitly mentioned to him the pressing needs for timely reforms, but he passed it saying, 'Politics doesn't run on a straight line, Vivek.'

I am also reminded of the time, while on our tour to Belgrade to participate in the Nonaligned Summit when Crown Prince Dipendra himself had remarked, 'Vivek, you talk in black and white, but political courses move along a gray line.'

In fact, the Panchayat was characterized by factious politics operated by fractious power seekers in the Royal Palace itself to have their own ways. Against this backdrop, other ambitious cliques, internal or external, would obviously play within the divisions of the factions themselves. Owing to the oddity of the manners and actions of the Panchayat's pundits and its system managers, and due to the geostrategic location of the country itself, this type of politics was gradually losing its hold on the masses. Nepal increasingly experienced the natural burden of being a landlocked [mostly 'India-locked'] country. There were reports of Indian aircraft violating, unhindered, Nepal's airspace. Nepal's decision to purchase antiaircraft weapons from China provoked India to such an extent as to impose a stringent economic blockade on the country for 19 months. The border sealing played a covert role to incite the political parties of Nepal against the Panchayat System. In my view, no matter how weak and unsuccessful the mandarins of the Panchayat System were, they were shrewd enough to think that had Nepal, as a sovereign country, not resorted to obtaining the necessary weapons from China, things would have been otherwise. The Royal Palace rightly took it as a signal leading to a mass movement against the Panchayat System, and not the strategic plan of the political parties. Therefore, the King preferred to decide to work out a deal with the political parties rather than kneeling down to external hegemony.

The political changes of 1990 created a new turn in the balance of power and power-sharing in the country. The King asked the Nepali Congress leader Ganesh Man Singh, the commander of the mass movement, to become Prime Minister. Singh, displaying a rare sense of wisdom and sacrifice, proposed Krishna Prasad Bhattarai, another Nepali Congress leader, for the post.

The resultant Interim Government accomplished its principal mandate of writing and promulgating a Constitution, on November 9, 1990, its Preamble guaranteeing constitutional monarchy, multiparty democracy, parliamentary system, an independent judiciary, and the protection of human rights in Nepal. The King believed that democracy would reach the grassroots to bring changes in the quality of life of the masses.

The Nepali Congress and the CPN (UML) were the main contestants in the general elections that were to follow, as per the new Constitution's other outstanding mandate. The Rashtriya Prajantra Party (RPP), composed of erstwhile *Pancha*-s, those pro-Panchayat leaders, did not fare well in the elections in the new environment. The Nepali Congress was led by veterans like Ganesh Man Singh, Krishna Prasad Bhattarai, and Girija Prasad Koirala.

Ganesh Man Singh set another example of statesmanship and sacrifices with his preference for remaining as a 'watchdog' of democracy, by staying away from power politics. With such grand gestures made by Singh, Krishna Prasad Bhattarai, the second leader in the party hierarchy, was expected to head the government for the second time around in the newly elected government. Bhattarai was believed to possess the ability to settle many intricate issues with inclusive politics. The Royal Palace also took him favorably, with a wish to see him as Prime Minister for the second time, as Nepal required such a man of understanding and political acumen to settle political discords.

In the general elections of 1992, the Nepali Congress fielded Krishna Prasad Bhattarai to contest the elections from Constituency No. 1 of Kathmandu as its prospective candidate for Prime Minister if the party won a majority of votes. From the same constituency, the General Secretary of the CPN (UML), Madan Bhandari, known for his eloquence and nationalist image, contested against Bhattarai. Thus, the elections at the particular constituency had become the talk of the town, and a sight to see for people at home and abroad. The Royal Palace wished the likes of Bhattarai and Koirala to win the elections so that democracy could take a smooth course ahead.

It is difficult to believe but it is true, that a certain faction within the Nepali Congress Party itself covertly campaigned to have Bhattarai lose in the elections because by doing so, a particular group could reap the dividends. To the best of my memory, the Royal Palace got the wind of the conspiracy against Bhattarai and alerted him about the developments. But the latter took it lightly, and ultimately, as was speculated, he lost the elections.

Girija Prasad Koirala covertly played his role to cause that loss, a typical pattern of repetitive political conspiracy so prevalent in Nepal. Thus, out of the 205 electoral berths, the Nepali Congress Party won a simple majority of votes, with 113 seats. The CPN (UML) became the second largest party to play the role of the Opposition Party in the Parliament.

Girija Prasad Koirala, notorious for his brand of biased and vengeful politics, did not only manage to have Bhattarai removed from the contest of Prime Minister – to become one himself – but also adopted a policy of the same vindictive tactics, particularly against the Communist forces.

As Prime Minister, his self-centered manner of working and controversial decisions contributed to damage his own image and reputation. He bulldozed, singlehandedly, with a complete disregard for the politics of accommodation, the norms of democracy, and mutual understanding. He introduced extreme politicization in the bureaucracy and completely ignored the Nepali Army, considering it as a 'property of the King.' As if that was not enough, he drastically reduced the Army's budget and recruitment. He seemed to encourage corruption rather than control it. He ordered to have withdrawn the cases against the most corrupt, such as the former IGP, DB Lama, and Colonel Bharat Gurung, who were convicted during the Panchayat System; he even had these men inducted into the Congress Party with honor. Protesting Koirala's decisions, with special reference to DB Lama, the Minister for Law and Justice, Taranath Ranabhat, resigned. Instead of rectifying his unilateral manner of working, PM Koirala, in fact, sidelined all those who opposed him. Always in favor of having his own coterie, he practiced unchecked nepotism and favoritism. As a result, corruption and poor governance took deep roots. While on a tour to India, he willfully and secretly signed the Tanakpur Agreement on a large hydropower project on the Mahalkali River bordering Nepal and India for harnessing the river for multiple uses of water. This misdeed caused much controversy, even disrupting the proceedings of the Parliament for several days. The Royal Palace also raised its eyebrows over the Tanakpur issue, a controversy big enough to damage any trust in PM Koirala because the accord was related to Nepal's sovereignty. The Royal Palace thereafter secretly sought leaders with a clean image and nationalistic zeal from within the Nepali Congress and the CPN (UML), presumably to form an alternative force to work together in which the then Speaker of the House, Damannath Dhungana, played a pivotal role. Madan Bhandari, General Secretary of the CPN (UML), also played a decisive role in making PM Koirala lose his face in the Parliament. The Tanakpur Case reached the Supreme Court which declared the agreement as against the norms of the Constitution.

Already at the height of popularity for his eloquence and his triumph in the general elections against Krishna Prasad Bhattarai, Madan Bhandari's image and stature grew. At the 5th General Convention, in 1993, he, as the CPN UML General Secretary, played a key role in having passed the party's policy of 'People's Multiparty Socialism'.This policy paper addressed his party's pragmatic awareness of the new political equations in democratic Nepal.

His proposition received an overwhelming response at both home and abroad, so much so that *Newsweek* published an interview under the caption, 'Karl Marx Lives in Nepal'.

Presumptively, the emergence of Bhandari as a nationalistic leader of international acclaim caused headaches to a particular external power. The role he played in the Tanakpur Case somehow bore upon Indian interests, sending ripples all over. Against this backdrop, Madan Bhandari, along with Jeev Raj Ashrit, Chief of the Organization Department of the CPN (UML), was killed in the Dasdhunga Accident on May 16, 1993. The fatal vehicle mishap on a highway gave rise to speculations, that it was a murder, mysteriously engineered, although it was described as an accident.

Certain people speculated that PM Koirala, The Minister for Home Affairs, Sher Bahadur Deuba, and the then IGP Moti Lal Bohara must have got wind of the planned murder, under the cover of an accident. Other people also talked of some strangers and their mysterious movement around the CPN (UML) Central Office at Bag Bazaar for a few days before the accident. Others wondered whether the Army had its plainclothesmen planted in the area to inspect the Party's activities. There was no grain of truth in the latter speculation as there was no need for the Army to do so.

The party called for a mass demonstration, obstructing the proceedings in the Parliament over the question of the mishap. Several 'Probing Committees' were formed. By then, a couple of UML leaders had headed the Ministry for Home Affairs. Bidhya Devi Bhandari, the widow of Madan Bhandari, was twice made Minister, and later became the Defense Minister. The mysterious murder of Amar Lama, the driver of the ill-fated vehicle and the only survivor of and witness to the accident, has made the incident all the more mysterious, dumping it into a cold case file. The CPN (UML) government itself did not find any clue to the accident.

In my opinion, fingers point to the possibility of external and internal power players' hand behind what was cold murder but it was given the cover of a vehicle accident.

The First Communist Government in Nepal

PM Girija Prasad Koirala, despite his image as a seasoned politician, could not provide satisfactory leadership to the country, and Nepali politics, from what was expected of it, slid from bad to worse. He could not bring together the factional groups within his own party, let alone those of others. Fearful of a Vote of No Confidence against him in the House of Representatives, and expecting to secure a majority anyway, he dissolved the Parliament in June 1994 for a fresh mandate, with elections slated for November 13, 1994. A case was filed at the Supreme Court (SC) demanding the reinstatement of the Parliament. The court, however, gave its verdict in favor of the Prime Minister for his decision to dissolve the Parliament, citing his action as his constitutional prerogative.

The elections were going to cost dearly to the Nepali Congress Party, as it was virtually split into two factions: one led by PM Koirala, and the other jointly by KP Bhattarai and Ganesh Man Singh. In many constituencies, Congress candidates themselves stood as rival contestants. The Royal Palace, as usual, wished nationalistic leaders with a clean image from the Nepali Congress and the CPN (UML) to win the elections.

It is said that two captains sink a ship. And even a small leak causes the sinking of a great one. The same happened with the Nepali Congress Party. The CPN (UML) emerged as the largest party, but without a majority. The Rashtriya Prajatantra Party (RPP) opened its new account as the third largest party in the elections. The Nepali Congress Party tried its hardest to form a coalition government, stitching together several other parties. But the political momentum moved towards the possibility of forming a Communist government, something that some power blocks in and outside Nepal deeply abhorred. One could sense unsuccessful attempts being made to avert it from happening.

It was from this very period that the diplomats of some embassies, contrary to their protocol, started overtly poking their respective noses in the politics of Nepal, visiting the leaders in their houses and offering their unsolicited advice. The hardsell and negotiation rounds delayed the formation of the government.

It must have been Saturday, November 3, 1994, when the King (Birendra) called me and expressed his concerns over the delay of the formation of the government, and about the political environment in the country. As his aide-de-camp (ADC), I updated him on the political deadlock created over the question of whether or not a Communist leader should head the government. I offered my advice that there was no harm in having a Communist as Prime Minister since the contender, Man Mohan Adhikari, was a seasoned and suave politician. Although a new experience in Nepal, it would convey good messages to strengthen democracy.

Having heard me out, the King instructed me to confidentially call Lokendra Bahadur Chand, the leader of RPP, to visit the Royal Palace without delay. I did so. There followed the King's consultations with him. I was not present at the meeting, but later I gathered that the RPP agreed to support the CPN (UML) in the Parliament without joining the government. Thus, the floor was open for a Communist-led government to be in power in the Himalayan Kingdom of Nepal. It was since then that the rapport between the Royal Palace and the CPN (UML) leaders improved. Before long, the Communist government unveiled its populist programs and received good responses in the country.

Presumably frantic at its popularity and populist programs, the sidelined parties tabled a vote of no confidence motion against the government in the Parliament. The Prime Minister, on his part, recommended to the King to have the Parliament dissolved for holding midterm elections, and the King, too, in pursuance of an earlier precedent, readily assented.

Again, the dissolution move was contested at the Supreme Court. On July 28, 1995, the Supreme Court gave its verdict against the dissolution of the Parliament, with its judgment reading that the Prime Minister of a minority government could not enforce this prerogative.

The SC's verdict triggered widespread criticism. The SC Chief Justice, Vishwo Nath Prasad Upadhyaya, with a slender majority in the full bench, had issued the verdict, and this judgment, taken as 'historic', would forever blur the future course of modern politics in Nepal.

The verdict reinstated the Parliament and opened the floor for the political parties to form a government. However, they seemed focused more on power politics than finding solutions to the deadlock. It was from this very reinstatement of the Parliament that unsavory and dirty practices entered Nepal's parliamentary politics. In order to pull together a majority, parliamentarians were lured with money, wine and women. It was a sordid abuse of democracy and the parliamentary system itself.

To the best of my knowledge and memory, the Royal Palace wished the government to complete its full term. It did not happen. The frequent toppling of governments became the political reality in Nepal.

Six Governments in Five Years!

In the five years that followed the mid-term elections of 1994 to the general elections of 1999, the politics of Nepal dived to the lowest levels of vile practices in the entire history of the nation. Entire parliamentary exercises were employed in the making and unmaking of governments. Sher Bahadur Deuba, Lokendra Bahadur Chand, and Surya Bahadur Thapa respectively headed the governments following the collapse of the minority CPN (UML) government. None of the three leaders' governments completed one year.

The Deuba-led government, backed by the CPN (UML), signed the Mahakali Treaty with India, and it also drew nationwide criticism. Madan Bhandari had become a popular politician and a national leader because of his vehement opposition to the Tanakpur Treaty, which related to the same river, Mahakali.This time around, the same CPN (UML), which had so strongly opposed the Tanakpur Agreement, took a U-turn to sign the Mahakali Treaty later, after Bhandari's demise. Some leaders of the Party opposed the decision of signing the treaty, leading it to a formal split into two factions – as CPN (UML) and CPN (ML).

The next government, with support from the CPN (ML), was formed by Girija Prasad Koirala, as Prime Minister. This was his second premiership. But the government tumbled in its first fifth month of its full tenure of five years.

Another government, again led by Koirala, was formed with backing from the mother faction of the CPN (UML).

Not a single government survived to complete its full mandate of five years. They failed to deliver peace and good governance. If the leaders of the political parties had displayed a degree of statesmanship to let the governments run full steam, Nepali politics could have taken the right course ahead. This did not happen because their leaders jostled for power.

Bam Dev Gautam, an influential leader who had proved his mettle as an emerging leader of the CPN (UML) by strongly opposing the Mahakali Treaty, became Deputy Prime Minister (DPM) in the Lokendra Bahadur Chand Government, in the beginning of 1997. He had built an image of a statesman by that time. He was also in the good books of the Royal Palace for his bold stance against the treaty which was more in the interests of India than those of Nepal.

But his image began to wear thin as he also introduced another form of politics. This he did in the Nepal Police by replacing the incumbent IGP, Achyut Krishna Kharel, by appointing Dhruba Bahadur Pradhan in his place. He also won the CPN UML a good number of seats in the local elections by his rampant abuse of the national coffer and using the police for his political maneuverings. That the 'Bam Dev Style' of electioneering set a license or precedent for other political parties to win in the elections through other nefarious ways and imaginative means, and the Home Ministry became the most coveted portfolio in the successive Cabinets.

It had barely been seven months in office when the Lokendra Bahadur Chand government was toppled down, as the parliamentarians deserted their camps for other greener pastures. Next followed Surya Bahadur Thapa's government which removed IGP Dhruba Bahadur Pradhan and brought Achyut Krishna Kharel back in the IGP's office. This was a mere tip of the iceberg of invalidating the decisions made by the former governments, setting another set of unsound precedents.

PM Surya Bahadur Thapa also contemplated the dissolution of the Parliament for a fresh mandate, though he could not achieve his mission. He instead stepped down, leaving the space yet again open for former PM Koirala to play the same old slimy politics of nepotism and favoritism.

As if these notorieties were not enough, there followed, like never before in the history of Nepal, other political excesses as were seen during the tenure of this government, because in the newly fangled name of globalization and privatization, many smoothly running public corporations were sold to individuals at throwaway prices by the Koirala Government of this time. That seriously undermined the opportunities for economic growth and employment in Nepal.

Tactically, PM Koirala announced the general elections of 2000. But being aware of likely defeat, he cleverly fielded Krishna Prasad Bhattarai as the prospective candidate for Prime Minister. True enough, the party won 114 seats out of the total 205, a good majority of votes in the Parliament. The CPN (UML) slid down to the second position while the CPN (ML) won no seat.

IGP Achyut Krishna Kharel had reportedly tried to settle his old accounts with Bam Dev Gautam by having caused his defeat in the elections. Interestingly enough, it was also heard that the Royal Palace had covertly played its card to have nationalistic leaders of the likes of Bam Dev Gautam win the elections. Gautam, however, lost, and he attributed the cause of his party's defeat to India's game plan.

When it comes to the general elections of 2000, one incident springs up in my mind. It concerns Bharat Mohan Adhikari. He was in the lead by a slender margin in the counting of votes in Biratnagar. He sensed a conspiracy in the ballot counting and so-called his relatives in Kathmandu to ask me to help him, as the rigging was going on at the vote counting area. I called Lt. Col. Ramindra Chhetri, the Commanding Officer of the Kali Bahadur Battalion in Itahari to have some soldiers under his command visit the counting center to avert any such alleged manipulations.

That one incident, though just a tip of the iceberg, illustrates the range of tricks and unfair means employed in Nepal's parliamentary politics.

Krishna Prasad Bhattarai won the elections this time around and became Prime Minister. But from his very first day in office, the usual internal squabbles began in the Nepali Congress Party. This particular tussle took place when GP Koirala plotted to have Bhattarai removed to become Prime Minister himself.

This manoeuvre was contrived at the very time when Bhattarai was preparing his groundworks to initiate dialogue with the Maoists. However, worried that by way of embarking on the said dialogue with the Maoists, Bhattarai would prolong his hold on to power, Koirala incited some Nepali Congressmen to mount pressures on the PM to resign. Bhattarai was thus pushed to the corner by his own party members to resign, unwittingly, and in this lamentable instance, he was betrayed by his own longtime party associate, GP Koirala.

Back again in office as PM, Koirala adopted tough measures to weaken the Maoists by the use of force rather than dialogue or any form of peacemaking overtures. The Maoists, too, on their part, and as per their tactical move, announced their 'No dialogue with the Koirala Government' stance, and launched a horrible attack at Dunai, the administrative headquarters of the Dolpa District in October 2000, causing heavy casualties of policemen and a great loss of property.

The Minister of Home Affairs, Govinda Raj Joshi, offered his resignation by citing the reason being the noncooperation of the Army at such a critical time and with such pressing needs. This stand infuriated the Royal Palace and the Army. Subsequently, IGP Achyut Krishna Kharel also resigned in that connection. Beleaguered by pressing circumstances, the PM was seeking to mobilize the Army to quash the Maoists.

The Royal Palace and the Royal Nepal Army (RNA) Headquarters were not averse to army mobilization. But both institutions wished to see constitutional procedures in place from the National Security Council (NSC) and a political consensus for a declaration of the State of Emergency. PM Koirala perceived it as dillydallying by the Royal Palace and the Army.

During this very lingering backdrop, the Royal Palace Massacre took place, on Friday, June 1, 2001. By that time, the Maoists, with their tactical strategy and sway of influence, seemed to be heading towards the very center of power politics. Their reasoned refusal to hold dialogues with the Koirala Government but their readiness to talk with any other team indicated the shimmering changes in the power politics of Nepal.

Koirala's Resignation: Holeri a Mere Excuse

When it comes to the recent political history of Nepal, the Holeri Incident – having taken place in the first six months of King Gyanendra's accession to the throne – deserves its due narrative.

Apparently, PM Koirala was not able to deliver any of the promises he had made – tranquility, good governance, and security in the country – while campaigning to remove PM Bhattarai. Even his party, the Nepali Congress, and the Opposition Party, the CPN (UML), did not like the way he had so unceremoniously removed Bhattarai from office. On his part, Bhattarai was secretly campaigning for Sher Bahadur Deuba.

While the game plans for power politics were in full gears, Maoist activities were also getting more intense across the country.

At the NSC meeting held on June 4, 2001, however, the PM and the CoAS of the Nepal Army voiced their sharp differences over the question of procedural instruments for the mobilization of the Army against the Maoists. The CoAS reiterated the need for political consensus, the declaration of a State of Emergency, and the activation of the NSC as preconditions for the mobilization of the Army, as enshrined in the Constitution, whereas the PM was intent on mobilizing the armed forces without delay. For whatever reasons, which remained obscure, the PM was seeking a shortcut in his purpose. But sensing difficulties in the process, he and some Ministers in the Cabinet attributed the delay to the unwillingness and noncooperation of the Army.

The Royal Palace and the Army always saw that the Maoist phenomenon was largely due to the weakening socioeconomic conditions and poor governance. The Maoist-affected areas needed help with employment opportunities for the youths to generate income through development projects, with the Army providing security. The Army developed a plan called the Internal Security and Development Plan (ISDP) and presented it to the government.

The ISDP went into effect in some Maoist-influenced areas. But the decision unanimously made was not directed as 'action' against the Maoists, but rather as an effort to exposing their misdeeds to the people. In the NSC meeting at the PM's office on July 8, 2001, the Army sought the government's clear concept and position on the Maoists, whether they were 'terrorists', or just 'rebels'. In order for the Army to be mobilized against them, it underscored the immediate need of a law related to the policy, thereby declaring the Maoists as terrorists engaged in radical activities and a ban on the entire sister organizations of the party. The CoAS explained the groundwork necessary for it to go into action against the Maoists as terrorists.

PM Koirala, on July 10, 2001, during consultations held in his office with top leaders, such as Surya Bahadur Thapa, Pashupati Shumsher Rana, and Dr. Prakash Chandra Lohani of the RPP, Jog Mehar Shrestha, Balaram Gharti Magar, Rabindranath Sharma, and Bam Dev Gautam of the CPN (ML), and Govinda Raj Joshi of the Nepali Congress, hinted at his mood to resign due to extreme factional politics within his own party, and the environment of noncooperation from all sides. The Royal Palace, too, was not happy with him for his failure to deliver peace and security.

Happily, the prevailing factors did nothing to belittle his greatness because he had displayed his fortitude and farsightedness to avoid any likely anarchy and turmoil following the Royal Palace Massacre.

However, at a time when the PM was increasingly getting singled out in vindictive politics, the Maoists launched a sudden attack on the Holeri Police Post of Rolpa District, and seized arms and ammunition and abducted as many as 72 policemen. At such a fresh spate of events and seeing it as the pressing need of the hour, the PM asked the CoAS to immediately deploy the Army to the beleaguered area for the release and rescue of the kidnapped policemen. The COAS gave his verbal consent to the PM but also sought the formal decision of the NSC. The RNA Headquarters instructed the No 4 Brigade in Surkhet to mobilize its troops to the affected area to rescue the abducted Police personnel rather than take military actions that might jeopardize the safety of the hapless prisoners. The No 4 Brigade Commander, Brig. Gen Bajra Gurung, was given the command to start an 'operation' in the area. But due to the delay in the enforcement of the action, the objective could not materialize.

Citing noncooperation of the Army as the reason, DPM Ram Chandra Poudel offered his resignation on July 13.

Meanwhile, all the three available helicopters that ferried the troops from Kathmandu to Dang were fired on by the Maoists. One chopper did an emergency landing in the Dang barracks. Pilot Binayak Bahadur Singh, Lt. Rudra Bahadur Poudel, and Private Nara Bahadur Khatri on board the chopper were wounded by shots fired by the Maoists.

It was the Maoists' first firing at the Army.

Tipped off by intelligence that the Maoists were heading to Nuwagaon where they presumably held the abducted policemen, army platoons were directed to it with orders to set an ambush on the Maoists.

The other two choppers ferrying two platoons were also hit by two shots in Salyan, and thus they were also grounded. There lurked a high risk for the platoon in Nuwagaon to be outnumbered and surrounded by the Maoists.

Every attempt made to send more platoons to the area failed due to inclement weather and the difficult terrains of the mountains. A company from Kapurkot marched to the area where a stand-off was going on but without firing from either side. The company linked up with the platoon.

After a couple of days, the abducted policemen were released, and they came in contact with the security forces.

On July 14, 2001, PM Koirala secretly flew to Biratnagar, his hometown, which he visited as and when he sensed a political problem. Back in Kathmandu the next day, he reportedly met former PM Bhattarai who advised him to first create an environment of unity in the party to move thence. In response to his request for help and his tentative decision to resign, Bhattarai reportedly suggested to Koirala that it would be a good idea for the two elderly NC leaders to stay in the Party as its guardians, thus leaving the space open for the young generation to take up the Party's leadership. By this proposition, Bhattarai seemed to refer to Sher Bahadur Deuba.

Koirala sounded rather worried in the meeting about the likely motion of no confidence vote against him in the Parliament. Feeling beleaguered, Koirala ultimately submitted his resignation on July 19, 2001.

There was the usual interplay of several partisan and political factors behind Koirala's resignation; the Holeri Incident was merely an excuse, a face-saving façade. If ethics was anything to go by, Koirala should have resigned much earlier, immediately following the Royal Palace Massacre.

Following Koirala's sudden resignation, politics took another turn, giving rise to speculations as to what might follow. Several game plans got underway once more.

According to casual information received, Koirala was mulling over Chakra Prasad Bastola as the next Prime Minister. The Central Committee of the Nepali Congress met to discuss the fresh political developments, and also formed a three-member election committee to choose the leader of the Party's Parliamentary Committee. The election committee comprised seasoned leaders like Bashu Rishal, KB Gurung, and Laxman Prasad Ghimire. Koirala had Sushil Koirala in his mind as an alternative to Chakra Prasad Bastola. The meeting discussed several possibilities and alternatives, and ultimately Sher Bahadur Deuba appeared to be the most competent candidate for the post. Because he was comfortable in every respect, be it his relations with the West, or with the Royal Palace, and with India, he, therefore, appeared to be a consensus candidate. The Royal Palace had already given him its green signal for the said position. He had also personally called me to ask my friends in the Congress Party to support him. I asked Narayan Singh Pun to lobby for him. The CoAS and other Generals also used their channels to create a suitable environment for Deuba.

As a result of all these campaigns, Deuba became the leader of the Nepali Congress Party, with 72 votes, leaving Sushil Koirala behind with 40 votes, out of the total of 113. This win opened for him the space to become the Prime Minister of Nepal for the third time.

The Late Night Dissolution of the Parliament

Not a single government in Nepal, since the very neo-democratic renaissance of 1990 and thereafter, was able to complete its full electorally mandated term of five years. Each was a victim of dirty political game plans of all sorts. The internal strifes and splits in the political parties themselves were another classic spectacle. For instance, even the Nepali Congress, the oldest and the biggest political party in Nepal, saw virtual splits within it, the balance of power swinging from Bhattarai to Koirala and then from Koirala to Deuba.

Soon after the assumption of prime ministership, Deuba formally initiated dialogue with the Maoists. The Maoists joined the peace talks as their tactical move to foment differences and create misunderstanding among the different political parties, while also to give an impression that they were not aligned with the Royal Palace. Under the cover of the peace talks, they also cultivated the leaders of the political parties, met the intelligentsia, and held discussions with the civil society activists, all in order to read their minds and gauze the opinions of the masses. They also used the hiatuses the talks created to reinforce their strength in weapons and training.

PM Deuba sought to extend the State of Emergency. But his party itself vetoed it. Therein was brewing up another game plan, and it was aimed at him. The Parliament, too, did not endorse his proposal.

Therefore, while the cancellation of the extension of the State of Emergency added complications to the maintenance of security, the size of Deuba's own Cabinet swelled up in his bid to keep the government going by accommodating as many Ministerial wannabees as possible in his government. The cumulative effect was the erosion of the people's trust in his government. Not surprisingly, rumors flew of a likely motion of a Vote of No Confidence against the Deuba Government in the Parliament, and it was allegedly floated by his own party, the Nepali Congress, this time in league with the CPN (UML). In fact, as was expected, from his very first day in office, a game plan to remove Deuba had already been hatched in the Nepali Congress.

It may be surmised that in a private meeting, King Gyanendra had assured Deuba of every possible help from the Royal Palace and the Army once he proposed to dissolve the Parliament. A tacit understanding was reached that the PM would also continue in office until the elections. I knew that the King, in his audience to the American, Chinese, and Indian Ambassadors, said it would be a good idea to dissolve the Parliament to break the ongoing impasse to give the country normalcy. The ambassadors were also known to have advised the PM to do so.

Accordingly, a letter from Prime Minister Deuba recommending the dissolution of the Parliament and declaring the date of the next elections was submitted to the Royal Palace for approval.

That very night, the King, in accordance with the recommendation of the PM, readily announced the dissolution of the Parliament with the date of midterm elections also announced. The National News Agency asked to remain on standby, aired the news late at night.

Following the unexpected turn of events, the Nepali Congress Party met and suspended PM Deuba even from its ordinary membership.

As he had received an encouraging nod of help from the King, he formed a new Congress Party of his own and laid claim on the election symbol of the party, the 'Tree' [*rukh*]. The case reached the court which gave its verdict in favor of the original side.

The October 4 Move

When all else appeared to have failed and nothing seemed to work to bring the country out of the doldrums, anarchy, and directionlessness, a situation brewed itself for a decisive move.

King Gyanendra took a fresh initiative to start consultations with political party leaders, intellectuals, journalists, civil society activists, and diplomats at Nirmal Niwas, his residence where he lived for a year following the Royal Palace Massacre. In the consultations, they expressed their worry over the steadily increasing Maoist insurgency, extortion, widespread killing, and poor governance.

However, the political leaders and their parties maintained double standards. I could sense that the King was anxious and contemplating to take an action in the interest of the country and the people. In the process, he met many leaders of all political parties, intellectuals, and others from all walks of life.

Although a bit hesitant earlier, in the days that followed after the Royal Palace Tragedy, the King increasingly gained his confidence and acquired a thorough knowledge of the real state of affairs in the country and its statecraft. In an interview, he expressed his desire to play the role of a 'constructive and visible King', a clear indication of his mindset.

There always was an interest group in the Royal Palace, at one time active and at other time dormant but persistently inciting the King to do something rather than merely watching and letting things get worse.

I counselled the king that whatever steps he takes should be taken cautiously and in pursuant to the Constitution. It was a time when the monarchy was under a question mark. Given the volatile and vulnerable situation, I cautioned the King that those with their intent to fish in troubled waters would rise to weaken the monarchy to serve their own interests.

I have no knowledge of any team of advisors formed in the Royal Palace except a couple of the King's close confidants, such as Sharad Chandra Shah, Prabhakar Shumsher Rana, Prabhu SJB Rana, Birendra Bahadur Shah, and the former CoAS, Gen. Satchit SJB Rana.

A few months earlier, the King had asked me to form a core team to work out a plan of a particular nature. Accordingly, the King granted audience to the heads of the security forces to give them clear instructions to stay alert and in a standby position, as he was going to make a move at the auspicious time of 10.45 o'clock on the night of October 4.

We, the four Secretaries of the Royal Palace, collectively prepared and discussed the contents of the draft of the Royal Address, word by word and sentence by sentence.

The Royal Address termed PM Deuba incompetent of holding the elections on the stipulated date for which he had recommended to the King to announce the dissolution of the Parliament in the first place. Instead, he postponed it for one further year.

The address also proclaimed the dismissal of Sher Bahadur Deuba as Prime Minister and the assumption of all executive authority by the King himself until another provision, of an elected government, would be in place. In his address to the nation, he further announced the postponement of the elections scheduled for November 12, the formation of a small and efficient Cabinet within five days with Nepalis of a clean image and without any intention of contesting elections. The King also called upon the political parties to propose such a candidate as Prime Minister.

After due hearing of the case for the reinstatement of the Parliament, the Supreme Court gave its verdict in line with the legality of its dissolution. It made space for the elections to be held.

But the President of the Nepali Congress and former Prime Minister, Girija Prasad Koirala, took a tough stand on the reinstatement of the Parliament. But, given the Maoist violence escalating all over the country, and the security situation being vulnerable, and political rallies for and against the royal takeover being held across the country, the environment for holding elections did not seem to be possible. Even the all-party meeting called by the Election Commission turned out to be ineffective in creating conducive environments for holding the elections as announced.

Meanwhile, the security forces were taking every detail of the security situation across the country. Prior to the October 4 move, all the security chiefs had met at the RNA Headquarters to make sure that there would be no threat from any quarter against the royal move. I had issued special instructions to all security organs concerned to strengthen their positions in the Kathmandu Valley.

Against the backdrop of the failure of the political parties to forward a consensus candidate to head the government, and the deadline for the formation of a government inching closer to expire, and with no consensus candidate in sight, the King chose Lokendra Bahadur Chand as Prime Minister.

Chand had upon himself the twin tasks of reviving the dialogue with the Maoists and creating a suitable environment to take the other political parties along with the government.

However, the Nepali Congress President and former Prime Minister, Girija Prasad Koirala, and the General Secretary of the CPN (UML), Madhav Kumar Nepal, seemed unwilling to oblige. IGP Pradip SJB Rana was informally persuading the CPN (UML) leaders, particularly Madhav Kumar Nepal and KP Oli, to join the government with the portfolios of either Deputy Prime Minister or Minister for Home Affairs. The King cancelled the slot for the Ministry of Home Affairs because Dharma Bahadur Thapa had already been given the portfolio.

Citing this as their reason, the CPN (UML) refused to join the government. Oli, however, showed his inclination to come aboard; but since not enough time was left with the King, that could not happen.

When the political parties declined to join the government, neutral individuals with a clean image and clear vision were asked to petition for the position of prime minister.

Keshav Sthapit preferred to help the government as the Mayor of the Kathmandu Metropolitan City with his revolutionary ideas to lure Maoist combatants with a package of interest-free loans to go abroad for employment. The likes of Kamal Chaulagain of the CPN (UML) were willing to join the government even at the expense of splintering the party, provided they were offered the position of the DPM and other key portfolios.

Interestingly enough, many politicians of almost all the major parties petitioned the King for the position of prime minister. Many others simply aspired for any ministerial berths.

My friend, Dr. Duman Thapa, who kept me posted on the developments in the country, informed me that the non-partisan Leftist leader, Padma Ratna Tuladhar, was offering to approach the Geneva-based Center for Humanitarian Dialogue (CHD) willing to broker peace negotiations in Nepal.

With the King's approval, I met the CHD group at the Shangri-La Hotel. I learned that the CHD, with a good track record of conflict management in different parts of the world, had approached the Maoist leadership for a peace initiative. Mr. Tuladhar had also approached the Prime Minister on the subject.

In an intensive talk, Tuladhar asked me to obtain the King's permission to proceed with the groundworks for the said peace negotiations. He explained the tentative modality and plan of the talk and added that the Danish Embassy in Kathmandu would bear all the expenses of flying the government and Maoist negotiation teams, with Tuladhar as the facilitator.

The groups would be secretly flown out of Nepal in a private aircraft to an undisclosed place in Switzerland with the assurance that the Maoist delegation would be given their safe conduct and passage to their addresses in Nepal in case the peace talks failed.

When I briefed the King with the details of the meeting, he mentioned of similar efforts being made at the Prime Minister's level that had actually moved one step further, to the preparation of the selection of the members respectively of the government and the Maoists.

The Prime Minister himself, from the very day in office, was making every effort to find a key person to establish a channel of communication with the Maoists. As part of the preparations, he had established contact with the Maoist leader Haribol Gajurel through a mediator, one Madhav Prasad Marahattha from Gorkha.

In the second round of meeting on November 10, the Maoist leaders were learned to have set a list of tough preconditions for dialogue:

1. Immediately remove of the 'terrorist' tag on the Maoists;
2. Reverse all the decisions of the erstwhile governments pertaining to the Maoists' so-called 'terrorist' activities;
3. Issue the formal call for dialogue from both sides;
4. Make public the names, status, and whereabouts of those detained Maoist cadres; and
5. Lay down the arms, and announce ceasefire from both sides.

When I briefed these fresh developments to the King, he instructed me to seek the opinion of the CoAS, too.

The CoAS sounded positive to hear about the developments but expressed his cautions as follows:

1. All Maoist activities must stop and cease making objectionable comments against the monarchy and the Army;
2. Make public the details on the detainees of the conflict from both sides;
3. No objections should be made by the Maoists to the regular management and transport of weapons and equipment of the security forces;
4. The regular functions of the security services will continue without targeting the Maoist combatants; and
5. The Maoists' extortion and torture of people must be stopped forthwith.

A Bracing Peace Endeavor

In the Cabinet expansion of November 19, 2002, such small-party politicians and non-partisan individuals as Narayan Singh Pun, Devi Prasad Ojha, Dr. Upendra Devkota, Deepak Gyanwali, Kamal Chaulagain, and Gore Bahadur Khapangi were inducted into the government.

At a meeting with me in my house, Narayan Singh Pun offered to initiate talks, as he claimed to have a channel of contact with the Maoists. I told him that it would be too early to tell him anything as a similar approach was going on in another form, and there were also many foreign agencies from Ireland, Italy, and South Africa involved in the initiative.

It occurred to me that there was a grain of truth in Pun's claim of having a grapevine for the reason that he had a long experience of service in the Army, and he could also arrange many things in confidential manners.

I had also known that the Maoists were desperate to join the mainstream politics with a face-saving and safe-landing mechanism in place for their purpose. So I had a long meeting with Minister Pun to work out a tentative plan.

In a briefing to the King, I updated him on the new developments. The King instructed me to first consult with the PM, and proceed only if he agreed.

Coincidentally, Sudhir Shah, my first cousin and an Undersecretary at the Ministry of Home Affairs, had been appointed Personal Assistant (PA) to the Prime Minister at Baluwatar, the Prime Minister's official residence. It proved to be instrumental in arranging things smoothly.

Consequently, the PM and I met to discuss how the government-Maoist talk could substantially be steered. As I was discussing the plan of action, I requested the PM if I could also call Minister Pun to the meeting.

Accordingly, the Minister of Physical Planning and Construction, Narayan Singh Pun, was summoned to the three-man meeting. We discussed the pros and cons of the prospective two levels of talks with the Maoists. The idea of the first option of talks through the mediation of the CHD looked good, but it rather appeared comparatively farfetched, given the cumbersome process of having to arrange an airplane from and to Switzerland, issuance of passports to the Maoist representatives, and bringing them back to their safe zones in Nepal after the talk.

The PM, therefore, gave Minister Pun his nod to reopen his channel to the Maoists to initiate the envisaged peace talks.

He later called to inform me that the talks with the Maoists were progressing smoothly. He added that they had suddenly asked him in the very beginning of the talk to convince them that he was the authorized representative of the Royal Palace to proceed with the talks.

As there were also other peace poseurs moonlighting as negotiators, I asked Pun to remind the Maoists that the Royal Palace could not delegate authority just as it pleased because it was the government's prerogative.

'So,' I said, 'At the most, I can issue you my letter as Military Secretary to the King attesting your role.'

The King was fully aware of the situation concerning the talks that Minister Pun was conducting with the Maoists.

Minister Pun spoke with the Maoists to this effect, and they agreed to the logic. So we drafted a 'To Whom It May Concern' letter and I signed it as Military Secretary. I also expressly asked him to bring the letter back to me once he showed it to the Maoists as proof because it was top confidential. It was not without risks to write such a letter without the knowledge of the King. Minister Pun said the Maoists took it positively and returned the letter after reading the content. I arranged his audience with the King for a briefing on the developments.

A couple of days later, the Maoists sought an official letter from the government to create a congenial environment of trust for the talks to proceed further. The PM wrote a formal letter to the Maoists, with the authority for Minister Pun to steer the dialogue to its logical end. In all these developments, only the King, the PM, and I knew what was going on. The entire formalities were conducted in a completely confidential manner.

Just when the government-Maoist talk was progressing, the second in command of the Indian intelligence agency, RAW (Research & Analysis Wing), JK Sinha came to Kathmandu, ostensibly on a personal visit. But he met the leaders of the parties and the officers of the security forces. In the meetings, he expressed his concerns over the length of time taken for the end of the ongoing situation. He also mooted the possibility of the mainstream political parties joining hands with the Maoists against the monarchy, and wondered about the likely leadership leading such a joint movement. He showed his interest in the agenda of the peace talks, and wished to know about the relations between the Maoists and the

Royal Palace. He also dwelt on KP Sharma Oli's ouster or stay in the CPN (UML). He spread other feelers as well.

It seemed to me that Sinha overstepped his limits to feel the pulse of the relations and coordination between the security officials and the political parties. He also recommended the replacement of PM Lokendra Bahadur Chand by Surya Bahadur Thapa or Rabindra Nath Sharma as PM Chand was considered too close to the Royal Palace.

He met the Chief of the NID and pledged help in the installation of telephone tapping equipment there. It sounded that the mission of the RAW deputy chief was directed at weakening the relations among the political parties and the security agencies, thereby to foil the peace talks in order to create an environment of the indispensability of Indian mediation in order to serve its interests. He expressed his concerns about the status of the government-Maoist peace talk through the mediation of the CHD.

On the ongoing peace negotiations with the Maoists, the talk team's facilitator/Minister Pun updated the King with the developments. Sometimes I arranged his visits, at other times the Principal Secretary, Pashupati Bhakta Maharjan, did so. The talk was progressing in utmost secrecy, with only four people knowing about it: the King, the PM, me, and Pun.

One evening, after Minister Pun finished briefing the King on the progress of the talks, the King and I discussed the legal aspects that could emerge. Because we were not competent enough in such matters, the King then decided to call Principal Secretary Pashupati Bhakta Maharjan in to include him in our deliberations. We discussed legal matters and their implications.

It was the first time that Principal Secretary Maharjan came to learn of our secret project and the progress it was making.

The very next day, Minister Pun called me, wishing to meet me immediately, to discuss a serious matter. I met him at my place, taking time off from my office. Sounding serious, he quoted the Maoists as saying that the former Minister of Foreign Affairs, Ramesh Nath Pandey, wrote to them to the effect that Minister Pun was not competent enough to be a party to such a talk, as the King only trusted him (Pandey). In the letter, he also claimed that the King listened to him, and therefore begged for the authority to negotiate with the Maoists be given to him.

The Maoists' disclosure came as a surprise, and we thought over the ways to deal with it. It occurred to me that the move by Pandey – in league with Principal Secretary Maharjan – was likely set to demolishing the ongoing negotiations.

We had no inkling that Maharjan, being no less than the Principal Secretary to the King, would go to such an extent of leaking such classified matters to his close advisor and mentor which Pandey was. But Pandey's questionable character was known to many anyway.

The news of the ongoing confidential negotiations reaching the media was by no means good and proper in form. Thus the peace process experienced complications. I asked Minister Pun to talk to the PM about it, and when I briefed the King on what was transpiring, he also sounded surprised. But he promptly instructed me not to lose hope at such an occasional misstep, and I conveyed the King's message to Minister Pun to carry on his mission.

After much discussion on the fundamentals and the modality of the peace talks, the code of conduct in a foreseen ceasefire, and the names of the possible interlocutors to be selected for the peace talks, the process reached its final form.

As the king was busy preparing for the fast appraoching wedding of Princess Prerana, the King's only daughter, the peace process could not progress as quickly as it should have.

But Minister Pun was in regular contact with the Maoist leaders, such as Krishna Bahadur Mahara and Ram Bahadur Thapa. They had formed their talk team headed by Dr. Baburam Bhattarai.

Amidst all this, the murder of the IGP of the APF, Krishna Mohan Shrestha, and his wife by the Maoists threatened to disrupt the painstakingly built atmosphere of trust and talk. However, with due seriousness and restraint, both sides displayed a sense of proportion through the announcement of a ceasefire to create a congenial ambience for talk although many in the security forces held doubts about the Maoists' actual intentions.

Minister Pun was delegated to maintain an air of trust with a thorough observance of the code of conduct by both the government and the Maoists. The Maoist top leaders, 'Prachanda' and Dr. Baburam Bhattarai, sought a meeting with the King himself to better read his intentions and plans ahead. Minister Pun conveyed me that the Maoists' will accept monarchy if the King agreed to the elections to a Constituent Assembly. The mainstream political parties shied away from the idea and forgot about the Constituent Assembly elections.

Minister Pun had prepared a document with 15 points to be discussed in the meeting between the King and the Maoists if and when it occurred. The King consulted me about the contents therein. We amended the draft.

The clauses in the document sounded far more progressive and advanced in attitude than the Maoists would expect. In a briefing, I explained to the King that since by then the delegate members were already appointed for the state-Maoist negotiations, we needed to finalize the points of issues to be negotiated with the Maoists. He instructed me to note every detail they wanted to discuss, and brief him accordingly.

After reading the document, the Maoists looked most eager to meet the King. I don't know if it was a rumor or a fact that a certain place in Balkot of Bhaktapur was chosen by Minister Pun as the venue for such a meeting.

I told Minister Pun that before the Maoists would meet the King, I should see them first and brief the King on it. The King agreed to the idea and Pun likewise arranged my meeting with the Maoist leaders, Krishna Bahadur Mahara and Ram Bahadur Thapa, at his Karnali Airlines office at Sinamangal, next to the Tribhuvan International Airport.

As befitted my position and responsibilities, I discussed several issues with the leaders. I explained the King's wish for peace and harmony in the country and asked them to contribute to the same goals as a common minimum meeting point.

They reiterated their wish to talk to the King about everything, in full. Ram Bahadur Thapa did all the talking. I asked him to take into consideration the country, its societies, geopolitics and Nepal's location as a buffer.

They asked me to convey to the King their demand for dissolving the present Cabinet to form an interim government to hold the elections to the Constituent Assembly.

It now seems to me that had the then government accommodated the Maoists' old preconditions, many of the problems of those times would have been resolved in one single sweep, and the situations facing the country today would not have grown as complicated as they are now.

The mainstream political parties always feared the elections to the Constituent Assembly. Therefore, they pushed the Maoist propositions under the carpet. They shortsightedly read it as a move to turn Nepal into a republic.

I recall the moment when the King wondered, 'Vivek, what's so wrong in announcing the elections to the Constituent Assembly?'

I think that if the leadership had been accommodative about forming an all-party government to hold the elections to a constituent assembly, the country would never have to pay such heavy prices. It would have brought an end to the Maoist conflict, and the politics of the country would also have taken a definitely better course.

The government formed yet another team to negotiate with the Maoists, this one headed by DPM Badri Prasad Mandal, and having Ramesh Nath Pandey, Narayan Singh Pun, Dr. Upendra Devkota, and Anuradha Koirala as members. The new team was asked to prepare its groundworks which Minister Pun had done earlier.

At the unexpected turn of the situations, Pun must have felt bitter being a mere member because he was entrusted with a limited scope of authority to deliver results. With Ramesh Nath Pandey prevailing over the team, complications of personality clashes and an environment of distrust took roots in the members, rendering the team unable to speed up its mandated functions.

Just when the government negotiation team was set to go to the meeting being held at Shankar Hotel, Minister Pun called to inform me of yet another demand made by the Maoists.

The demand, utterly unacceptable, called for the restriction of the Army within the perimeter of five kilometers from the barracks during the peace negotiation.

That morning, the King had gone to the Dakshinkali Bhagwati Temple for his annual puja, so I could not contact him immediately. But I contacted the PM and the CoAS to advise them that a national army could not be confined to the barracks and told to remain within a certain boundary. Army units can go anywhere without any hindrance.

No wonder I was shocked when the CoAS, General Pyar Jung Thapa, told me that it was reasonable to restrict the army not just within five km, but even 20 km, of the barracks! I reminded him that it was against the norms of the National Army.

The new demand created confusion for the members of the government's negotiating team. The PM sounded perturbed by the new demand. I advised him that it would be wise to consult the CoAS as a member of the NSC and also duly convey the news of the new turn of events to the King.

I called the CoAS to ask him not to accept the Maoist demand of restricting the space and the movement of the Army. I knew that the PM had also talked to him regarding the matter. I also heard that the CoAS gave his nod to restrict the Army to a perimeter of 20 kilometers.

I could not believe my ears!

In a press meet after the first round of dialogue, Krishna Bahadur Mahara, a member of the Maoist talk team, mentioned the tacit understanding with the reference to the five-kilometer quarantine. Minister Pun maintained his silence on the matter, as if to prove that 'Silence is half consent.'

The security chiefs held a meeting with the government talk team. The CoAS expressed his objection to the understanding to restrict the Army within the perimeter of five kilometers. Maj. Gen Rookmangud Katawal termed the understanding 'erroneous.'

However, Ramesh Nath Pandey, a member of the negotiation team and a Minister in the King's Cabinet, remarked that since everything had been already agreed on, there was no point in speculating otherwise or magnifying the issue.

Minister Pun never bothered to explain as to who and why such an objectionable consent was given.

Perhaps Prime Minister Chand, a man known for his quiet, amicable and easygoing nature, experienced complications at the new turn of events, and he had no authority to select the members of the state's negotiation team of his own free will.

As expected, the talks could not make any headway, and as a result of the standstill, Prime Minister Lokendra Bahadur Chand resigned on June 2, 2003.

Following the PM's resignation, the prestige of the office of the Prime Minister further lost its glory and dignity when the Royal Palace itself called for applications from all interested Nepali nationals for the position, as if it were a berth to be petitioned for, as for a job application.

People from the likes of the CPN (UML) General Secretary, Madhav Kumar Nepal, down to every Ram, Shyam, and Gopal submitted their applications to the King. With his papers filed, Madhav Kumar Nepal became a laughingstock in town.

When I shared my idea of the prospective PM in Taranath Ranabhat, the Speaker of House of Representatives, the King said he already had Surya Bahadur Thapa in his mind for the opposition. In his opinion, Thapa could bring the CPN (UML), or the Nepali Congress, or both parties, to join the government.

At this precise time, Surya Bahadur Thapa had an audience with the King and had gone to India for a few days. When he came back, he met the King again to assure him that political parties would join the government formed by him.

We, the Secretaries at the Royal Palace, did not like the idea and so opted to keep quiet. With blessings and pledges of help from key players, and presumably he also expressed his own confidence, he became PM, in accordance with Article 35(2) of the Constitution but without the House of Representatives in place.

Around the same time, five political parties, embittered at being quietly segregated from power politics, were constantly staging daily protests and agitations at Ratna Park. However, the responses and participants' turnouts were slender, with poor turnout.

With the major parties spoiling for a fight, it was obvious that the government was feeling jittery. The PM, however, spared no efforts to earn the support of the Nepali Congress which had softly pledged to support the government. But, later on, they began calling for the reinstatement of the Parliament.

The King had given the PM, at the moment of his assumption of the post, three mandates: to take all the parties on board the government, to resume the Peace Process, and hand the authority over to the elected representatives.

No matter how much he tried, PM Thapa, however, could not accomplish any of the tasks. Nor was he able to coax the parties into joining his government. The Peace Process remained like an unclaimed baggage, and Maoist violence escalated again.

At one briefing, I reminded the King that the current state of affairs reflected a condition of statelessness, with the situation deteriorating every day. I also added that the lackluster performance of the government would, in turn, cause the erosion of his own image.

Although the PM formed yet another team of facilitators and negotiators comprising leaders of the RPP, with Dr. Prakash Chandra Lohani and Kamal Thapa in it, the Maoists did not show much enthusiasm for it.

In the audience granted to the negotiators, the King authorized them to concede the demand for Constituent Assembly elections with certain preconditions, among them the chief being the settlement of the problems to ensure peace in the country.

However, in the meeting with the two political party leaders – Girija Prasad Koirala of the Nepali Congress, and Madhav Kumar Nepal of the CPN UML – both opposed the idea of the elections to the Constituent Assembly. They feared that such elections would wipe out the achievements of the 1990 political changes.

Additionally, too, because the new talk team had a fixed agenda with preconditions and without any room for flexibility in the peace talks, the Hapure negotiation round fizzled out. It seemed that there was a sinister design from abroad to scuttle the Hapure peace talk.

Given the failure to deliver any of his commitments, Surya Bahadur Thapa was ethically expected to resign as Prime Minister, but he displayed no such virtue.

I had advised the King to have Narayan Singh Pun, former Minister and facilitator of the initial government-Maoist talk team, included in the Cabinet as well as in the government's negotiation team because he had already proven his worth by achieving a measure of progress in the previous peace overtures with the Maoists. He had developed his own working relations and trust with the Maoists. But his inclusion did not happen, for obscure reasons.

I heard that the PM, on November 6, 2003, briefed the King on the transfer of some DIGs (Deputy Inspector General of Nepal Police). I did not know about the other items discussed in the meeting.

The next day, quite early in the morning, and all of a sudden, the King asked for my resignation. I later heard a rumor that the PM had complained against me as interfering in the transfers of the said police officers. I sensed that some power coteries in the Royal Palace and outside of it were not comfortable with my being the Military Secretary to the King.

With the failure of the Hapure peace talk, the Maoists resumed their violence, including targeting the RNA. They shot Colonels Kiran Bahadur Basnet and Ramindra Chhetri at their doorsteps one morning, killing Basnet and injuring Chhetri.

The government, too, on its part, massively mobilized the Army and the Police against the Maoists. The agitation of the major political parties also reached its highest scale. Left with no alternative, the PM resigned in June 2004.

The Speaker of the House, Taranath Ranabhat, allegedly and already attired in formal Nepali apparel, was on his toes to the prospective appointment as the next PM. But when the Royal Palace's Principal Secretary, Pashupati Bhakta Maharjan, called him to resign his portfolio of Speaker as a precondition, and when Ranabhat refused to do so, his chance of becoming PM vaporized.

Sher Bahadur Deuba, only recently labeled 'incompetent' more than once by King Gyanendra in his royal takeover declaration, was again called upon to become PM.

Before long, the King sacked the PM and took over the entire state powers in his own hands, on February 1, 2005.

King Gyanendra's advisors gave wrong counsels to him. And by so listening, accordingly, he moved anticlockwise, more so in the changed contexts of the 21st century, in assuming the active role of Chairman of the Cabinet in addition to being King.

In fact, it was the consultants to the King who actually sowed the seeds of republicanism in Nepal in the truest sense of the term, and thus consigned the monarchy itself to oblivion.

Thus, another fateful game was played out in the Nepali history's sequels of the Basnets conspiring to annihilate the Pandes, and the Thapas eliminating the Basnets, and in turn, the Ranas wiping out certain Thapas, and similar other power grabs taking place in the political history of Nepal in which the power-hungry cliques swung into actions against each other in their deadly mortal game for thrones. Nepalis were and are being thus destined to witness the everlasting subsequences of conspiracies and political game plans for power.

I wonder as to how long yet we must wait for the country to shake off this culture of dirty political schemes to give space to reliable and accountable governing systems to serve the Nepali peoples' interests through peaceful political processes and for the progress of Nepal.

CHAPTER: TWO

PAGES OF REALITY: THE GENERAL'S JOURNALS

The Undercurrents that Culminated in the Present States of Affairs in Nepal

The Office of the Military Secretary was one of the key agencies in the Royal Palace. The overall security, in general, and the entire administration in the country, communication and coordination with the government Ministries and the Departments thereof, and resolving security-related problems, in particular, fell under the scope and the ambit of the office I served at.

This part of the book in your hand charts out the annotations of the events taken place during my tenure of service at the Royal Palace (Narayanhiti Raj Durbar) of Nepal in Kathmandu for more than three years, from April 2000 to November 2003.

Because the book is based on the entries in my diaries, I don't claim it to be a replica of the entire conditions and events of the time and period. It represents only those incidents that caught my attention, and which came to my notice and knowledge during my years of service in the Army and the office of the Military Secretary.

I've tried my utmost best to do justice to the events to present them in their 'as was where was' manner without any kinds of fabrication, additions or deletions.

I've already said that the book isn't written to please or displease anyone; it's merely my honest effort to record the events as historical documents.

The Incidents that Unfolded in the Year 2000

After the general elections of 2000, the rifts within the Nepali Congress widened, bringing on the surface sinister schemes of power politics, catapulting one clique to power, and another interest group pulling it down, to be themselves toppled very shortly and eventually by the same old or the other clique.

By that time, too, the Maoists, though weakened, still frightened the Police Force. The political leadership, ignorant or acting as such, sought to mobilize the Army against the Maoists, just like it was in the deployment of the police, without fulfilling the due constitutional procedures.

There also appeared an environment of confusion and misunderstanding between the King and the government over the question of the Army's mobilization to address the Maoists. The leaders of the Nepali Congress then in power, and unable to mobilize the Army, as expected, resorted to their blame game, even to the extent of saying, 'There's a sinister nexus between the Royal Palace and the Maoists, for they are operated by the Royal Palace itself.'

King Birendra read the Maoist problem as the byproduct of socio-economic and political discriminations and deprivations that demanded to be addressed politically. According to him, government/s must give serious thought in that direction, rather than setting one Nepali to kill another Nepali. Security operations should go side by side with development packages, and delivery of services, and with good governance. Killing was no solution to the problem.

The mobilization of the Army wasn't the same thing as the previous deployment of the Nepal Police. It required the government to promulgate the necessary Acts of the Parliament for the purpose.

The Army also required training, sophisticated weapons and logistics to fight the Maoists. The Maoists cashed on the delays and confusion within the Royal Palace and the government, and thereby increased their strength to the scale of even attacking Dunai of Dolpa District.

The APF (Armed Police Force) was formed as an alternative paramilitary force. There was also a pressing need for the activation of the NSC (National Security Council).

By that time, criticisms of the poor security system in the land, the power-hungry attitudes of the political parties and leaders, the recurrent political deadlocks, poor governance, and other chronic ills were being voiced louder every single day. Pleads of various kinds from several quarters were made to the King to make a timely political move to alleviate the situation. He indeed seemed to be contemplating on the ways and means in his hands to do just that.

Wednesday, April 19, 2000

Today is my first day in the office of the Principal Military Secretary. In my first briefing to His Majesty King Birendra Bir Bikram Shah Dev at 11 a.m, I'm also set to give him my short introduction.

I expressed my sincere gratitude and firm commitment to fulfill my responsibilities to the King and to the Institution of the Crown for having entrusted to me the position of Military Secretary.

As planned, I presented my brief introduction that included my birth, family background, address, schooling, recruitment in the Army, my sincerity to duty, my love for my profession, respect and loyalty to the monarchy, and my straightforward nature of expression, etc. Having heard me, he said, 'That's precisely why I've appointed you Military Secretary.'

My briefing began. In reply to my first item, he instructed me to send five non-combatant 'pipa' men to be at the service of His Royal Highness Prince Paras at Nirmal Niwas.

I then briefed the King on the Australian Ambassador's request to confer upon Crown Prince Dipendra the title of 'High Altitude Chamber'. I said it would be decent for the Crown Prince to accept it and send it over to the High Altitude Training Center of the Army.

The British Foreign Minister, Robin Cook, is visiting Nepal from April 20. He's scheduled to meet the Minister of Home Affairs to express the British Prime Minister's concerns over the objectionable activities of some ex-British Gurkhas resorting to extortion on their members.

The civil servants concerned are seeking the King's permission for staging a demonstration of Disaster Management in his presence.

The Army is ready to submit to His Majesty their concept paper on the establishment of the National Security Council, covering the need for an office and other essentials.

The Army has obtained the information that 'Prachanda' – the nom de guerre of the Maoist supreme leader, Pushpa Kamal Dahal – is a member of the Revolutionary International Movement (RIM), and Mohan Baidhya is receiving medical treatment in Siliguri in India. It's also known that they are planning to attack a government office somewhere to commemorate Lenin's Anniversary on April 22.

I then briefed the King on the need for the maintenance of the communication system in the office of the Principal Military Secretary from the funds held in the office's bank accounts. He granted his permission to go ahead.

My next briefing to him concerned the CPN (UML) General Secretary, Madhav Kumar Nepal's recent visit to Delhi under the cover of medical 'treatment', the costs of which are to be borne by the Government of India. He said it wasn't new for the leftist and democratic leaders of Nepal.

Barely one year after his election to the House of Representatives Narayan Singh Pun has resigned his membership in the Nepali Congress. He's allegedly fed up with its leaders' unfair power politics. He wants to found his own party.

Monday, April 24, 2000

The Crown Prince is attending the basic course of parachute jumping with keen interest. He's to take the first of the seven jumps in the dropping zone on the banks of the Manohara River on Wednesday, May 3. The course will be followed by a Wing Parade where he'll be conferred on with the Para Wing by the CoAS in the Royal Palace itself. The King commanded me to arrange for Their Majesties to watch the first parachute jump at the site.

There's information of the Local Development Officer (LDO), Dirgha Bahadur Pokharel, who is charged of corruption, being smeared with soot and garlanded with shoes by the cadres of the CPN (UML) of the Kalikot District.

It's learned that there's Maoist investment in three schools: the Gyankunja Higher Secondary School, the Fluorescent Boarding School, and the Greenland Children's Academy.

The King instructed me to prepare a work plan on the possibility of using the ex-army's organization in various fields, and the modality of assistance to be given to them from the Welfare Fund of the Army.

He also instructed me to update him on the developments of disaster management so far.

He also expressed his worries and concerns over the discrimination against the lower-caste [now called Dalit] population and for their unhindered entry to and worship at the temples and shrines across Nepal. His instruction to me was to tell the government to seriously work to that effect before it would be too late.

Sunday, April 30, 2000

Today, the CoAS shared his experiences with me about the decision of the first ever meeting of this kind in the office of the PM. There was no provision for the CoAS to attend the meeting before this. It discussed the ongoing reconnaissance of the Army deployment with one Company each in the districts of Salyan, Kalikot and Jajarkot, with adequate budgetary allocation. Other members at the meeting emphasized on the transfer of quite a large number of soldiers to the APF. The CoAS said it would be a good idea to leave the decision to the soldiers themselves as it concerns their free will and choice. The meeting reached a decision to form a task force led by the DPM to prepare a detailed work plan to that end. The meeting also discussed the development of positive attitude for help and of good faith from the government, political parties and the civil society to the Army. The CoAS asked me to schedule an audience with the King for a briefing on the organization of the NSC.

The King granted time to inaugurate the conference of the World Hindu Federation on May 9, 2000 in Pokhara.

There's information of a likely protest demonstration by the INF [Indigenous Nationalities Forum], an indigenous ethnic group.

Tuesday, May 2, 2000

The Honorary ADC, Maj. Gen. Bharat Keshar Simha, has sought a leave with the permission from the King to go to Nagpur of India in response to an invitation from there. According to a source of information, he's scheduled to visit Ashok Singhal of the Bharatiya Janata Party, and the latter is to visit Kathmandu soon at a future date.

The Army has submitted to the King the proposal for the policy and program for the foundation of the NSC complex, seeking an appointment for a briefing.

There's news of as many as 40,000 Sri Lankan army troops being surrounded by the LTTE rebels in the Jaffna Peninsula of Sri Lanka. The Sri Lankan Government is reported to have requested India for help in the rescue and return of the soldiers. But India has pledged only non-military help.

The Pakistani participants in the NDC (National Defence College) Course, currently on a visit to Nepal, called on the Secretary of the

Ministry of Defense and discussed mutual goodwill, and the problems of India, Pakistan and China in the SAARC [South Asian Association for Regional Cooperation].

The Secretary briefed the visitors on Nepal's contributions to the UN Peace Keeping Missions in several parts of the world. He expressed Nepal's desire to expand the UN Peace Keeping Mission Training Center at Panchkhal as a standard training center for South Asia, and requested Pakistan to help the plan.

Thursday, May 4, 2000

The long-awaited and much-talked-about concept paper for the foundation and operation of the National Security Council (NSC) Office prepared by the Army has been forwarded to the King. The operation of the office is expected to recommend to the PM the solutions and offer advice on the programs of national interests and security-related issues.

The CoAS, Gen. Prajwallala SJB Rana, and the DGMO [Director General Military Operations], Pradip Pratap Bom Malla, were present when the DMO [Director Military Operations], Col Dilip Rayamajhi, the main architect of the concept paper, briefed the King on it. There followed a discussion on the reforms to be made therein. The Army proposed for the structural foundation of the Center for Research and Plan of the National Security Bureau, and the National Security Policy Implementation Committee under the NSC for the sake of clarity and efficiency in work. The King said the structure of the office should be small, have specific objectives, scopes and functions, and efficient work performance, without any executive authority therein. The NSC is to frame a research-based work plan for the overall national security and advise the PM on ways to tackle the problems.

The King remarked that an apt name for the office could be the 'Directorate of National Security Council'. He gave me instruction to proceed with the work in coordination with the architect of the scheme, Col Dilip Rayamajhi.

There's a report of suicide committed by Rudra Prasad Dahal, an Undersecretary serving at the Royal Palace Service.

Thursday, May 11, 2000

The IGP, Achyut Krishna Kharel, telephoned me to say he's hurt by the CoAS's rudeness in their meetings on security matters. He objected to the CoAS's unbecoming manners. When I briefed the King on this, he said, 'Prajwalla still has the hubris of the Rana ruling family.' He instructed me to ask IGP Kharel not to mind it.

His Majesty said it would be a good idea to have a new Army Act rather than just making amendments to the one in existence.

He also instructed me to talk to the CoAS about the due process of budgetary allocation for the purchase of the RJ-Aircraft.

Tuesday, May 16, 2000

There's a provision for my office that there should be a Lt. Gen. as the Principal Military Secretary, and a Major Gen. as Military Secretary, and the rest of the staff to be appointed from the administration department of the Royal Palace.

Since 1996, there's been the practice of having a Lieutenant Colonel from the Royal Nepal Army Headquarters in a deputation to the Royal Palace as its Liaison Officer. Likewise, Lt. Col Rana Dhoj Limbu in 1996 and Lt. Col Dambar Singh Gurung in 1998 served at the post as per Army regulations. Today, Lt. Col. Kumar Budhathoki came to relieve Lt. Col Gurung.

A process of disposing of the Army's obsolete weapons, and with a reserve of 25% of it in savings, and the purchase of more sophisticated weapons with the money made from the sale, has gathered momentum.

There's a fresh report of a clash between the Hindu and Muslim communities in Birgunj. Similarly, there have been occasional frictions between the Christian and Buddhist communities in Dhading, too. It's by no means a good omen for a country long known for its religious harmony.

Wednesday, May 17, 2000

The preliminary report on the formation of a new security force has been forwarded to me for comments and suggestions, if any, by the member of the study commission, the former IGP, Moti Lal Bohara. I'll study it thoroughly to brief the King on it. The concept of having such a force isn't to replace or weaken the Army. Therefore, it needs open discussions in the NSC. I advised His Majesty to discuss the concept of the report with the PM when the latter calls on him. He instructed me to seek suggestions from the CoAS, too.

The PM said he was going through the report, and will ask the CoAS to submit it to the King. He added that henceforth, he'll discuss security issues with the King prior to forwarding them to the Cabinet for deliberations and decisions.

The PM sounded suspicious over the concept of the NSC Secretariat as if it looks more in the interest of the Army.

The King instructed me to ask the CoAS to brief on the details of the report to the PM.

Monday, May 22, 2000

India's Foreign Secretary, Lalit Man Singh, is in Kathmandu to consult with some key persons over several matters about the current situation in Nepal. This calls for a detailed analysis.

The Army has prepared its workplan, known as the Internal Security & Development Project (ISDP), for resolving the Maoists problem, and facilitating a procedure of mobilization of the Army. It's a long-felt need, a common practice in the world, and a fitting model for Nepal. According to the work plan, the Army shall reach the people with development packages and win their trust with services according to their needs and interests. The work plan is hoped to create an environment of working together for the common good in the local areas, to give an impression that the Army comes to help the people when so needed.

Thursday, May 25, 2000

As planned, at a special function held inside the Tribhuvan Sadan of the Royal Palace, the CoAS conferred the Para Wing on the Crown Prince on his successful completion of the Para Basic Course. The program was followed by dinner.

There's information of the Maoists holding a mass meeting in Rolpa with people from the four villages [VDCs, or Village Development Committees]. It's reported that they have a policy of imposing a ban on the teaching of Sanskrit, seizing of the questions papers and obstructing the annual examinations in the schools which are to begin soon. They are said to occupy the land of the displaced to start collective farming there.

It's also learned that the Maoists have redoubled the strength of their militia, expanding from squads to platoons and from platoons to battalions, each squad containing 25, and each platoon with 120 combatants.

They have also reportedly launched a five-point campaign to expand the Maoist Party to every nook and corner, and to redouble the strength of their combatants to the scale to confront the Army, to best exploit the media for publicity, to reenergize and strengthen the Party further, and to infiltrate the government security forces.

There's is a fresh report of Prachanda's return from his participation in the meeting of the RIM in London.

Another source has it that the Maoists are preparing for their party meeting in New Delhi on January 1.

It's also been reported that the Maoists' next move is to cripple the Kathmandu Valley from four sides with every move within their capacity.

Another intelligence report reveals that the Maoist Communist Center's (MCC) training camp in the Chakrata forests of Dumariya in the District of Hazaribag of Bihar in India has been training Nepali Maoists from time to time.

The Saudi Ambassador to Bangladesh, currently on a visit to Nepal, is learned to have contributed cash assistance to some madrasas here.

An Indian Hindu fundamentalist organization, with its newly formed Universal Security Guard of Nepal, and by inducting Nepalis in the [Nepal-India] borders, is learned to be active to curb the activities of Pakistan's ISI (Inter-Services Intelligence).

The Secretary at the Ministry of Defense, Keshavraj Rajbhandari, called me about his wish to visit Pakistan, a plan earlier postponed following the change of government.

He also informed about India mounting its pressure to obtain Nepal's permission to set up and operate its pension camps for the ex-Gurkha soldiers of the Indian Army in ten more places whereas 14 of them are already in operation.

The Secretary also asked me to arrange his audience with the King.

Friday, May 26, 2000

IGP Achyut Krishna Kharel asked me to arrange his audience with the King for a very important briefing about weapons for the Police. He's also eager to update the King on the current situations.

Brig. Gen. Gajendra Bahadur Limbu has been nominated for the National Defense University Course (NDC) in China. Earlier, the CoAS wanted Colonel Dilip Rayamajhi for the seat.

The concept paper on the structural foundation of the NSC Office calls for yet another round of detailed discussion on the respective concepts of the Security Council, and the Security Council Secretariat.

Thursday, June 1, 2000

The incessant torrential rains in Betun of East Timor have claimed the lives of 43 people and displaced over 20,000 others. The Nepalese Mission has been requested for relief packages for the flood victims.

According to a tipoff, the Maoists are menacingly intensifying their cash extortion from people in many areas.

The ex-British Gurkha Soldiers' Organization Nepal (GESON) has intimidated the paymasters at the British Gurkha pension offices to extort money from the retired soldiers who came to collect their pension.

In response to the CoAS's briefing on the fresh spate of Maoist insurgency, the King gave instruction to strengthen the security in the National Parks and Reserves.

His Majesty also asked him to follow the due procedure by way of discussing at the NSC the purchase of weapons for the Nepal Police, and to talk to the PM about the capitalization and utilization of the Army Welfare Fund for the development of the Army.

He also sounded worried over the occasional corruption scandals in the Army, and asked the CoAS to address it.

The CoAS briefed the King on the need for weapons for the Army in case of its mobilization any time against the ever-increasing Maoist insurgency.

Monday, June 5, 2000

The Public Accounts Committee (PAC) of the Parliament has pointed out the weaknesses of the Army in the proper utilization of its resources, its needs to control of corruption in the higher ranks, and its long-term plans.

A conference of the top-level security officials from the Royal Palace Service, the Brigade at the Royal Palace down to the Battalions and Companies, and including the ADC Office, has been organized to discuss the problems facing the security forces and the strategies to solve them.

According to a source, Gajendra Narayan Singh and Hreedayesh Tripathi of the Nepal Sadbhavana Party (NSP) visited Brajesh Mishra, ostensibly on a personal visit to Nepal, at Hotel Yak & Yeti. Mishra is the Principal Secretary and National Security Adviser to the Prime Minister of India.

Wednesday, June 7, 2000

At a briefing to the King, the PM shared his groundwork of collecting suggestions about the concept of the structural foundation of the NSC from the Chief Secretary and the Secretary at the Ministry of Defense, and also consulting with the CoAS to build the infrastructure to mobilize the Army with a limited role of facilitation in all the districts with a view to improving the security situation.

In his audience with the King, Brajesh Mishra expressed his concerns about the increasing activities of [Pakistan's] ISI in Nepal, its status of relations with the Maoists, Nepal's tilt to China, and the Citizenship Bill pending in the Parliament.

He expressed India's pledge of support Nepal and also asked to specify the types of weapons and helicopters needed by Nepal, and the modality of the consignment of the materiel to Nepal.

The King commanded me to take Mishra's concerns most seriously.

Monday June 12, 2000

Heavy rains have caused the loss of life and property in the far western Nepal. The government has dispatched the army for rescue operations and the distribution of relief materials to the flood victims.

The embankments from the Indian side caused inundation and surge of water over the land in four villages of the Banke District.

There's a tipoff on the abduction of the District Education Officer

(DEO) of Rolpa by the Maoists. Every attempt has been made for the immediate release of the abducted.

In a bid to replace the obsolete SLRs (Self-Loading Rifle), the Army is testing the models and qualities of some sample weapons from different countries. Two sample sets of the G-36 from the Heckler & Koch Company of Germany are arriving in Kathmandu in a day or two.

Thursday, June 22, 2000

The Public Accounts Committee (PAC) of the House of Representatives has issued a directive to the government to have the purchase of an Avro RJ-100 airplane for the Army cancelled. I briefed the King about it.

In their tactical move, the Maoists seem to be launching their sixth round of program from June 15 to utilize the peace talk to their advantage to expand their organization, raise funds and provide orientation to their cadres.

By spreading rumors that Nepal is intent on breaking the embankments constructed by India, playing ignorant of ISI's transaction of counterfeit Indian currency notes and investment in the madrasas, India is promoting anti-Nepal activities around the borders. It's imperative for Nepal's political leaders to talk to their Indian counterparts to refute such false allegations.

The Tamil Nadu State Government of India supports the separatist movement of the LTTE rebels in Sri Lanka whereas India's Central Government is opposed to it. It's feared that a similar divisive fate may strike India, too. One can see Sri Lanka's deft play of the India card.

Monday, June 26, 2000

In connection with his health checkup, the King is scheduled to fly to London on June 30.

The CoAS briefed him on the endorsement of the concept paper prepared by the NSC meeting, with the Nepali Congress taking it in a positive light, and happy about the good turn of events for the mobilization of the army in action against the Maoists.

He also briefed the King on the Maoists' seeking to meet him through a channel of the Army. He added that it would be a good idea to know their views through a senior military officer. The CoAS added that the preparation of the Army's website is reaching its completion.

Thursday, July 6, 2000

An in-law of the Royal Family, Kumar Mohan Bahadur Shahi is learned to have quietly met PM Koirala at his residence. The purpose and content of the meeting have been kept confidential. I briefed the King on it.

There's a tipoff of a large number of Maoist rebels moving over to the east, presumably to receive some financial assistance from the Gorkha National Liberation Front of Kharsang in Darjeeling.

The Nepal Police is demanding for weapons such as SMG with the Army to fight the Maoists.

An Indian team of representatives headed by Kamal Pandey, the Secretary for the Home Affairs, and comprising the Deputy Chief of Mission at the Indian Embassy, the Deputy Chief of Indian Intelligence Agency, RAW, the Secretary General of the Nepal Desk, and four DIGs has come to Kathmandu to raise the issue of the control of terrorism and unlawful activities on the frontiers, the treaty of extradition, and the ISI's transaction in Indian counterfeit currency notes.

The representatives also inquired about Nepal's needs and expectations of the types of help it seeks from India to address the Maoist problem.

The Secretary at the Ministry of Home Affairs of Nepal asked India to disallow Nepali Maoists' activities from Indian soil.

A conference of the Brigade Commanders has been called on at the Army Headquarters.

I've briefed the King on the murmurs of dissatisfaction and occasional comments heard over the promotion of some army officers by the CoAS, Gen Prajwalla SJB Rana.

Earlier, he had been given a free hand, initially for his first year, in recommending promotions for convenience in his work. Currently, the CoAS has breached the norms by his recommendations for Col. Bajra Gurung for promotion to Brigadier General, thus bypassing the more deserving candidate, Col Ujjawal Bikram Rana.

Unhappy about the breach of hierarchy, Col. Rana has already petitioned the King for a stay on the promotion.

In turn, the King asked me to find the truth of the matter.

I forwarded Col Rana's letter to the CoAS for acknowledgment.

I also advised the King that a tactful way would be to study the files on Cols. Bajra Gurung, Ujjawal Bikram Rana, and Prahlad Singh Gurung to take a decision on the basis of merits and length of service.

He gave me his instructions to proceed as per the norms and system.

From the point of view of merits, length of service and efficiency, Col Ujjawal Bikram Rana appeared to be the best-qualified candidate for the promotion.

I briefed the King accordingly.

Later on, I heard that the CoAS talked to the King to give him a free hand for the placement of the right officers in the right posts for his convenience of work.

And thus, the promotion of Col. Rana just didn't materialize.

Thursday, July 13, 2000

The Nationalist Youth Front organized a four-hour discourse in Kathmandu in which the speakers presented a bleak picture of Nepal, it being increasingly crippled by the Maoist insurgency and the lurking threat of a civil war, deep-rooted corruption, anarchy, statelessness, and such unhealthy conditions. They questioned the lack of accountability of the House of Representatives, the indifference to or the silent endorsement of the Citizenship Bill, and the inundation of the Nepali lands caused by the construction of unilateral embankments by India on its side of the borders.

According to a confidential source of information, a top-level Maoist leader clandestinely met Dhirendra Shah [the King's third and youngest brother] at Sanepa on July 5.

The police have tightened security in the Kathmandu Valley. The Maoists are learned to be in a mood to initiate the peace talks once PM Deuba is back from America.

With three million Rupees made from the sale of its obsolete weapons, the Army must reach a quick decision on the purchase of the best model from among the sample weapons.

Tuesday, July 18, 2000

Asha Kaji Basukala, the Central Committee Member of the Nepal Peasants and Workers' Party, publicly pledged the Maoists of financial assistance should such a need arise. The Maoists are set to massively raise funds from people of all walks of life for the establishment of local-level people's government. According to a tipoff, they are receiving every moral and material support from the GNLF of India and the LTTE of Sri Lanka.

There's a complaint against Her Royal Highness Helen Shah taking her car right upto the airplane at the Tribhuvan International Airport. The King instructed me to make arrangements for the members of the Royal Family to use the airport's existing VVIP Lounge for their entries and exits.

Police Sub Inspector Dil Bahadur GC is reported killed and two other policemen have sustained serious injuries in an ambush set by the Maoists in Chakratirtha of Lamjung District.

The King today instructed me, taking it as an issue of immediate importance, to consider the status of the recruitment and the numbers of lower-caste [Dalit] soldiers in the Army and to enforce a directive to end all forms of discrimination, if any existing in any form. High ranking army officers sharing the same food at the same table boosts the courage and morale of the Dalit soldiers.

Tuesday, July 25, 2000

Sources say that PM Koirala quietly met the former Royal ADC, Maj. Gen Tara Bahadur Thapa at his own residence at Maharajgunj. Their closed-door talk touched upon the procurement of arms for the Army and other issues.

The Secretary at the Ministry of Defense, Keshavraj Rajbhandari, called me to inform that the Royal Palace has already consented to the budget allocation for the procurement of arms, requesting for a quick approval from the Ministry. He says such matters call for clear-cut norms, rather than following the same old practices, to ensure transparency.

The ex-NID Chief, Haribabu Choudhary, visited me in my office to share fresh developments in the country. He said he's neither associated with nor a member of any political party, but of course, he's slightly tilted to the Nepali Congress, and now would like to serve the Royal Palace if it finds him worthy. He added that because he belongs to the Tharu community, the people there seek his advice, and more importantly, have a good channel of contact in the Indian Embassy. He sounded desirous to have an audience with the King if ever such an opportunity presents itself.

Member of Parliament (MP) Narayan Singh Pun, currently visiting Europe for personal purposes, called me to share that he's going to meet General Sam Cowan. 'PM Koirala asked me to advise him on the NSC and it's in this connection that I'm going to learn about the structure of the NSC,' he said.

Since the start of the Maoist insurgency, a total of 200 police personnel, including a few officers, have lost their lives. Two policemen have been suspended and sixteen others have deserted. There's a massive murmur of discontent over the insufficiency of weaponry and adequate training among the police personnel. It calls for a timely rethinking to prevent a likely mutiny in the police. As part of a package for the police force, the NA has started training a commando unit at the Jungle Warfare School in Amlekhgunj from today.

In their overt strategy of physical attacks on political party leaders, the Maoists have assaulted district-level Congress leaders such as Lal Chand Bohara (in Gulmi), Dandapani Timilsena (in Dailekth), Bhakta Bahadur Basnet (in Bhojpur), and Chandra Narayan Yadav (in Siraha).

Thursday, July 27, 2000

The Maoists have reportedly established their so-called 'People's Government' in four hilly districts of Midwestern Nepal. Political parties and the government have been confined to the district headquarters.

Discussions are underway at the Royal Palace on the modality of the role and move of the King in this situation of extreme turmoil, taking the mainstream parties in confidence.

The need for the mobilization of the Army against the Maoists has been more of a necessity than a matter of discussion.

The police-army rapport needs further mending between the two forces to work together to bring peace.

Friday, August 4, 2000

PM Koirala went to India last Monday on a goodwill visit. What matters were discussed or if any agreement was signed during the visit is unclear.

Sources say the Maoists have called for a ban on the teaching of Sanskrit, singing of the national anthem, and charging fees to students in the community schools.

The Police Headquarters have suspended seven fugitive Inspectors.

The King asked me to look into the details of the amount received from the UN, its transfer to the principal account, and the generation of interest from the Army Welfare Fund, as well as the volume of budget thereof.

He further asked me to capitalize the principal amount by the purchase of Treasury Bills and Security Bonds to generate more income which should be invested in the infrastructure development of the Army.

Misunderstanding and differences have developed between the CoAS and me over policy issues relating to the Army. The King, too, has knowledge of it. I've mentioned my position on honesty, dutifulness, disinterestedness and meritocracy when it comes to the recommendations for promotion, all to create an environment of respect and a system in the Army.

The personal assistant to the CoAS, Col. Ranadhoj Limbu, visited me at the Office of the Principal Military Secretary to narrow down the differences between the CoAS and me through regular dialogues and discussions, a good way to remove misunderstanding. I agreed to meet the CoAS on a day every third week to build an environment of trust and confidence.

Monday, August 7, 2000

As planned, we met. But the environment did not turn out to be congenial, as the CoAS talked rudely, assuming his air of privilege and rank. I was taken by surprise by his way of talking. But I maintained my composure and politely expressed my side of my presentation.

Later, while briefing, and when I shared with the King the contents of the discussion, he only smiled and remained silent.

IGP Achyut Krishna Kharel called me to say that singer Pravin Gurung was run over and killed on the spot by a Land Cruiser on Sunday night, just a little away from the South Gate of the Royal Palace. The police couldn't stop the vehicle as it drove too fast and entered the Nirmal Niwas. I briefed the King on the incident. Later, it came to our knowledge that the vehicle belonged to Prince Paras Shah, son of His Royal Highness

Prince Gyanendra. The King sounded disturbed. He told me to deal with the entire episode with due process of the law.

Tuesday, August 8, 2000

I consulted with the IGP about the accident. The one who drove the vehicle must own up to the responsibility to give the impression that law governs everything. I also consulted with Prakash Bikram Shah, ADC to His Royal Highness Gyanendra. He said it would be a good idea to offer a relief package to the family of the deceased first and find a fall guy to own up to committing the crime. The relief to the family must include an assurance of free education to the children of the deceased.

Because bad news travels fast, the news of the incident has spread like wildfire. The print and electronic media have covered it. The media is attempting to smear the accident with political colors. Some members of the Parliament have already raised the issue in the House of Representatives. The royalist RPP seems to be maintaining its double standards in this issue. The King also asked me to keep IGP Kharel at arm's length for his known double standards. The student wing of the CPN (ML) shouted objectionable slogans against the Royal Palace, demanding immediate action against Paras. They obstructed the roads in several places, bringing the traffic to a standstill.

Thursday, August 10, 2000

It's been found out that PM Koirala quietly visited Maj. Gen. Tara Bahadur Thapa, presumably to talk about the purchase of weapons, particularly the RJ aircraft, for the Army.

Dr. Bhekh Bahadur Thapa, Ambassador of Nepal to India, currently in Kathmandu, in a meeting with me, expressed his worries about the erosion of Nepal's pristine image, owing to the incompetence of PM Koirala, his listening to and working with his coterie, unable even to take his entire party in confidence. He also complained that the NC leader, Chakra Prasad Bastola, has planted anti-monarchy sentiments in some officials at the South Block [the foreign ministry of India in Delhi].

New Delhi is persuaded to think that democracy in Nepal suffers due to the Royal Palace, and the Palace peoples' fondness for leftist and rightist extremes.

Former PM Krishna Prasad Bhattarai had once rightly said that the Koiralas can go to any extreme when it comes to power.

CoAS, Prajwalla SJB Rana's decision to pay his official visit to China first [before India] has raised concerns in India. Given the situation, it's imperative for Nepal to remove India's fears to mend relations with it. The royalists themselves must take initiative to this end.

A fissure in the Nepali Congress has once again surfaced, this time it being the tabling of the Motion of No Confidence against PM Koirala.

Preemptively, Koirala has prevailed on IGP Achyut Krishna Kharel to prepare a case against NC leader Khum Bahadur Khadka on corruption charges.

The CoAS, in his speech at the Staff College, has reiterated the need for the activation of the NSC to keep the Maoists at bay.

Friday, August 11, 2000

There's been a torrent of negative comments against His Royal Highness Prince Paras following the death of Pravin Gurung caused by the car he was driving.

So far, there's been no response from the King, either, but the newspapers are sensationalizing it with political coloring, and the political parties are joining the demonstrations.

The RPP seems to be quiet on this issue while still continuing to instigate other parties to that front.

It calls for a wise and careful move as to how to deal with the public outrage by, say, the King asking for the file to study the case, or even better, consult with the PM.

'It seems to me that IGP Achyut Krishna Kharel is double-dealing with the Palace and the anti-Palace forces,' the King said, calling for his activities to be monitored.

He then touched upon many current issues, ranging from the Maoists' formation of Company-strong combatant forces and the preparedness of the state security forces to fight it, the creation of an alternative reserve force should there be the mobilization of the Army.

The King's next concern was with the Amnesty International's reference to the Maoists using children as combatants. This inference amounts to the acceptance of the Maoists as a force. He then instructed me to see to it that Amnesty International finds no reasons to raise its concerns regarding human rights violations in the post-Army mobilization period.

Monday, August 14, 2000

Today, two Indian Parliament Members and the Indian Ambassador to Nepal visited the former PM Krishna Prasad Bhattarai at his place in Bhainsepati to patch up the differences between Bhattarai and Girija Prasad Koirala in a bid to unite the Nepali Congress Party as one block, not just in form but also in spirit and substance, in the context of the present situations in Nepal.

The Indian side is also known to have advised the Nepali Congress to cash in on the car accident and death caused by Paras to their benefits in order to weaken the Royal Palace and the monarchy, and to hit the iron while it's hot.

Both Bhattarai and Koirala are said to have agreed to beam the message of unity in the Nepali Congress Party.

Thursday, August 17, 2000

Col. Khadga Bahadur Gurung of the Royal Guards Military Police, in a personal meeting with me, shared his intention to resign. He, however, didn't disclose his reason. I asked him to reconsider his decision.

The rift within the ruling Nepali Congress Party circle is deepening, thereby threatening it to split.

The Royal Palace must develop a strategy to have the concept of the NSC passed in this opportune environment of a likely coalition government.

The Army also must proceed with the purchase and arrangement of logistics necessary to fight the Maoists.

The King today asked me on the status of the preparedness of the Army to open and operate a bank with the Army Welfare Fund as a partner and raise it even higher by expanding its scope to purchase and make provision for medicines to the Army Hospital.

A driver called Khadga Bahadur Bhujel has been found willing to own up to the crime of the vehicle accident caused by Prince Paras and surrender to the police as a way to put the case into the legal procedure.

An arrangement to airlift the ailing Yogi Narahari Nath from Nepalgunj has been made.

The CoAS asked me to make plans for the Crown Prince to inaugurate the Army Golf Course and the club building, completed to the east of the Guhyeshwori Temple at Pashupati. When briefed on it, the King sounded angry, as he has never been happy about it ever since the plan for its design and construction were broached. He said it wasn't under the priority of immediate importance.

The Crown Prince has given September 8 as the date for the inauguration.

Thursday, August 24, 2000

The Secretary at the Ministry of Defense, Keshavraj Rajbhandari, called me to brief the King about his Indian counterpart's invitation to him for an official visit to India. He said it would be an opportune occasion to talk to the Indian authorities about the need for weapons and logistics for the Army.

As was referred to earlier, Col. Khadga Bahadur Gurung resigned his post. I asked him to submit his resignation to the Military Secretary's Department of the Army Headquarters for it to go into the process.

As per the wish of the King and according to the need of the times, the concept to operate the entire Army in a corps system is gradually taking shape. The Corps Concept is to include such levels as the RNA Headquarters Policy Level, the MOD, Human Resources, and Coordination with the Royal Palace, Training and Procurement of Military Hardware, and the like. The Corps Commander is to operate the formations and units of the Army, a concept thereby to assist and lessen the workload of the CoAS.

The King instructed me to attend to every detail of the concept paper on the development of the Army according to the Corps Concept before it reaches its final form. He wanted me to make changes in it if so required.

He asked me to go with the Principal Secretary to the Tribhuvan International Airport to receive and escort His Excellency the Prime Minister of Japan, Yoshiro Mori.

Monday, August 28, 2000

The Kantipur Media Group has obtained permission to install its antenna tower for the Kantipur FM radio station's broadcast on the heights of Bhendetar near Dharan which is supposed to facilitate clearer broadcast to the Terai districts. Its effects on the national transmission policy are yet to be assessed and analyzed.

The government is considering buying a helicopter for the Nepal Police. As it's likely to incur huge expenses to build the necessary infrastructure, the Police operating it in coordination with the Army will certainly cut costs.

Thursday, August 31, 2000

A decision has been taken to place Col. Keshavraj Mahat as Chief of the Royal Guard Military Police.

The King expressed his anxieties at the Secretary of the Ministry of Home Affairs, Padam Prasad Pokharel, clandestinely supporting the Free Tibet movement in Nepal.

He also sounded angry over the idea of forming two parallel police forces: one under the Ministry of Home Affairs, and the other under the Police Headquarters.

The APF should be organized under a separate Act.

Today, the King visited the United Nations Mission's Peace Keeping Training Center at Panchkhal to inspect it and also give a good sendoff to the contingent of the Bhairavnath Battalion. Present on the occasion were the PM and the Secretary of the Ministry of Defense.

He instructed me to see that those going on the peacekeeping mission have the required standard weapons.

He also told me to pay my attention to the purchase of the land adjoining the training center, fence it with barbed wire, and make arrangement for the compensation to be paid to the farmers in case of crop damages during training.

He talked to the PM and the CoAS about the need for constructing dikes and tree plantation as safety measures to prevent landfall and landslides at the site of the training center.

He also gave clear instructions to the CoAS to make specific standards of time and duration of the training for all ranks.

Tuesday, September 19, 2000

The Cabinet, at the recommendation of the security forces, has put a ban on torch rallies throughout the kingdom.

There's information of several cadres of the CPN (ML) and the CPN (UML) of different districts joining the Maoists.

There have been complaints from the common people of torture and extortion from both the police and the Maoists. People find themselves trapped in between. The victims are noncombatant civilians, either branded as Maoists, or informants to the police.

Sharad Chandra Shah called me to arrange for Dr. Tulsi Giri's audience with the King. He also told me to consult with the CoAS about the Maoists' mysterious recent activities and their likely infiltration of the sensitive places of the government and the Army.

It's also heard that their recent plan is to confront the Army.

He further asked me to put in place the old system wherein Brigadiers Generals and those in the higher ranks in the Army need offer coins to the monarch during an audience with him.

The Maoists have planted pointed bamboo stakes on the ground in Rukum and Rolpa districts to prevent Army helicopters from landing.

There's fresh information of the Dinanath faction of the CPN (Mashal) merging with the Maoist Party.

The Secretary at the Ministry of Defense called me to inform of the cancellation of his visit to India and mentioned his wish to visit Pakistan instead.

He also tipped off that the leaders are inciting the people against the Royal Palace and the Army. It may amount to dividing the Royal Palace and the Army to one side and the political parties to another.

I also knew from him that the PM has exposed all the details of the master plan of the National Security Council about the formation of the APF.

Former PM Bhattarai has said that the Army wasn't positive about its mobilization against the Maoists during his premiership.

Thursday, September 21, 2000

Sources close to the PM say that he's in a mood to reshuffle the Cabinet after the Dashain Festivals.

He has also accused the former Minister Khum Bahadur Khadka of paying money to the Maoists.

According to confidential sources, Ravindra Nath Sharma of the RPP sponsored an open discourse at the Reporters' Club on issues related to the controversial Citizenship Bill.

The King quoted the Sri Lankan Ambassador to Nepal as saying that the LTTE is training Maoist guerillas in the hills of Nepal. He asked me to convey his directive to the Army to see if what was alleged is true.

He also wanted to know about the current status of the master plan for the operation of the Army according to the Corps Concept.

He then instructed me to develop ethical, professional and physical standards of training courses for the cadets at the Military Academy at Kharipati.

Sunday, September 24, 2000

The King visited the Royal Nepal Military Academy at Kharipati.

The CoAS, Gen Prajwalla SJB Rana, updated him on every detail of the progress and development works being done at the Academy.

The King realized the insufficiency of space, accommodation and budgetary allotment thereof. Therefore, he expressly commanded the CoAS to take the necessary measures to expand the area by possibly purchasing the land adjoining the Academy.

He inspected the Academy and also observed the training.

Keen to learn about the quality of the trainers and the training, the King instructed the Commandant of the Academy to maintain the Academy's curriculum at par with that of a university.

Monday, September 25, 2000

The Maoists launched a massive attack last night on Dunai of Dolpa District. The District Administration Office and the Police Post were reduced to ashes and rubbles. Fourteen policemen were killed, and all the weapons were lost to the Maoists. One bank was robbed of all the cash and valuables it held. The additional police force mobilized from Nepalgunj couldn't reach there as the Maoists had blasted the bridge on the Bheri River.

The Minister of Communications, Jay Prakash Prasad Gupta,

criticized the army for non-cooperation after the government structures right under its very nose were attacked brutally. He accused the Royal Palace of operating the Maoists and threatened to spill everything out the instant he is out of the government. He accused the King of seeking to take over by amending the Constitution. He also said that should a situation of confrontation arise, the leaders may have to go into exile, but still public uprisings and agitations will continue.

It's indecent for a Cabinet Minister to make such comments to the press.

The Army is set to place a squad at the entry point of Dolpa District from tomorrow.

Tuesday, September 26, 2000

According to a tipoff, a man called Arjun delivered a letter from 'Prachanda', the Maoist Chairman, to Narayan Man Bijukchhe, Chairman of the Nepal Peasants and Workers' Party.

The PM, causing more speculations, quietly visited the former ADC, Maj. Gen Tara Bahadur Thapa.

The King sounded worried at the sporadic negative comments made against the Army by Ministers. He said the government would do well to discuss and find out the loopholes responsible for the Dunai attack. He said had the government adopted its preparedness in time to activate the NSC, the country wouldn't have had to suffer so much. What happened speaks of a sheer lack of coordination between the security forces and also between the government and the security forces. It would be wise on the part of the government and the security forces to do their own soul-searching and go ahead with a concrete analysis of the operations to be planned against the Maoists.

The PM visited the King to brief him on Dunai. The PM seems to be in a mood to seek the resignation of the Minister of Home Affairs, the Secretary at the Ministry of Home Affairs, and the IGP for their nonperformance or incompetence to effectively defend Dunai.

The King advised the PM to place Army squads at the entry routes to Dolpa District and proceed with operations against the Maoists. He also asked the PM to consult the CoAS in this regard.

The Regional Police Chief, DIG Rajendra Bahadur Singh, and Zonal

Police Chief, SSP Gyanendra Raj Rai, have been suspended for failing to maintain adequate security and preparedness to avert the Dunai incident. They have been replaced by SSPs Amar Singh Shah and Deepak Singh Thangden.

Saturday, September 30, 2000

Following the Maoist attack on Dunai, the Minister of Home Affairs, Govinda Raj Joshi, resigned, accusing the Army of not making arms available to the Police.

IGP Achyut Krishna Kharel, at a private meeting at his residence, told me that the Communists in Nepal, particularly the CPN (UML) and the Maoists, must never forge unity; else they will emerge as a very strong power.

He recounted his strained relations with the CoAS. He also talked about the lack of weapons and logistics for the Nepal Police. He also expressed his dissatisfaction with the role of the Army for not immediately coming to help the Police as a backup force. If the Army had cordoned off the area, the Maoists could've been wiped out in a single sweep there.

Having heard him, I said, 'In both the Police and the Army, we're serving the country to the best of our abilities and resources. It's no time for trading accusations and counter-accusations in the security forces. There have already been negative comments and accusations on the Army made by some Ministers and others.'

I added, 'You, as the IGP, didn't provide me with any details on Dunai. I briefed the King with whatever details about the incident I managed to gather from other sources. There's been poor coordination or a lack of exchange of information between the Principal Military Secretary's Office, the Home Ministry, and the Nepal Police Headquarters. I wonder why the successive governments don't consider mobilizing the Army constitutionally, even when there have been enough discussions on the need for the NSC. It seems the government and the Police still don't understand the nature and scope of the Army. Each successive government has sought to move the Army against the Maoists without considering to make constitutional provisions for it.'

IGP Kharel looked touched by my remarks, and he promised to host a dinner for the political parties to create an environment of trust and mutual help.

Sunday, October 1, 2000

The suspended DIG, Rajendra Bahadur Singh, has defended himself as innocent. He said he had been away for a few hours, with the approval of leave from the AIG of the Police Headquarters.

Newspapers report that IGP Kharel is likely to resign soon.

The King has granted time to see Dr. Tulsi Giri some time in November.

The former CoAS, Gen Satchit SJB Rana, now Royal Nepalese Ambassador to Myanmar, called me to say that he wishes to be transferred from Myanmar to Canada or Australia as Nepal's envoy. A little later, he informed me that he's back in town from Myanmar and requested me to arrange an audience with the King.

Monday, October 2, 2000

Amidst much hue and cry following the Dunai attack, the PM has appointed Mahesh Acharya as Defense Minister. The PM is in a mood to remove the IGP to appoint him ambassador to Myanmar and transfer the Secretary at the Ministry of Home Affairs.

The Defense Minister appears to be busy in consultations regarding the mobilization of the Army as per the law of the land. It requires consultations with the Army and the Police for the formation of the APF.

According to information received recently, the Maoists have looted Rs 300,000 and two handguns from the Nepal Bank Branch in Jiri of Dolakha District.

Tuesday, October 3, 2000

IGP Kharel has been replaced by AIGP Pradip SJB Rana as Acting IGP. He has consulted with the CoAS about the strategies of actions against the Maoists. The two heads of the security forces have agreed to go ahead together, with mutual understanding and cooperation, giving no room to friction or lapses.

The King has expressed his concerns about the observance of law behind the imposition of curfew in any part of the country.

Monday, October 16, 2000

The Military Attaché of the British Embassy called to inform me of the Colonel Commandant of the British Gurkhas, General Sir Samuel Cowan's upcoming visit to Nepal, asking me to make arrangements for a reception on his arrival.

The Army is scheduled to brief the newly appointed Minister of Defense on Thursday about the security situations of the country.

Thorough tests of the sample weapons are progressing under a team of technicians in the Army. The Army must consider the assemblage of the rifles' sealing, telescope and magazine along with the weapons before the final process of placing orders on the most suitable weapon system is undertaken.

Tuesday, October 24, 2000

There's been a provision for the medical treatment of the employees of the Royal Palace at the Birendra Military Hospital which also bears the expenses of the patients. It would be a good idea, therefore, to consider organizing a trust fund at the Royal Palace itself to pay for its staff's medical treatment.

The Hospital has sent its development plans for the expansion of its maternity and cardiology units.

Some Principal Staff Officers (PSOs) have expressed their strong wish to meet the King and these Generals have asked me to make such an arrangement. I briefed the King on this and advised that it would be a good idea to host a dinner for the CoAS and the PSOs together to instill a spirit of camaraderie among them. The King took my idea positively.

Maj. Gen Chitra Bahadur Gurung is scheduled to leave for New York for the UN's Department of Peacekeeping Office (DPKO). It would be a good idea to appoint someone with an acting promotion in such an appointment and send him on deputation.

Information has been received of the theft of valuables worth millions of Rupees from the Tripura Sundari Temple in Dolpa.

The Supreme Court's full bench, in response to a writ, has cancelled the privilege of pension for lawmakers.

The Minister of Defense has been briefed on the preparation of the purchase of the most essential weapons, the construction of barracks and other structures, and the introduction of the Corps Concept in the Army. On the occasion, the CoAS also briefed him on the operations mounted by the Army and the Police against the Maoists so far.

Thursday, November 2, 2000

In a regular briefing to the King, the CoAS quoted the PM as asking him (the CoAS) to brief him (the King) on the decisions reached in the NSC meeting. In fact, given the seriousness of the matter, it requires the PM to brief the King on the TOR and operation of the NSC. The PM's reluctance to do so reflects his lack of seriousness in it.

The CoAS shared that the government wants the Army to place a platoon in some district headquarters. But it doesn't go well with the Army's norms, as only a company at the most can be posted.

A fresh recruitment drive in the Army has been essential for the deployment of soldiers to such remote posts.

The King instructed me to obtain information about the structure of the Maoist operational command.

Sunday, November 5, 2000

The King, with the Minister of Defense, the CoAS, and other high ranking Army officers, visited Nagarkot for his inspection of the Army Training School there. The briefing on the nature of training given was followed by lunch. It was then the King expressed his concerns, emphasizing on the availability of adequate space and the curricular contents of the training. He gave instruction to the CoAS to spend from the Army Welfare Fund to pay for the construction of the Center's physical infrastructure, and turning to the Defense Minister, he asked him to allocate the necessary budget from the government for the purpose.

The Koirala Government has given a green signal to resume talks with the Maoists. It also has displayed its flexibility by releasing Maoist leaders Dinanath Gautam and Dinesh Sharma.

Thursday, November 9, 2000

The Nepali Congress leaders such as the former PM, Sher Bahadur Deuba, the former Minister, Bijaya Kumar Gachhadar and the others joined a mass meeting in Janakpur.

The DPM, Ramchandra Poudel, highlighted on the current state of affairs in the country to journalists.

Sources close to the Nepal Police said that the AIG, Ramkaji Bantawa, is just back in Kathmandu from his secret visit to the officials of the Ministry of Home Affairs of India.

There's yet another piece of breaking news of the former PM, Lokendra Bahadur Chand, having visited Dr. Tulsi Giri in Bangalore in India.

The newly appointed IGP, Pradip SJB Rana, had an audience with the King to brief him on the current strength of the Nepal Police: a total of 51,374 personnel, and not enough to maintain law and order in the country. He said that something went amiss earlier in the proper analysis of the Maoist problem and in taking the Police as a cure to all ills. He emphasized on the need for the enforcement of the Terrorist Act to fight the Maoists. He also mentioned the poor information network system in the entire security forces.

It's very likely that India will support the Maoists once the Army is mobilized against them in Nepal. We should be serious about this hypothesis.

The King said the government and the leaders must know that the function of the Nepal Police isn't to go in direct confrontation with the rebels. But just the reverse is the ground reality. The training and the types of weapons with which the Police operate don't enable them to fight the terrorists. So the Police must work in coordination with the Army.

Friday, November 10, 2000

The Maoists are learned to have strategically dropped the idea of holding the peace talks with the government for the present to concentrate on their objectives in the east to exhaust the government to bring it down to its knees. They also seem to have halted their strategy to clash with the Army but to study the regimen of its training in order to beat it with its own stick.

The Nepali Congress leaders, PM Koirala, Khum Bahadur Khadka, Bijaya Kumar Gachhadar, and Sher Bahadur Deuba respectively are working out their strategies to have their upper hand in the upcoming General Convention of the Party.

The Brig. Gen of the Army's Directorate of Military Intelligence

(DMI), Rookmangud Katawal, shared with me his experiences of his recent visit to China. He said he found that the Chinese Government felt more comfortable with the King than with Nepal's political leaders, as they tilted more to India. China is focused to keep its ties with the Middle East for its oil reserves. Nepal must open its route in the north to link itself with Tibet, and cash in on the opportunity for its benefits. The Chinese Army holds the Royal Nepal Army in high esteem. He said he also watched the demonstrations of the military operations while in China.

Tuesday, November 14, 2000

Today, the King instructed me to consider to moving the company-level strength of the Army first to the frontline of military operations in case the Army is mobilized.

He also told me to include the detailed plan in writing when it comes to having the budgetary allocations made for the Army's logistics.

His other concern has been to involve the local-level Army commander as and when the district security committee makes the operation plan.

Similarly, the CoAS should take care to inform the Chief District Officer (CDO) concerned of the move and assume the control of the Army.

He asked me to duly inform the administrative departments and the security units of the districts where an ISDP project is to be launched.

The Military Attaché at the British Embassy has said that he has no knowledge of the British Army creating a Sikh regiment, just like the Gurkha Rifles.

Thursday, November 16, 2000

The King granted his audience to Maj. Gen. Chitra Bahadur Gurung at the Nagarjun Bungalow. He advised the outgoing officer to work diligently as the first senior Nepali officer at the UN Headquarters in New York.

In my briefing to the King, he quoted Maj Gen Gurung as saying that the workload of the CoAS is heavy and those of the PSOs light.

The concept of the ISDP is good, but it should think of getting the support of all the sectors of the country. The current operation plan of the Army desperately needs reforms.

Friday, November 17, 2000

The King ordered me to maintain foolproof security, and also have a female doctor posted for timely services, and have the barracks and its outlying areas cleaned at Tiger Mountain in Chitwan during the visit of the Princess Royal, Anne, daughter of the Queen of Britain.

Sources say that Chhatra Subba, Sonam and Keshang are visiting the poor and underprivileged people in Chitwan for conversion to Christianity. Their other plans have been to establish a Baptist Church at Bharatpur, organize a conference of Christian youths in the Midwest Region, and their conversion campaign to be carried by 2005 in many of the underprivileged areas.

Tuesday, November 21, 2000

In his regular briefing to the King, the NID Chief, Devi Ram Sharma, described the insufficiency in budget and equipment in his office to collect, analyze and disseminate information. Provision of incentives would also encourage the staff to work harder and better. No requisite fund is allocated for the SSP-level officer placed in New Delhi is a case in point.

Several I/NGOs operating in Nepal for many years now need monitoring of their activities.

The Intelligence Department hasn't been able to monitor the activities of the opposition parties.

The NID suffers from budgetary insufficiency and bloating recruitment of political cadres. Unlike the practice elsewhere, the NID is under the Ministry of Home Affairs; in principle, it should be under the office of the PM, or alternatively, under the NSC.

Having heard this, the King commanded me to look into an efficient dissemination system of information, and to maintain transparency in budgetary expenditures.

Wednesday, November 22, 2000

When Haribabu Choudhary, the former Chief of the NID, called on me, he quoted Govinda Raj Joshi's displeasure at being made a scapegoat for the Dunai fiasco. He asked me to take an initiative to remove the confusion of the Royal Palace and of the Army that the Dunai incident was the result of laxity on Joshi's watch.

Thursday, November 30, 2000

The King affixed his Royal Seal of Approval (Lal Mohar) to the Regulations of the Army Welfare Fund.

Sources have it that the Maoists have issued their threats to schools to refund the admission fees charged to students, ban the singing of the National Anthem, and the teaching of Sanskrit. They are also extorting and intimidating private school managements.

The CoAS briefed the King on the rift deepening between the two high-command leaderships of the Congress Party, threatening its split.

He also informed the King of the Indian CoAS's scheduled goodwill visit to Nepal from December 19 to 23 at the invitation of the RNA. The CoAS also asked the King's for his audience to the guest.

Monday, December 4, 2000

Now the procedure for the formation of the APF has finally got momentum. Enough thought has gone into the scope, scale and modality of its operations.

PM Koirala, the DPM and the Minister of Defense are believed to have discussed the current security situation of the country.

The former Indian Prime Minister, IK Gujral, was wishing to meet the CoAS to discuss Nepal's purchase of weapons from India. But he had to immediately return to India due to his health complications.

Tuesday, December 5, 2000

It'll be proper to supply weaponry to the APF from the Army's armory. The Army also needs replacement of its old weapons with sophisticated ones without delay.

Sharad Chandra Shah called me to help Rajendra Gurung, his business partner, to have the contract of essential supplies to the Army. I told him to talk to the CoAS, a good way to avoid it.

The ISDP concept, now under intensive discussion in the Army, deserves the same attention from the political fronts too; sooner its implementation the better.

Friday, December 8, 2000

Today's issue of the *Saptahik Bimarsha* Weekly has covered a newsstory. Accordingly, the DPM, Ram Chandra Poudel, also holding the Home Affairs portfolio, accuses the Royal Palace of playing the Maoist card against the government. The weekly further writes that the Maoists operated by the Royal Palace will do its own harm, more so because Communists can never be friends of the Royal Palace, a truth the Royal Palace is yet to learn.

Stricken by the fear psychosis spread by the Maoist Revolutionary Students Wing, over 40,000 schools, both public and private, have closed down indefinitely.

Saturday, December 9, 2000

The Minister of Physical Planning and Construction, Narayan Singh Pun, updated me on the fresh information of five or so Congressmen, including himself, joining the Deuba camp. It indicates that Deuba's side tilts the balance more in the Congress Party.

Pun also said there are complications in the solution of the Maoist problem owing to the ill-defined role of the Royal Palace and India.

The present PM, GP Koirala and the former PM, Sher Bahadur Deuba, visited the former PM, KP Bhattarai, separately at his residence. The reasons for the meetings and the contents of the discussions aren't known.

The CoAS and the Minister of Defense have discussed the Maoist problems and the current security situations.

Sunday, December 10, 2000

Today, the King, along with the Defense Minister, and the Secretary of the Ministry of Defense, visited the Command and Staff College at Shivapuri. The CoAS received them there.

The Commandant of the College briefed the King on the activities and the future plans of the Staff College. He added that the College's infrastructure development will soon complete. He also stated the objectives of the College as to provide training and equip officers with knowledge to enable them to take the responsibilities of higher ranks.

The King advised the Minister of Defense to allocate sufficient budget, possibly from other friendly countries, for the professional development and the construction of quarters for the resident officers.

He also instructed the CoAS to see if some funds from the Army Welfare Fund could be managed for the needful infrastructure development at the training center.

He further advised the CoAS to include Disaster Management in the curriculum of the College.

Wednesday, December 13, 2000

The British Defense Minister, Geoffrey Hoon, currently on visit to Nepal, advised PM Koirala to mobilize the Army to fight the Maoists. Against the backdrop of India's pressure on Britain not to sell any weapons to Nepal, he told the PM to talk to India about Nepal's interest to purchase the HK G36 rifles from the German company, Heckler & Koch.

Applauding Nepal's contributions to peacekeeping in the world, he asked the PM to send a battalion of the RNA to Sierra Leone for the UN Peacekeeping Mission in the next dispatch.

In response to the British Defense Minister's request, the PM asked him to provide the Army with military equipment for its operation in Sierra Leone, and for the infrastructure development of the UN Peacekeeping Training Center at Panchkhal.

He also asked Britain to provide pension to the Nepali soldiers who served in the British Army during World War II.

His next request to Britain was to monitor the activities of the RIM in London with references to Nepal.

Tuesday, December 19, 2000

There's information of Prince Nirajan having had a motor accident at Maharajgunj. He's reported to have returned home after minor treatment.

Sources say there's been a tacit understanding between the Maoists and the CPN (UML) not to disrupt each other's mass meetings.

Friday, December 22, 2000

The King granted audience to the Indian Air Chief Marshal, AY Tipnish.

The guest reportedly advised the King to form an air force effective enough to fight the Maoists in a mountainous country like Nepal. He also said that MI-17 helicopters would be more effective than the Super Puma.

Meanwhile, a Nepal Army representative has been officially invited to observe the Air Fire Power Demonstration in India.

In his reply, the King expressed his pleasure in meeting the Indian visitor and thanked him for his support and cooperation.

He also expressed confidence of further Indian help and support in the days ahead, particularly in training for the pilots of the RNA.

Monday, December 25, 2000

Today, the PM, the DPM, high ranking security officers, and the Secretary of Defense discussed the current security situation in the country. Sounding angry and frantic, the PM expressed his dissatisfaction over the inaction even when Maoists leaders, according to information, are taking shelter in the houses of some high and mighty in Kathmandu.

He also disclosed that former Panchas [of the previous Panchayat Politics] and other revisionist forces are trying to cash in on the volatile situation in the nation. This will cause harm to the country and the monarchy as well.

In his turn, the DPM expressed his displeasure that the NID, once so efficient, during the Panchayat era, was now underperforming.

Devi Ram Sharma, the NID Chief, explained that insufficient budget allocated to his department was the reason of its gradual decline in competence. He also described the recurrent interference of political parties with the NID as another reason of the low performance thereof. He also noted the difficulty in work due to understaffing and inadequate incentives to those in the intelligence department. He also made his dissatisfaction clear at the way the political parties and leadership tend to use the professional institution as a mere tool to prolong the tenure of the government or to topple it.

IGP Pradip SJB Rana deplored the successive governments' indifference to strengthening the intelligence department by not allocating adequate budget for its infrastructure development, training and the purchase of essential equipment. He said that despite the lack of weapons and training, the Police contained the Maoists for five years. He added his worries at the reluctance of the government to pass the necessary laws to authorize an immediate mobilization of the Army against the Maoists even when the concept paper to that end had been submitted to the government ten years earlier.

PM Koirala met the former Minister of Foreign Affairs of India in Janakpur at a memorial program held to express condolences at the demise of the leader of the Nepali Congress, Ram Narayan Mishra.

Tuesday, December 26, 2000

There have been angry demonstrations and protest marches following the Indian filmstar Hritik Roshan's alleged comment on Nepal being a part of India. Four schoolchildren lost their lives in a police firing to control an angry crowd at Chabahil. Demonstrators shouted slogans demanding action against those responsible for the firing, the resignation of the Minister of Home Affairs, a public apology made by the Indian Embassy, and a total ban on Bollywood movies in cinemas in Nepal.

It's also come about that the student wings of their respective principal parties are marching together to hand a letter of their protest over to the Indian Embassy in Kathmandu.

There's reporting of demonstrators having smashed the shops of some Marwari entrepreneurs. It's likely that mobs may attack Indian visitors, too.

The Indian Embassy has demanded that the Nepal Government be concerned with the safety of their citizens in Nepal.

Monday, January 1, 2001

PM Koirala is known, again, to have quietly visited the former ADC, Maj. Gen Tara Bahadur Thapa, at his residence in Maharajgunj. Also present on the occasion were the former CoAS, Gen. Dharmapal Bar Singh Thapa and businessman Rohini Thapaliya.

Knowledgeable sources say that both the Deuba and Koirala camps are lobbying to ensure its win in the General Convention of the Nepali Congress held in Pokhara. Lately, there's been much mudslinging between the two camps to bag the coveted post of the Presidency of the Party. Sources say the Koirala camp's balance appears to be heavier.

There's been a baseless rumor of the arrest of the Maoist leader, Dr. Baburam Bhattarai.

The King is set to go to the Diyalo Bungalow of Bharatpur, Chitwan, for rest.

Monday, January 22, 2001

The candidacy filed by Sujata Koirala for the position of Central Committee Member in the Nepali Congress Party has been annulled on the grounds of the immaturity of her active membership at the party base. She has threatened some Congressmen with action in the Party.

The former Nepal Students Union leader, Bir Bahadur Balayar, with his supporters, is heard to have shouted slogans in support of the former PM, KP Bhattarai.

At the NC Party's General Convention, 13 loyalists from the Koirala Panel and six of the Deuba Panel have been elected to the Party's Central Committee.

Some CPN (UML) leaders have left for the site to inspect the Purnagiri Dam being constructed by India.

According to information received, the Maoists have torched the police post near the Tripurasundari Temple in Dolpa District.

Wednesday, January 31, 2001

During an audience with the King, the PM advised him to grant regular audience to the IGP.

He also asked the King to stamp the Royal Seal of Approval to the Citizenship Bill, duly endorsed by the Parliament, and now at the Royal Palace for the King's assent.

The King, in turn, has consulted with the Secretaries in the Royal Palace about the future implications of the Citizenship Bill. After an in-depth discussion on the matter, we, the Secretaries of the Royal Palace, advised him with four alternatives – first, His Majesty, the government, and the leaders of the political parties to squarely discuss the Bill; second, to send it back to the Parliament with suggestions for amendments; third, to seek the view of the Supreme Court; and the fourth, forthwith accord the Royal Seal of Approval to the Bill.

In view of the future implications and with the respect of the people of Nepal, we reached a decision as the best option: to seek the views of the Supreme Court. Accordingly, the Principal Secretary was asked to send the Bill to the Supreme Court for its deliberations.

The King has expressed his desire to elevate the former CoAS, General Dharmapal Bar Singh Thapa to Field Marshal of the Army soon. So he has asked me to tell the Principal Secretary to talk to the PM in this regard. He has also asked me to take the measurements of Thapa's hand for the Field Marshal's baton, and make all other preparations for his uniform and insignia.

Thursday, February 1, 2001

Following the King's recent approval of the Ordinance, the APF has been officially formed, with Krishna Mohan Shrestha as its IGP. On the occasion of his assumption of office, IGP Shrestha told me to convey his thanks and gratitude to the King for his trust in him and the opportunity of service given him.

Applications have been called from interested Army personnel for their transfer to the APF. As many as 119 armymen have applied for it.

The King has asked me to follow up the on progress of the reforms he had emphasized on during his inspection visits of the Army Training School.

As per his instruction, I also consulted with the security heads about the Citizenship Bill. They unanimously opined that should the Bill receive the status of Law, it would entail a lot of complications. Unforeseen problems are likely to emerge as it would open the way for non-Nepalis to easily procure Nepali citizenship from the Terai region, eventually leading Nepal to suffer the fate of Sikkim.

The King expressed his concerns about the upcoming visit of the Defense Minister of China to Nepal. I told him that the intention behind his visit may be to obtain fresh information about the situation in Nepal amidst the Maoist violence, Nepal's desire of help from China, and the expansion of relations between the national armies of the two countries.

The King instructed me to tell the CoAS to circulate instructions in the Army to make proper arrangement for the dispatch of doctors and paramedical equipment for treatment from sections down to the unit levels.

Wednesday, February 7, 2001

The King today told me about China's invitation to him as a chief guest at the Boao Conference in the Henan Province, and asked me to consult with the COAS to prepare a list of equipment Nepal would expect from China in case of the Army's mobilization against the Maoists so that he could discuss the matters with the President of China.

He also instructed me to learn the process for Nepal to write a keynote paper for presentation at the Asia Forum.

The Maoists made an unsuccessful attempt on the life of the Chief Justice, Keshav Prasad Upadhyaya, with a landmine blast in Surkhet last Saturday. The blast claimed the lives of five people on board the vehicle. The King has expressed his concerns about the condemnable act.

He has started fresh consultations over the Sixth Amendment to the Citizenship Bill, 1964, now forwarded by the Parliament to the Royal Palace for the King's Royal Seal of Approval.

Thursday, February 8, 2001

An intensive discussion is underway on the operations of the Army under the Corps Concept, with the provision for all the units under Corps Commanders and the Headquarters to make the policy for fresh recruitment, training, procurements, and coordination with the Royal Palace, implementation of the directive of the government, and the plan of strategic importance.

The King has been duly briefed on it.

Present at the briefing were the Crown Prince, the Secretary of the Ministry of Defense, and PSO Generals.

A number of major reforms in the Concept before its implementation has been envisaged, particularly about the locations of the Corps Headquarters.

Thursday, February 15, 2001

The King asked me to study and advise him on the proper modality and process of the appointment of the new Field Marshal in keeping up the spirit and norms of the RNA.

It's learned that the morale of the Police is low. The trend of its officers deployed in Maoist-affected areas trying to secure the blessing of the political power centers for their transfer to safer Kathmandu has been on the rise.

Col. Hemendra Dhoj Khadka, appointed Military Attaché of Nepal to India, expressed his extreme displeasure at the cancellation of his appointment at the last moment. Col. Bikas Bikram Shah has replaced him at the post while Khadka has been transferred to the RNA Headquarters. He sounded hurt and humiliated with a wish to resign soon.

Wednesday, February 21, 2001

The Minister of Defense, Mahesh Acharya, hosted a dinner for the visiting Chinese Defense Minister, Gen Chai Haosain, who is on his first official visit to Nepal. Minister Acharya, speaking on the occasion, recalled the age-old multidimensional Nepal-China relations. He applauded the efforts of the PLA of China for making the country prosperous.

In the same breath, he admired the RNA for its efforts to maintain peace at home and in the world through its participation in the UN Peacekeeping Missions.

An exchange of reciprocal visits by high-ranking army officers of both countries would contribute to further strengthening of Sino-Nepal ties.

The Defense Minister further expressed his pleasure at having his Chinese counterpart visit the birthplace of Lord Buddha, Lumbini. He said Nepal was eagerly looking forward to the King's upcoming official visit to China.

In response, the Chinese counterpart expressed his happiness at his visit to Nepal at the opportune time of the Army Day.

On behalf of the People's Liberation Army (PLA) of the People's Republic of China (PRC), he expressed his confidence that the King's upcoming visit to China in the New Millennium will further strengthen Nepal-China ties.

Friday, February 23, 2001

The King granted his audience to the Defense Minister of the People's Republic of China. The Chinese Defense Minister expressed his extreme pleasure at visiting Nepal and seeing its beauty. He recalled the commencement of visits to China by the late King, Mahendra, 45 years ago, which proved to be bedrock in Nepal-China relations. He fondly recalled the introduction of the One China Policy of Nepal by the late King, which Nepal is consistently pursuing ever since.

On the occasion, he pledged the RNA of help with weapons or logistics according to its requirements, and also expressed his confidence that Nepal will develop further under the leadership of the King.

The King appreciated China's continued support for Nepal's sovereignty and its territorial integrity. He said China has always been a good neighbor and friend of Nepal. He added he was eagerly looking forward to his imminent visit to China, and the opportunity to confer with Chinese leaders and discuss matters of mutual benefits and share experiences of development and cooperation.

He lauded China's deep understanding of Nepal's strategic importance due to its geographical location in South Asia, and the placing of high value on this fact. He further wished the official guest every success in his visit and a pleasant stay in Nepal.

The King also asked him to convey Nepal's best wishes to the President of China, and his eagerness to meet him soon.

I briefed the King on the agenda to be raised and discussed with the Chinese leaders during his China visit. I said the priority areas could be about the opening of a transit route to China from the northern border of each development region of Nepal, the construction of an Outer Ring Road in Kathmandu, and securing China's support for Nepal's security and help in logistics and communication equipment for the Army.

The Maoist insurgency is escalating, making it increasingly likely to mobilize the Army. It's becoming urgent to search for consensus for the modality of the Army's mobilization. The political parties, however, don't bother to appreciate the fact that the Army is the ultimate backup force, the last resort – a common practice in the world.

The pressures of pleading and advice from diplomats, civil society activists and wellwishers alike to make a timely move to avert further damages and destruction in the country have been mounting on the King with each passing day.

The timing of his official visit to China is very important, considering these factors.

The RPP leader, Dr Prakash Chandra Lohani, shared with me PM Koirala's remarks about the King. He quoted Koirala as saying that the recent visit of the Chinese Defense Minister to Nepal, the fresh remarks made by the Chinese Ambassador, and the King's upcoming visit to China were all a rehearsal for a 'Grand Design'. Dr. Lohani added that Koirala sounded displeased with the King.

Thursday, March 1, 2001

PM Koirala reportedly visited the former Prince (ex-His Royal Highness) Dhirendra Shah at a time when the latter is said to have a channel of communication with the Maoist leaders. The contents of their talk are yet unknown.

There's speculation of the PM's resignation. He seems interested to visit France soon.

Thursday, March 8, 2001

Krishna Mohan Shrestha, the IGP of the newly formed APF, was received in audience by the King soon after the latter's return from his official visit to China. The King expressed his concerns about the Ordinance on APF with its command headed by the IGP, and asked him to take the initiative to have it endorsed once the House convenes.

The newly promoted Brig. Gen Dr. Gautam Ratna Shakya, was also received in audience by the King.

There's information about one Dhan Bahadur Khatri bragging his association with the Royal Palace. Actually, he's the one brought to Kathmandu and educated in the best schools on the expenses of Their Majesties who, on their tour of Okhaldhunga in 1988, were moved by his pleas for support for his study. The abuse of his privilege came to light when a Police Sub-Inspector called the ADC Office to report that Dhan Bahadur had promised him to secure a transfer to a better workplace.

Tuesday, March 13, 2001

The King granted audience to Kofi Annan, the UN Secretary General, currently on a visit to Nepal. The Secretary General lauded the efforts of the Royal Nepal Army in the UN Peacekeeping Missions. He expressed his view to have the RNA in the UNIFIL during its mandate with the UN.

The King and the Secretary General also discussed the prospect of further developing the UN Peacekeeping Missions Training Center in Panchkhal.

The Indian media have carried objectionable and baseless news coverage about the Chinese Defense Minister's visit to Nepal and the King's official visit to China as though the two countries are set to sign an important agreement.

Wednesday, March 14, 2001

Soon after the briefing by the PM, the King commanded me to arrange a meeting of the Defense Minister and the security heads in order to have them discuss the current security situation of the country and to brief him on the reports thereof.

He also referred to the PM's complaint against some Army officers' involvement in wildlife poaching. He said it would be wise to put it to an investigation to bring the guilty to book according to Army rules and regulations.

Today, the King expressed his surprise at PM Koirala's recommendation for the elevation of the former CoAS, Gen Dharmapal Bar Singh Thapa, to Field Marshal while the King has already decided to do so, on April 14, the New Year's Day of the Bikram Era. He told me to put the process of the elevation on hold until another decision.

Thursday, March 15, 2001

The tehalka.com of India has gone viral with its live coverage on the BJP Chairman, Bangaru Laxman, who allegedly received a bribe in the purchase deal of defense supplies with West End Arms. He's shown accepting the bribe in the live sting coverage on the Indian TV channels. One Major General, one Brigadier, and one Colonel of the Indian Army, with the Joint Secretary at the Ministry of Defense and other officials, altogether five individuals, have been accused as being involved in the corruption scandal. Jaya Jetley of the Samata Party, a close friend of the Defense Minister, George Fernandes, is also reportedly involved in the scam. It's also been revealed that the two have shared the same house for the last 20 years.

There's a tipoff that the Maoists have extorted Rs 300,000 from the Hattiban Resort at Pharping.

The former Indian Ambassador to Nepal, KV Rajan, is currently in Kathmandu. The objective of his visit has been kept secret.

The King also expressed his serious concerns and asked me to find out and brief him whether or not the Ministry of Foreign Affairs has had any knowledge of the visits of some Ambassadors to the districts of Nepal.

The work on the installation of the 'Command Network' for the Army by a Singapore-based Motorola Company is progressing. The first phase of the work is already completed. Some Generals in the Army question the price and quality of the communication sets of the company, asking for reconsideration of the decision made for it, whereas the CoAS, Gen. Prajwalla Rana, defends the deal as the best and the cheapest.

His Majesty's Government is set to endorse the Ordinances of immediate importance in the Parliament despite the CPN (UML)'s obstruction of the House.

PM Koirala, again, visited the former ADC, Maj. Gen Tara Bahadur Thapa, today. I briefed the King on the frequency of such visits and meetings between them. The King seemed aware of it but remained quiet as I briefed him.

The local media reported a rumor about that the Maoists will likely target the NC leader Govinda Raj Joshi.

My briefing to the King covered the call by the UN on Nepal to despatch a battalion of the RNA to the Peacekeeping Mission in Sierra Leone.

I also touched upon the fresh recruitment of female soldiers by ten percent, a common practice worldwide. He appreciated the idea and instructed me to coordinate with the CoAS on the ranking system for the recruitment.

Sunday, March 18, 2001

The King, with the Minister of Defense and the Ministry's Secretary, visited the Jungle Warfare School in Amlekhgunj for inspection. They were received by the CoAS, the CGS, the DMT and the Commandant there. In a briefing to the King, the Commandant described the school as the best for jungle warfare course, but still in need of infrastructure development.

The King keenly listened to the briefing and instantly commanded me to note down all the pointers discussed and send one copy of it to him, the CoAS and others, to see to it how much has gone into the implementation and how much of it is still in progress.

He also gave me instruction to see to it that the teams of the Army as Customs Patrols don't turn to corruption and aren't targeted by the Maoists.

Wednesday, March 28, 2001

The suspended Governor of the Nepal Rashtra Bank, Dr. Tilak Rawal, called me to update that he, at last, got justice from the Supreme Court as it gave its judgment in his favor.

At a briefing, I explained to the King the additional quota made available by the Military Staff College Course in America to the Army Headquarters, and the preparations of sending Maj. Raju Kumar Karki, ADC to the Crown Prince Dipendra, to it.

The briefing also covered the issues of climate change, its impacts, and the necessary preparedness for electric power generation, rainfall, and the availability of drinking water.

Tuesday, April 3, 2001

The Maoists have launched a massive attack on the Police Bases of Rukum and Dolakha districts simultaneously, leaving behind 35 policemen dead and several others critically injured.

Against this backdrop, the King expressed his grave concerns over the incident and instantly commanded me to find out the structure of the Maoist combat squads, command control system, their training centers and strength.

He also instructed me to convey to the CoAS to adopt transparency when it came to the purchase of logistics for the Army.

He added that it was too early at this time to provide support weapons to the newly formed APF.

The government has finally endorsed the TOR of the NSC but without taking into consideration the aspects of its secrecy and sensitivity, as it's already been leaked out before the endorsement of such a decision.

The Maoist problem has grown too stupendous for the Police alone to control it. The situation is developing itself to necessitate the rapid mobilization of the Army.

Thursday, April 5, 2001

Ramesh Nath Pandey, the Royal nominee to the National Assembly, has publicized himself as the key person to approach the Maoists. The King has expressed his displeasure at such posturing as it would only tarnish the image of the Royal Palace.

The Principal Secretary, Pashupati Bhakta Maharjan, dismissed it as 'insignificant'.

Monday, April 9, 2001

According to unofficial information, the Maoists have formed a battalion in the Sishne area, and two members from the RIM are training the cadres.

The policemen at the Pashupati Aryaghat, while performing the last rites for their fellow policemen killed in the Rukum and Dailekh attacks, sounded very angry and shouted slogans against the PM and the IGP. It's an expression of sheer dissatisfaction, or a bad sign, or an alarm of mutiny in the Police if it isn't addressed soon enough. It calls for an immediate mobilization of the Army against the Maoists.

The CPN (UML) has formed a task force headed by Jhalanath Khanal who's seeking an appointment with the CoAS to discuss pressing matters.

Sources say a Vote of No Confidence motion against PM Koirala is likely to be tabled in the House of Representatives.

The King, at the request of the PM, has agreed to a meeting on Wednesday to discuss the ISDP plans and procedures.

With the experiences seen and the reforms made to the shortcomings in the ISDP while being carried out in the District of Gorkha, it's been decided to replicate the same in the districts of Salyan, Pyuthan, Rolpa, Rukum, Jajarkot, and Kalikot.

Wednesday, April 11, 2001

None of the successive governments down the years since the start of the so-called People's War of the Maoists in 1996 has been able to evolve any substantial measures to settle the problem. By now, a large number of policemen have lost their lives in action against the Maoists, and government infrastructures worth millions of Rupees have been destroyed by them. The Maoists have also been able to double their forces to form combatants of Company strength.

The King, the PM, the DPM, the Minister of Home Affairs, the Minister of Defense, the Chief Secretary, and the Secretaries of the Ministries, the Secretaries at the Royal Palace, myself included as Military Secretary, and the security heads in detail discussed the current state of affairs at the Royal Palace.

The Chief of the NID presented a detailed report on the police operations and Maoist activities from the beginning of the Maoist problem. His report outlined the Maoists having redoubled their strength and expanding their influence over as many as 50 districts.

The NID, with its networks and resources, has been doing its utmost to collect and disseminate confidential information to the security agencies.

He expressed his displeasure at the successive governments' apathy to formulate stringent measures to nip the problem in the bud. He also accused the political parties of abusing the NID as a recruitment depot for their cadres.

The CoAS, Gen. Prajwalla SJB Rana, said that Army posts have been built in 20 more districts where all the security forces will be operated under the central command of the Army.

He emphasized on the need for a clear mandate and the simplification of the legality of Army operations to move in coordination with the ISDP, with its model experiment in Gorkha District in mind. He also underscored the need for the political parties as well as the people to support the Army to fight the Maoists and attain success to that end.

IGP Pradip SJB Rana admitted to some flaws on the part of the Police but made his displeasure at the lack of even common priority and coordination among the Police, the Army, and the government. The successive governments always held that the Maoist problem was simply a law and order issue. If the government had understood the nature of the trouble, the situation wouldn't have worsened to such a low point as it has reached now. He concluded that there's no alternative to the Army's deployment.

The King said that up to the present time, different political parties have had different understanding and approach to see the Maoist issue. The first and the foremost is that there should be a uniform approach to and identification of the goal in the name of national interest. The political parties and the state machinery must deal with the Maoists with a common approach as to whether they are terrorists or a political force. The government and the security forces should move ahead with two alternatives – operation, and dialogue; it's so because the main objective is to bring them to the political mainstream. Similarly, the government should consider a proper arrangement of arms and essential supplies of logistics to the security forces when in operation. The intelligence bureau must enhance its level of performance by insulating itself from political influences. He gave instruction to work with team spirit and perfect coordination to achieve the goal of national interest. He also said, we all must discuss the timeframe of the operation of the security forces, how the Army is to move out, and how and when it'll go back to the barracks. The security forces must remember that non-combatant civilians shouldn't be the victims during their operations. The government must give assurance of its total support to the security forces.

All the participants in the discussion expressed thanks and gratitude to the King for his instructions and guidance.

Thursday, April 12, 2001

The Nepali Congress Party in the Parliament has sought PM Koirala's resignation for failing to deliver what he had promised.

It's been reported that Lt. Gen Krishna Narayan Singh Thapa met the PM today. Thapa allegedly requested the PM for an ambassadorship or the position of security advisor to the PM.

It's also been known that the press advisor to the Prime Minister of India, Baleshwor Agrawal, visited PM Koirala. Agrawal supposedly expressed India's concerns about the King's visit to China and the Nepal visit by the Chinese Defense Minister.

The Indian state is overtly interested in Nepal's internal matters, and this quidnunc is a most sensitive issue for Nepal.

2001: The Most Fateful Year in the History of Nepal

The Royal Palace Massacre of Friday, June 1, 2001 ranks as one of the most heartrending and horrendous incidents that even shook the world. The Maoists tried to cash in on the tragedy, signifying it as the end of the monarchy in Nepal. The year is characterized by a mix of events Nepal and the Nepalis witnessed and endured with fortitude. Against the backdrop of the shooting in the Royal Palace on the one hand and the Maoists bent upon capturing the state power on the other, it was a tough task for those working in the Royal Palace Services to provide a smooth continuity to the monarchy in Nepal. The occurrences and incidents in a series complicated the smooth passage of Nepali politics. The DPM, and ultimately the PM, too, resigned, citing the noncooperation of the Army as their reason. PM Koirala fired a salvo, terming the Royal Palace shooting as a rehearsal for a 'Grand Design.' Sher Bahadur Deuba replaced Koirala and tried to give a new turn to Nepali politics by initiating dialogue with the Maoists but the drumbeat of the peace talks did not last long. Internal rifts deepened within the Nepali Congress Party circles for their own vested power politics. The Maoists, watching for an opportunity in the inter- and intra-party feuds, reinforced their strength to the level of even attacking the Army barracks in Dang. It called for the declaration of a State of Emergency, and the movement of the Army from the barracks in operations against the rebels.

Monday, April 16, 2001

PM Koirala has breached a common protocol, that of notifying the Royal Palace in advance of his actions, by airing his communiqué from the radio and TV.

Students at Putali Sadak pelted stones at the PM's car. But due to the advance precaution of the Police in place, no damage has been reported.

The Police took under control 129 political leaders, including the General Secretary of the CPN (UML), Madhav Kumar Nepal, from their street demonstrations and were subsequently released that same evening.

Tuesday, April 17, 2001

Indian newspapers have carried the frontpage news of the appointment of the Chief Admiral, Sushil Kumar, as Chief of the Defense Staff with scope to supervise the Army, Navy and the Air Force of India.

In my briefing to the King today, we discussed the preparedness and precautions to be exercised in the implementation of the ISDP. He said the Army should try to bring home to the political leaders the core objectives of the ISDP to bring the Maoists into mainstream politics.

He also instructed me to see to it that the commanders of the security forces abided by the rule of law and their professional ethics.

Wednesday, April 18, 2001

The Royal Nepal Army has been deployed for Customs Patrolling on the request of the Ministry of Finance for already a few months now, and as a result, there's been a considerable increase in revenue collection, by 33%, and decrease in pilferage, theft and crime.

The Senior Superintendent of Police (SSP), Ashok Shrestha, has reported back to Nepal Police from his service of 16 years as ADC in the Royal Palace, his return being effective from April 14. The Principal Military Secretary's Office arranged a farewell program on his return to Nepal Police.

Tuesday, April 25, 2001

The remarks made by the newly appointed Lt. Gen, Pyar Jung Thapa, at the Staff College in Shivapuri have triggered mixed reactions, mostly negative, and particularly from the political parties.

The Nepal Students Union (NSU) and the Tarun Dal, the sister organizations of the Nepali Congress, have taken the call for cooperation in the ISDP in a positive light.

Questions have been raised over the CoAS's Rs 50,000 'tips' to the *Jana Astha Weekly* for publishing objectionable pieces against the monarchy.

The CoAS briefed the King on the detection of the correction of the dates of birth in the certificates of Brig. Gens Rookmangud Katawal and Balananda Sharma by the late King's special command. The CoAS explained the preparation to retire the two officers on the basis of the original dates of their birth. He said there was no supporting documentary proof to justify the special order for the change of their dates of birth in the records of the Military Secretary Branch at the Army Headquarters. The King instructed him to look for the documentary proofs in the office concerned first.

When I was in mourning for my deceased mother in the Pashupati Area, Brig. Gen Rookmangud Katawal visited me whereupon he complained of the manipulations of the documentary proof of his date of birth by the special command of the previous King from his file by the CoAS and the Adjutant General. He recalled his being brought up by the grace of the late King Mahendra, and tearfully pleaded me to help him in this regard.

I consoled him and promised to help by searching for the documentary proof from the RNA Headquarters or from the Office of the Principal Military Secretary.

By the time I was back in office, Katawal's file had been found, and it was placed on my table. The documents in the file showed that his date of birth was corrected with due process and official formalities and after the necessary scrutiny made of the documents from the concerned school and offices. The change of the date was duly effected in 1977 on the special orders of King Birendra.

Tuesday, May 15, 2001

The CoAS today called me to arrange a VVIP helicopter for the Mul Bhatta, the principal priest, of the Lord Pashupati Nath Temple, for his pilgrimage to Muktinath. When I briefed the King on this, he told me to leave it to the CoAS himself to arrange one.

A directive to immediately repair its sewage pipes has been sent to the Army barracks of Jomsom.

A report says there's been a substantial increase in the collection of revenues after the Army's Customs Patrols were pressed into service. But the Customs staff, by inventing every excuse, aren't cooperating with the Army.

According to another report, China is displeased with Nepal regarding the recently concluded visit of the PM, Zhou Rongji. The report says Nepal took the visit lightly by keeping it in low key without even the government delegating an accompanying Minister to his reception.

PM Koirala called CoAS Prajwalla SJB Rana for consultations but it's learned that the latter sent the Chief of the General Staff (CGS), Lt. Gen Pyar Jung Thapa, as his representative.

Monday, May 21, 2001

The NSC informally met in which the PM, the DPM, and the Minister for Home Affairs, the Minister of Defense, and the CoAS discussed the issue of setting up of the NSC and its construction.

The plan for the purchase of weapons for five years for the newly formed APF was also on the agenda. The members minutely discussed the measures to make the new security force competent with training and vehicles and to enhance its intelligence network.

The members also raised their concerns about India's pressures on the PM for Nepal to purchase the INSAS Rifles from India.

The King expressed his intention to gear up the corruption control drive about which he has twice cautioned PM Koirala for the appropriate measures to be adopted. It's believed that the Commission for the Investigation of the Abuse of Authority (CIAA) must be made competent enough to take stringent measures to fight corruption.

The security forces are told to stay alert to prevent any untoward event in the wake of the CPN (UML's) call for a nationwide closedown for three days, from the 26th to the 28th of this month. It's highly likely that the Maoists will try to infiltrate the UML to incite its cadres to any eventuality.

Thursday, May 24, 2001

The King recently inspected the Recruit Training Center at Trishuli, the RNA School at Nagarkot, the Jungle Warfare School in Amlekhgunj, and the Command and Staff College at Tokha. He's soon scheduled to inspect the High Altitude Mountain Warfare Training Center at Kaisang of Manang District.

The King has given his strong instructions to the CoAS for investigating and taking actions against those who leak out military secrets. The news printed in the *Jana Astha Weekly* relating to the decision made on the promotions in the Army is a case in point.

He also expressed his concerns over the implementation and monitoring of the ISDP carried out by the Army and people's participation in it. He then pointed out the need for coordination between the theoretical and practical aspects of the ISDP.

Next, he gave instructions to set out the priorities for the military equipment to be made available to Nepal by Britain.

Friday, May 25, 2001

PM Koirala quietly consulted with the key leaders of the Nepali Congress, the RPP, and the CPN (ML) yesterday evening.

There's news of the return of the helicopter which was hit by bullets fired by the Maoist militia while on a reconnaissance mission to Banfikot in Rukum.

The Act for the newly formed APF is being formulated. My advice for the APF included four-year tenure for the IGP, its coordination with the

NSC for the purchase of weapons and equipment, and clear provisions to be made for the transfer of security personnel to the APF from other security forces.

I also advised the King to talk about the criteria of age limits of all the ranks in the APF when the IGP next visits him.

Nepali Congress leaders, Ramchandra Poudel and Khum Bahadur Khadka, are reportedly inciting people on the scandal of the accident and death of Pravin Gurung.

Ramesh Nath Pandey, the Royal nominee to the Upper House, is set to file his candidacy for the Chairmanship of the National Assembly, and canoodling party leaders to help him. He's also visiting the PM frequently to have his son, Nischal Nath Pandey, appointed as Nepal's Deputy Chief of Mission at the UN in New York.

The Cabinet meeting held today has decided to call the House for the budget session at 11 a.m. this Monday.

The meeting has also decided to help Chakra Bahadur Thapa, a former employee of the Royal Palace, with the cash assistance of Rs 50,000.

Thursday, May 31, 2001

The IGP of the APF, Krishna Mohan Shrestha, briefed the King on the progress made in the newly formed paramilitary police force. He described the provisions made by the APF Service Commission for fresh recruitment. While appreciating the progress of works, the King instructed the IGP to insulate the force from politicization and to develop the mechanism of checks and balance between the APF and the Army. They also discussed the design of the APF emblem and its presentation by the King to the Force four days before the Dashain Festivals.

The King said that the APF was a long-felt need for a paramilitary formation with its own scopes of internal security, particularly of VVIPs, and the airports – quite different functions from the maintenance of law and order for which the police force is already there in the first place.

The IGP expressed his concerns about the scope of the APF in the present context of both the police and the army being operated for security. The two discussed the possibility of the deployment of the APF to maintain border security.

The King and the IGP then discussed the Maoist problem and the immediate need for the development of separate internal and external intelligence cells for it to enhance its performance.

The King also told the IGP to consider the provisions and privileges of dispatching the AFP personnel to the Peace Keeping Missions of the UN.

The King instructed me to appraise the concept of the foundation of the NSC Secretariat to the IGPs of Nepal Police and the APF.

As per the instruction of the King, Col. Dilip Rayamajhi and I are discussing the framework of the NSC Secretariat.

Friday, June 1, 2001

ADC Col. Sundar Pratap Rana told me about the schedule of the King's program to inaugurate the Rara Festival soon, and with instructions for me to arrange for security during his visit and stay there.

Their Majesties visited the Royal Priest, Nayan Raj Pandey, at his residence in Dilli Bazaar. Back from there, His Majesty called me to update him on the weather with the Company Commander and the Brigade Commander of the Rara area, and his stay at the No. 4 Brigade in Surkhet.

He sounded a bit upset while he talked.

Monday, June 18, 2001

In my first formal briefing to His Majesty King Gyanendra Bir Bikram Shah Dev, I advised him to have the Tribhuvan Sadan cleaned and then close it; transfer its previous ADCs to Shree Sadan and have Capt Pravin Khadka transferred from the Tribhuvan Sadan back to the Army Headquarters.

Next, I briefed him on the report on the Royal Palace Massacre, prepared by the Army task force, having reached its completion. The task force comprised the former Principal Military Secretary, Lt. Gen Shanta Kumar Malla, as the coordinator, and Brig. Gen. Tika Dhamala, Acting Brigade Commander Bharat Bahadur Rayamajhi, Colonels Kumar Budhathoki and Keshavraj Mahat as members.

I advised the King to grant an audience to the heads of the security forces, and grant an appointment to meet the newly promoted Maj. Gen, Rookmangud Katawal, and Brigs. Gen Dilip Rayamajhi, Bharat Bahadur Rayamajhi, and Bishnu Bahadur Gurung.

I briefed the King on the news of the arrest of one Panchayat Shrestha, aged 25, a local of Gorakhpur, having shouted objectionable slogans against him (the King) at the BP Chowk in Butwal. It's likely that he was set up by the Indian Intelligence Agency, RAW, as a feeler and to see the resultant responses to his deeds.

The next update of developments to the King was on the progress of the ISDP in Gorkha and other districts.

The task of providing SLR rifles to the APF currently being used by the Army and the purchase of and provision for more modern weapons to the Army is progressing.

The preparations for the dispatching of an Army battalion to Sierra Leone, as requested by the UN, are underway.

Similarly, the process of the sale of the obsolete weapons of the Army and the upgrading of the No. 17 Brigade to the Central Maintenance Center, are also underway.

The Customs Patrols formed by the Army at the request of the Finance Ministry is yielding good results.

Because the Nepal Police is operated under the mechanism of the Ministry of Home Affairs, and the Army under the Defense Ministry, the two have poor coordination and communication between them except in security-related matters and reports of sensitive events.

The newly formed APF is still operating under the provision of an Ordinance without an integral Act. The next sitting of the Parliament must pass it for the new force to receive its national status.

Concurrently, the AFP is in the process of creating15,000 posts from fresh recruitment, and accepting transfers from the Army and the Nepal Police to the APF is in the pipeline.

I updated the King on the current political situations, particularly that of the CPN (UML)'s 'Wait and See' approach in the wake of the new King's moves. The Party is also taking to the streets against the Koirala Government. And, on the other hand, the CPN (ML) seems to cash in on the current situations to its benefits.

At its recently concluded Central Committee Meeting held at Kirtipur, the Party has demanded a referendum to settle the question of 'monarchy or no monarchy', the conversion of the Narayanhiti Royal Palace into the proposed Birendra Museum, and the transfer of the late King Birendra's entire property to a Trust.

The recent interview of the CoAS to the high-level Investigation Panel of the Royal Palace Massacre has come under controversy, drawing widespread criticism. It has raised doubts about and fears of the status of the RNA, its relations with the Royal Palace, and the like.

I briefed the King on all these current issues.

I also advised the King to have his portrait placed in each Army barracks and government offices.

My next advice to him was about cash prizes or awards to be given to the medical staff who displayed their best performance of duty during the treatment of the severely wounded in the Royal Palace shooting.

He instructed me to work out the modality of awarding letters of appreciation to the doctors of the Birendra Military Hospital.

Wednesday, June 20, 2001

The King granted his audience to the newly promoted senior Army officers and took from them fresh reports on the overall security situation of the country. He keenly heard the report on military operations against the rebels. The King then expressed his wish to inspect the areas of the Royal Palace and the Secretariat there.

Monday, June 25, 2001

The Speaker of the House of Representatives, the PM, and the Leader of the Opposition Party expressed their best wishes to the new King and paid their most heartfelt respects to the deceased Royal Family members in the Massacre.

Thursday, June 28, 2001

Today, I described to the King the enhancement of the security mechanism and system to monitor the arrival and departure of the visitors at his private residence, Nirmal Niwas. I asked him for his comments on the system.

The Field Marshal, Nir SJB Rana, asked me to convey to the King the news of his impending flight to India for medical treatment.

At the request of the high-level Investigation Panel on the Royal Palace Massacre, the ADC Office collected the weapons discovered in the Tribhuvan Sadan.

It would be better if the modality of action to be taken against the ADC on duty at the time of shooting comes out at the earliest.

The Maoists attacked the Khilatpur Police Post, left two policemen dead, three others severely wounded, and whisked away 31 items of weapons.

In a similar attack in Melamchi of Sindhupalchowk, the Maoists took away Rs 750,000 and 12 guns from the Nepal Bank Limited.

The Army has commenced the procedure of the purchase of an MI-17 cargo helicopter and a STOL airplane.

The King has nominated Yankila Sherpa from Taplejung, Retired Brig. Gen Dipta Prakash Shah, Dr. Roopjyoti of Kathmandu, and Dev Narayan Choudhary from Sunsari to the four vacant seats in the National Assembly. The royal nominations have been received as balanced and inclusive of Nepal's geographic regions, ethnicities, and professions.

The Heckler & Koch Company, earlier interested to supply the RNA with their HK G36 Rifles, has declined to do so, citing the reason being the non-issuance of export license by the German Government.

The RNA is seeking a process to claim compensation from the company for the loss the RNA suffered due to the cancellation of the supply of those weapons. The failure of the deal is likely to dent the image and credibility of the RNA. It should pursue a reliable alternative for the purchase of weapons with due observance of transparency.

Adjutant General Sadip Bahadur Shah allegedly made critical remarks to his close ones against the performance of the ADC Department and the Royal Guards Military Police, demanding their replacement by a special Company of the Bhairavnath Battalion.

When briefed on it, the King dismissed it, saying, 'He still has the hauteur of the Chandra Shumsher lineage.'

Monday, July 2, 2001

I advised the King to consider withholding the release of Brig. Gen Dilip Rayamajhi for the NDC course in China as he's taking an important responsibility in the operations against the Maoists. The King appreciated my advice.

A bomb went off at the Balaju Textile Factory, leaving a passerby severely injured.

The National Planning Commission (NPC) is set to host a two-day seminar on the ISDP, on July 6 and 7. Secretaries at the concerned

Ministries, Regional Administrators, Chief District Officers, CGS, DGMO and local heads of the security forces are the participants to discuss the budgetary allocations required for the districts where the ISDP is in operation.

Tuesday, July 3, 2001

Today, IGP Pradip SJB Rana, in his first briefing to the King, mentioned the recent events of the Maoists vandalizing and littering temples to hurt the sentiments of the people.

Events justify that the Maoists take safe haven in India. Even so, Ministers in Nepal don't bother to speak up against it with their Indian counterparts.

Nepal should try diplomatic channels to convince China to exert moral pressures on India to solve the Maoist problem.

There's a crisis of trust between the Ministries and the Departments thereof.

The CPN (UML) is maintaining double standards in this regard.

It's better for the Royal Palace to maintain its relations with Mahesh Acharya, the high-caliber leader of the Nepali Congress, with good intentions and guts to call a spade a spade.

Friday, July 6, 2001

I was stunned to hear Suresh Acharya, the Undersecretary at the Press Secretariat, who happened to overhear the former ADC, Maj. Gen Tara Bahadur Thapa, in a drunken state venting out his fury at the Royal Palace and King Gyanendra. Talking to Dr. Janardan Subedi in a Tibetan restaurant in Lazimpat, Thapa reportedly claimed to know the person responsible for the Royal Palace Massacre and spoke threateningly to divulge every detail. He sounded bitter at the injustice meted out to him.

When I asked why he didn't react immediately, Acharya said he preferred to keep quiet, for the obvious reasons. He handed me a paper in which he had noted all the offensive things said by Thapa.

At a briefing, I handed the paper to the King. Ever since his retirement, Maj. Gen (Retd), Thapa has been close to PM Koirala. He has campaigned to influence the PM to appoint the former CoAS, Gen. Dharmapal Bar Singh Thapa as his security advisor while also lobbying to have the former

CoAS, Gen. Pyar Jung Thapa, nominated to the National Assembly. As for himself, he wants to be Royal Nepalese Ambassador to America.

In fact, it's been often heard that Maj. Gen Tara has started bragging too much. Furthermore, he's heard to have made reprehensible remarks against the members of the Royal Family. He's also displeased with the replacement of Col. Sundar Pratap Rana as ADC. He's furthermore known to have been in close contact with PM Koirala and the leader of the Nepali Congress, Shailaja Acharya, convincing them to call for a referendum on 'either monarchy or no monarchy'.

Monday, July 9, 2001

In my regular briefing, I updated the King about the scheduled meeting of the NSC, and the matters of budgetary allocation for the districts with ISDP in operation, which is to be raised by the CoAS. He's set to ask the political leadership for their support of the ISDP, or it would go nowhere. RPP leaders Pashupati SJB Rana and others have given negative comments on the report released by the Royal Palace Shooting Investigation Panel.

The King appreciated the video documentary on the operational activities of the Army. He asked me to have it aired on TV for the common people to see it.

I advised the King to invite the PSOs of the Army Headquarters for a dinner to enhance rapport. He sounded to be considering it.

Tuesday, July 10, 2001

Today was the first briefing by Devi Ram Sharma, Chief of the NID, to the King about the Maoist activities and the current security situation as seen through the scanner of his intelligence cell. He updated the King on the information collected and disseminated to the security forces about the operations against the Maoists.

He complained of political parties of abusing the NID, and constantly trying to do so by turning it into a job center for the recruitment for their cadres, to render it thereby devoid of skilled manpower to collect confidential information. He attributed this lack of competent manpower to political interference.

The King assured the chief that everything will turn out to be fine if they (at the NID) work to their best. Meanwhile, he promised to talk to the concerned authorities to rectify things gone wrong at the NID.

Thursday, July 12, 2001

Asked by the Birendra Military Hospital authorities as to what to do with the clothes of the deceased in the Royal Palace Massacre, the King instructed them to consult with the Investigation Panel first and then incinerate them.

The CoAS sounded dissatisfied over the reduced budgetary allotment for the Army by the government and expressed fears that it might affect the mobilization of the Army in operations against the Maoists.

He also complained of the government's unclear view on the Maoists, whether they are just political rebels or terrorists. The NSC's decision only can allow the Army to move out from the barracks. In order for the Army to move forward with its operations, the government must formally declare the Maoists as terrorists, ban their party, and give the Army a free hand to disarm them by launching military operations in any part of the country.

PM Koirala has visited the RPP leaders and the members of the Parliament for undisclosed reasons.

The King has appreciated my idea of having the Crown Prince tutored in the socio-political and economic conditions of the country.

The Maoists have attacked the police post at Holeri, left one policeman dead, and abducted 72 others who have been out of any form of contact. The No 4 Brigade of the Army stationed in Surkhet has been ordered to immediately start searching for the Maoists to rescue the kidnapped policemen.

The PM has called the CoAS for consultation on this matter.

The Maoists have also detonated bombs in Baglung, Salyan, Kalikot, and Baitadi districts.

Friday, July 13, 2001

The CPN (UML) has condemned the Nepali Congress government as incompetent. They, along with some Nepali Congress leaders of the ruling party itself, have alleged that a clandestine meeting has had taken place between the King and Maoist leader Mohan Baidhya. To that effect, the DPM and the Minister of Home Affairs, Ramchandra Poudel, made an irresponsible remark on the 'Palace-Maoist Nexus.'

Following the Maoist attack in Holeri and the abduction of a large number of policemen, the Army has been moved in search operations. The Maoists foiled the airlift of the soldiers as they fired several shots at the helicopter.

Former Chief of the NID, Hari Babu Choudhary, expressed his worries at the declining morale and performance of the Police. He offered to serve the country if the King entrusted him with any responsibilities that befit him.

Sunday, July 15, 2001

The inclement weather and difficult terrains of the mountains hindered immediate search operations against the Maoists and the rescue and release of the abducted. An additional security force has already marched and is expected to reach the affected area by tomorrow.

Monday, July 16, 2001

Today, the King called me to brief him on the details of the Holeri incident and the reasons behind the lapses of the security forces. I described the security lapses and inclement weather behind the cause. Hearing about it, he sounded worried at the laxity of the security forces and the lack of preparedness and coordination.

Amidst fresh developments, the PM reportedly discussed the present political complexities and requested the senior leader and former PM, KP Bhattarai for help. The two are expected to reach a consensus to hand the leadership of the Nepali Congrees Party over to the young generation, and for themselves to remain as guardian figures in it.

The DPM and Minister for Home Affairs, Ram Chandra Poudel, is learned to have visited the senior RPP leaders for help.

Tuesday, July 17, 2001

The PM is set to reshuffle the Cabinet in order to maintain a balance in the Congress Party. NC leader and former Minister, Khum Bahadur Khadka, has vehemently opposed the PM's intention of making Chakra

Prasad Bastola a prospective candidate for PM. Khadka says Bastola is much junior to him in the Nepali Congress in terms of his years of work and dedication.

Police Inspector Surendra Nepal is reported to have committed suicide. It's indicative of the mounting dissatisfaction in the psyche of the police personnel following the Holeri débâcle.

Senior Police officers are discussing the shimmering discontents in the security forces.

The Army has set up a base camp in Rolpa. The sick have been sent from the fields of operation to the hospital for treatment.

The helicopter that was shot by the Maoists has been flown back to Kathmandu after repairs.

The CoAS is seeking an audience with the King. He has reported to me that some leaders of the Nepali Congress have been consulting with him.

I offered my advice to the King on the design of the logo and his signature for use as the Royal Seal of Approval on formal documents. He commanded me to attend to this matter via the Chief of the Royal Household.

The practice of hunting and supply of wild fowls for the table and firewood for use in the Royal Palace kitchen needs immediate revision in the context of the growing emphasis on the conservation of wildlife and forest products and in view of the increasing use of modern fuels and other alternative resources. The King instructed me to notify all the members of the Royal Family before the new provisions are brought into use.

In the context of the government intending to mobilize the Army immediately without due constitutional procedure and decision from the NSC, the relations between the PM and the CoAS have been strained.

The King sounded tough about the PM in my briefing.

The CoAS has recommended Brig. Gen Dilip Rayamajhi for the NDU training quota received from China whereas the King prefers Brig. Gen Chhatra Man Singh Gurung for the slot.

Thursday, July 19, 2001

There's speculation of the PM resigning this evening under the grinding pressures of the obtaining situations. He's also learned to be in a mood to recommend Chakra Prasad Bastola to succeed him, anticipating India's support to this move.

Bastola has complained that the King is seeking to use the instrument of the State of Emergency as a precondition for the mobilization of the Army.

The Former CoAS, Gen. Dharmapal Bar Singh Thapa, is said to have remarked that, with the whole family of King Birendra being massacred in the Royal Palace, the new King didn't trust him, and he was seeking a channel to contact him. Speaking his mind with the King's Principal Personal Secretary, Sudan Prasad Pokharel, he said he would serve the country should the King entrust him with responsibilities, and he wasn't the type to bargain for posts and privileges.

The IGP of the APF, in his briefing to the King, disclosed the speculation that the PM's resignation would lead the Congress Party to a division.

He reported on the progress of the AFP's activities. The King gave him instructions to scrutinize the current situations, and take steps to standardize the quality of training when it comes to the expansion of the Force and initiate to create an environment to have the APF Act passed soon.

Friday, July 20, 2001

Lending credence to the already rife speculations, the PM indeed resigned and triggered further speculations of power balance within the Nepali Congress Party. The Central Working Committee of the Party has formed a four-member board to elect the leader of the Party in the Parliament. The election is scheduled for Sunday. Chakra Prasad Bastola, Sushil Koirala, and Ram Chandra Poudel are in the race for the post, giving rise to apparent factional politics and politics of personal gains.

The CPN (UML) made its position clear that neither Sushil Koirala nor Ram Chandra Poudel can strike a deal with the Maoists. The Party further observed that Sher Bahadur Deuba would be the best choice in these trying times.

Given the Royal Palace Massacre, the Holeri fiasco, and the complicated turn of events in recent times, the King had wished the PM to resign. He also wished to see Deuba as the leader of the Nepali Congress in the Parliament because the Royal Palace finds him comfortable to work with.

Saturday, July 21, 2001

I updated the King on Deuba's win as the leader of the Nepali Congress and advised him to have his activities and performances as PM closely observed for a certain period of time.

The new PM entrusted Durga Subedi to establish a channel of communication with the Maoists to resolve the problem. He's learned to have started the preparatory groundworks. In response to the PM's initiative of dialogue, the Maoists, on their part, demanded the scrapping of the ISDP, a formation of an all-party government, and the writing of a new Constitution. But they're learned to have displayed their flexibility to drop their demand for holding elections to the Constituent Assembly for the time being. Their other demands include the release of all Maoist detainees, making public the whereabouts of the disappeared, and the provision of compensation to be paid to the families of the killed and disappeared. In a tactical move, they seem to be cashing in on the dialogue to widely publicize the Royal Palace Massacre as a political conspiracy, to expand their military strength, and to further train their combatants and to raise more funds.

The Secretary of the Ministry of Home Affairs has asked the CoAS to continue providing security cover to the PM even after his resignation, taking into consideration the fragile situations prevailing in the country.

Khum Bahadur Khadka is known to be bargaining for the portfolio of the Ministry of Home Affairs while the ex-PM Koirala is lobbying for Amod Prasad Upadhyaya, Prakash Koirala, and Mahesh Acharya for the same portfolio.

Monday, July 23, 2001

The King appointed Sher Bahadur Deuba as Prime Minister.

All the four chiefs of the security forces discussed the current situations in their meeting at the RNA Headquarters. According to the IGP of the Nepal Police, the CoAS lightheartedly remarked that there's been a perceptible decline in Maoist activities. He attributed it to the recently formed government. The IGP shared his bitter experiences of the years when Deuba was the PM for the first time.

In response to the CoAS's proposal of constructing facilities at the Pashupati Area to house mourners for the families of RNA soldiers by using certain percentage of the Army Welfare Fund (AWF), the King said it would be a good idea to utilize the Fund as recommended by a committee comprising former CoASs, Generals, and other retired Army officers.

I informed the CoAS of the King's suggestion, but he didn't seem to take much notice of it. He said that such an idea entailed procedural hassles and complications in the enforcement of the plans and programs.

The King instructed me to look into the process of renaming the Military Hospital at Chhauni as The Birendra Military Hospital.

Tuesday, July 24, 2001

According to a report, a clandestine meeting between the former Member of the Parliament, Deep Kumar Upadhyaya, and 'Prachanda' took place at the residence of the former. It's speculated that Prachanda was escorted to Kathmandu all the way from Ayodhya in India. Also present at the meeting were the NC leader, Pradip Giri, and contact person, Durga Subedi.

Monday, July 30, 2001

With the view to making the RNA Act compatible with the spirit of the Constitution, 1990, the Ministry of Defense, with a due circulation of its public notice, has called for experts to offer their needful counsel for the amendment to the Constitution.

Earlier, I had called Col. BA Kumar of the legal department of the Army Headquarters, for consultations in this regard. We discussed the implications of the Ministry's move and concluded that it was by no means within the spirit of the Laws and Acts. It's the government itself that makes a new Act or amends it to be effective for the Army. The Office of the Attorney General can offer legal advice if so requested. The notice is seen to raise the morale of the Army.

The Royal Palace instructed the CoAS to convey to the PM the outright cancellation of the notice of the Defense Ministry. The PM and the CoAS met, discussed it and agreed to do so. The PM also asked the CoAS to help the government's peace initiative to resolve the Maoist problem. He expects the Army's cooperation to that end.

The PM, along with his spouse, Dr. Arzu Rana Deuba, called on Her Royal Highness Prekshya at her residence, Jeevan Kunj, in Maharajgunj.

The sister organizations of the Maoist Party have organized mass meetings in several places across the country. The speakers at the mass meetings have voiced that they accord high importance to the peace talk and wish that it reaches a logical conclusion. They reportedly made critical remarks against the monarchy.

It's reported that Maoist leaders have expressed their commitment to drop their anti-Indian slogans and terms such as 'Indian expansionism' henceforward. It's also learned that the Maoists may play both the India and China cards in this tactical policy.

A fresh report has it that the Maoists are having problems in feeding and maintaining their approximately 20,000 hardcore cadres.

Political analysts opine that closeness between Nepal's democratic and republican forces can pose threats to the monarchy. The analysts likewise recommend an alliance among the political forces to isolate the monarchy. They add that the Maoists must identify republican forces and join hands with them for the common good of the country.

The Students' Revolutionary Wing of the Maoist Party is gearing up for its upcoming national convention in August. As part of its preparations, it's seeking permission from the District Administration offices for holding its mass meetings and demonstrations. Authorities have issued directives to the district authorities to allow such mass meetings and demonstrations but within the framework of the laws. They have organized open mass meetings and have shouted anti-monarchy slogans.

The NID Chief has complained to the PM about the Secretary at the Ministry of Finance for his negative comments that the budgetary allocations for the security forces are ballooning.

The PM sounds hopeful of the government-Maoist peace talks. I advised the King on the mistake made by the Sri Lankan Government's dealings with the LTTE rebels, and this can be a useful lesson for Nepal while dealing with the Maoists.

He appreciated my precautionary ideas and asked me to think about having articles published to create positive images of the monarchy.

Thursday, August 2, 2001

Today I briefed the King on the fresh peace initiative of the government.

There's every possibility of the supply of arms and ammunition to the Maoists from Indian border areas. The Army and the APF engaged in their Customs Patrols can achieve both functions – monitoring the Customs points as well as the borderlines and entry points.

The King expressed his interest to know which security force can been assigned the duty of border security. I told him any security force can do that provided there's a provision for such an emplacement.

The timeframe of one year for the Army's Customs Patrolling has expired recently and needs an extension. I advised the King to talk to the PM in this regard.

Next, I updated the King on the progress in the process of the RNA's AWF and the Chinese Bank jointly operating a bank in Nepal, the proposal for which has already been forwarded to the Nepal Rastra Bank for approval. It has triggered mixed responses from several quarters. We have to bring home to all concerned the fact that the AWF will have its involvement in the proposed bank, not the Army. Because it's sure to have good future implications, it's good if the King talks to the PM in this regard.

For some years now, Brig. Gen Dilip Rayamajhi has been entrusted with the responsibility to deal with the Maoists. The process has progressed substantially.

There's been a correspondence between the Royal Palace and the Maoists. I drafted and signed the letter on behalf of the Royal Palace. Brig. Gen Rayamajhi asked me to keep it top secret; even the CoAS has no knowledge of it.

Monday, August 6, 2001

The Dhulikhel Municipality's Mayor, Bel Prasad Shrestha, and the Vice-Chancellor of the Kathmandu University [in Dhulikhel], Dr. Suresh Raj Sharma, in an informal meeting with me expressed their fears of any eventuality in a volatile situation such as this.

They fondly recalled the late King Birendra who graciously granted KU permission to open a Medical College with a clear provision for the

AWF to invest in it and grant the Military Hospital its affiliation. Sounding worried about the future of the ambitious plan in the changed context of there being a new King in Nepal, they asked me to arrange a time for them to explain everything to the King in an audience.

I promised them to do so and expressed my hope for everything to go as planned.

My next update to the King concerned the supply of logistics to the Army at the local bases and the extension of intelligence services to its best advantages.

The *Saptahik Bimarsha* weekly published an article written by Lt. Gen (Retd) Krishna Narayan Singh Thapa, bringing to the surface the weaknesses of the RNA. Sounding like a know-all, he offered his unsolicited prescription that the Army be placed under the Parliament.

The CoAS has talked severely to the writer and suggested to me never to invite him to any event the RNA hosts.

Lt. Gen.Thapa expressed his grievances of being a victim of the government and pleaded with me to talk to the King on his behalf. The King only smiled when he heard it.

Some analysts have raised their eyebrows over Dr. Baburam Bhattarai's article about the Maoists' dialogue with the government that's to begin soon.

I advised the King that it would be a good idea to take the ethnic leaderships of the likes of Gore Bahadur Khapangi, Bam Kumari Budha, and Narayan Singh Pun into confidence to convey a good message of inclusiveness the Royal Palace pursues.

Former PM Koirala has once again ranted his dissatisfaction over the RNA and PM Deuba. He accused PM Deuba, his wife Dr. Arzu Rana Deuba, and Maj. Gen Tara Bahadur Thapa of having conspired to cause his failure in the mobilization of the Army. He has issued a threat to put the proposal of 'monarchy or no monarchy' choice into a referendum.

A book entitled 'From Dolpa to Rolpa' by Maoist sympathizer and Indian journalist Ananda Sworoop has been released. The book claims the journalist's visits to several places and his records of the actions and activities all over for a long time. The book also discloses how, without being noticed by the security forces, he could do what he did.

Wednesday, August 8, 2001

The opposition parties have opposed the Ordinance on the APF as though it was brought up to suppress the voice of the populace. The Ordinance was put to the vote and was passed by both the Houses.

A subtle analysis of events and activities suggesting the Maoists' opposition to the Royal Palace is just a tactics to keep the RIM and COMPOSA in good humor. They seem to be gearing up for a grand mass meeting in Kathmandu to unveil their future course of actions.

The PM has included former Minister Narayan Singh Pun in his core group. As per the PM's request, I advised the King to grant him an audience to command him to prepare the modality of negotiations with the Maoists.

The Editor of the *Kantipur Daily*, Yubaraj Ghimire , has asked me to arrange an audience with the King.

The Nepal Peasants and Workers' Party Chairman, Narayan Man Bijhukchhe, has aired his observations to his close aide of the King's hardening attitude and shimmering move.

The Army Headquarters has sent a request to Queen Komal to accept the role of the Patron of the Army Officers' Wives' Club. There's a long tradition of the Army Officers' Club presenting the Queen its monogram and uniform. The King has accepted the proposal and gave instructions for the arrangement.

In the recently concluded elections, Rampreet Paswan of the CPN (UML) got elected as the Vice Chairman of the National Assembly, with a comfortable margin of 31 votes against his contestant from the Nepali Congress, who earned 26 votes.

The Royal Palace had encouraged all the members to cast their votes in favor of the CPN (UML) candidate.

Dipta Prakash Shah, a nominee of the King to the National Assembly, abstained from the elections, an act contrary to his duties and ethics. When asked why he did what he did, he said his conscience prevented him from casting his vote for a Communist.

Monday, August 13, 2001

Today, the CoAS breached the established customs of the Army by attending the dinner with his wife at the PM's official residence without prior information forwarded to the Royal Palace.

The *Mulyankan Monthly* magazine has published objectionable articles by Padma Ratna Tuladhar, Lila Mani Pokharel, and Shyam Shrestha against the monarchy.

The PM and the CoAS are scheduled to visit Hile of Dhankuta to set a foundation stone of the Hile-Bhojpur Road Construction Project by the PM.

At the appointed time, the King granted his audience to the editor of the *Kantipur Daily*, Yubaraj Ghimire .

The RNA has been asked to furnish the Public Accounts Committee (PAC) with the file regarding the purchase of weapons and helicopter in order to look into irregularities, if any.

The King asked me to convey to the CoAS to brief him on the work progress with the Embassy of Israel on the supply of communication equipment for intelligence gathering. Earlier, the King had asked the Israeli Ambassador for such help.

I advised the King that a Working Dinner with the security officers would boost up their morale.

Wednesday, August 15, 2001

Blaming the government for its indifference to the peace talks, the Maoists wish to talk directly to the King himself. They want the King to create a congenial environment for talks by making public the detained and disappeared Maoist cadres.

Apprehensive of another Kilo Sierra-2, the Maoists haven't made public the names of their facilitators and interlocutors for the peace talks.

As was decided earlier, the decision to purchase modern arms and ammunition with the money earned from the sale of the Army's obsolete weapons has gone into effect.

Regarding my request to the Master General of the Ordnance to inform the PM about the purchase decision, the CoAS sounded angry with me on the telephone.

Thursday, August 16, 2001

Today, the Maoists publicized the names of the member of their peace talk team. This time around, the government alone has taken the initiative for dialogue. The PM is set to unveil a special program in this regard during the session of the Parliament.

The King has given me special instructions. Since electoral democracy has proved to be a failure in the context of Nepal, we have to think of a suitable modality of the polity. He asked me to compile a list of the names and details of those with experiences of having worked in the Panchayat Policy and Evaluation Committee (PPEC).

There's no alternative to working for the people even at the costs of covert operations against a few political and Maoist leaders.

It's important for the Royal Palace to adhere to the tenets of democracy. The King asked me to work to that end with short- and long-term programs.

Friday, August 17, 2001

Sources say a clandestine meeting between the top mainstream Leftist leaders and Maoist leaders has concluded in Siliguri of India. The participants included Madhav Kumar Nepal of the CPN (UML), Bamdev Gautam of the CPN (ML), Chitra Bahadur KC of the CPN (Mashal), Lila Mani Pokharel of the CPN (United Leftist Front), and Narayan Man Bijhukchhe of the Nepal Workers and Peasants' Party. It's said that all of them favored republic, with Bijhukchhe seeking some kind of role for the monarchy within the republican setup for Nepal.

Sources say that Maoist leaders Prachanda, Baburam Bhattarai, and Mohan Baidhya are to come to Kathmandu soon.

Bamdev Gautam of the CPN (ML) is said to have a link with one [Swami] Brahmanand Udasin who allegedly funded the splinter of the CPN (UML) into the CPN (ML). It's very likely that he's the one who's supported by the RAW to introduce republicanism in Nepal. It's the same sadhu who arranged a meeting between Gautam and ADC Maj. Gen Tika Dhamala. It's again the same sadhu who opened a channel between Gautam and the RAW.

I was surprised today when the Royal Household Secretary, Vishwo Bikram Shah, asked me for my biodata, to be sent to the Indian Embassy, as those of the Master of Ceremony, Chiran Shamsher Thapa, the Principal Personal Secretary, Sudan Prasad Pokharel, and his own having been already sent. Refusing to oblige him, I also warned him to refrain from such activities.

I also learned that he had invited the Indian Ambassador to his daughter's wedding.

He breached the protocol of palace employees having to inform the Royal Palace prior to sending invitations to foreigners or officials. It hints at India's interest in the key officials serving the Royal Palace.

I briefed the King about all these. He instructed me to develop an effective intelligence cell in the security agencies to gather confidential information.

I described the Thai models of Royal Projects in the interest of the King's subjects, the people. I advised that it would be a good idea if something similar was replicated in Nepal, too.

The King hear me keenly and said that it would be good to send a team to Thailand to study the natures, frameworks, and implications of such Royal Thai Projects to see how the concept could be replicated in Nepal as well.

Monday, August 20, 2001

Owing to his deprecatory statements, there's been a gradual erosion of PM Deuba's image within the Nepali Congress Party. Consequently, a question hangs over the very future of the peace talks with the Maoists which the PM is harping on ever since he became Prime Minister.

The government's team for the state-Maoist peace talk should be able to convince the Maoists to accept monarchy, and shun violence; and in order to make it all happen, all the mainstream political parties also should have a common position that other political agendas are negotiable.

The Maoists declined the government's proposal of the No.4 Brigade in Surkhet or the Bhimphedi Plastic Industry as the possible venue for dialogue, as proposed by the PM, to ensure secrecy and security.

Today, the former PM, GP Koirala, surprised everyone with his remark, 'By now, it should've been clear to one and all that the Maoists are operated by King Gyanendra himself, Nirmal Niwas being their headquarters. The King's bent upon sabotaging the Nepali Congress and Nepal's democracy by using Deuba as a tool to achieve those ends.'

He slammed similar accusation at the CoAS, too.

Tuesday, August 21, 2001

The Military Attaché of Embassy of China to Nepal updated me on the news of the upcoming visit to Nepal by a delegation led by the Vice

Commander in Chief of Chengdu Province. I sensed difficulty in that as the CoAS has said because the visitor has a low protocol standing, he is not entitled to visit the King. It had so happened that the CoAS did not inform the Military Secretariat while setting the program, neither did he specify the exact status of the visitor. Finally, we scheduled the appointment for the visitor to meet the King.

The amount of unaudited expenditure in the army is burgeoning every passing year. The Army Headquarters is to seriously look into it.

Today, Chief of the Royal Household Sharada Prasad Pradhan proposed a fresh provision of monitoring the King's health by the Military Secretariat. I proposed the name of Deputy Military Secretary Bhupendra Bahadur Thapa with that responsibility.

Monday, August 27, 2001

Fissures, coupled with complications and confusions regarding the government-Maoist peace talks, have surfaced. The Royal Palace has instantly rejected the Maoists' proposal to offer lifelong presidency to the King, or the formation of a State Council. The Royal Palace should take into consideration the Maoists calling themselves nationalists and at the same time weakening the very institutions of the country by continuing to serve the interests of foreign powers. To prove their nationalism, they must forthwith stop their extortion and intimidation of the people.

Sources say that the Maoist leader, Krishna Bahadur Mahara, is currently in New Delhi, presumably to discuss the contents of the imminent dialogue with the government. His movement and activities need monitoring.

The top officers of the security forces have requested the PM to issue directives to the Maoists to immediately halt their anarchic activities.

The Home Minister and NC leader, Khum Bahadur Khadka, reportedly flew to India without notifying the authorities concerned.

Wednesday, August 29, 2001

Dilip Maharjan of Kirtipur reportedly received Maoist leaders Krishna Bahadur Mahara, Top Bahadur Rayamajhi, and Agni Prasad

Sapkota with flowers and garlands in his hometown. In a hastily organized mass meeting there, they stated that republicanism for Nepal was in the pipeline; it's only to be institutionalized.

Leaders of the mainstream political parties, such as KP Oli, Pradip Giri, and Narayan Man Bijhukchhe, refused to see them.

There's unofficial information on the Maoists operating FM radio broadcast stations in several districts, such as Baglung, Gulmi, and Myagdi, to name a few. The security forces have been ordered to jam the broadcasts.

The CoAS briefed the King on the current security situation. He also told the King about his request to the PM to issue strong directives to the Maoists to disband their so-called people's government, halt verbal assaults on the monarchy and lay down their arms.

The Army's been asked to provide security to the peace talks at Godawari, the venue of the government-Maoist dialogue.

The briefing also included the information of the Indian Admiral's arrival in Kathmandu on a six-day official visit.

The King commanded the CoAS to talk to the PM to gather support from the political parties for the ISDP for it to achieve its desired goals.

The King also asked the CoAS to talk to the PM about the equipment and logistics of immediate and long-term necessity for the Army.

Thursday, August 30, 2001

The much-publicized and belated peace talk finally took place at Godawari and concluded at around 1:30 PM. The Maoist leaders left for their safe houses immediately.

Prachanda has allegedly promised the General Secretary of the CPN (ML), Bamdev Gautam, a high post and profile in the Maoist Party if the latter joins them. This move has been speculated as a tactic to prevent CPN (ML)'s merger with the CPN (UML), the mother party. It was the same Gautam who, in 1998, had proposed the Terrorist and Subversive Act in the Parliament to quash the Maoists.

The Chinese Ambassador, in a meeting with the PM, reportedly said that China has had nothing to do with the Maoists in Nepal. Instead, they are operated by India, so much so that the Maoists and the Indian intelligence agency, RAW, have entered into a secret pact to put an end to the monarchy in Nepal.

Friday, August 31, 2001

The media has given unnecessary prominence to Lila Mani Pokharel's unfounded statements on the RNA and the Royal Palace. Such baseless utterances by political leaders don't contribute to the success of the ongoing peace talks.

IGP Pradip SJB Rana has raised his eyebrows at some Army Generals' contact with the Maoist leaders.

The Maoists are doing their utmost to utilize the time allowed by the peace talks to visit the leaders of the mainstream political parties in trying to woo them to their political agenda. They have assured them of an interim government to be headed by the former Speaker of the House, Damannath Dhungana. The NC is likely to agree to this proposal.

Monday, September 3, 2001

It would be a good idea if the security forces could use the ex-army and policemen to collect information about the Maoists' activities.

The King today granted his audience to the delegation from the Chengdu Province of China to discuss bilateral cooperation.

The NSC meeting is expected to reach a decision on the establishment and functioning of its Secretariat complex.

Sources say that IGP Pradip SJB Rana quietly attended a private dinner with the PM to discuss the current security situation. He requested the PM to take the initiative to mobilize the Army, as Nepal Police alone isn't strong enough to deal with the Maoists.

The PM reportedly instructed the IGP to have the phone of Ramesh Nath Pandey tapped, as his activities have appeared suspicious in recent days.

The Minister of Communications, Jay Prakash Prasad Gupta, has reportedly instructed his Ministry officials to air discourses on republicanism on the TV. The Director of NTV (Nepal Television) needs to be interrogated in this regard.

The CoAS hosted a dinner in honor of the Chinese military delegation. The Chinese delegation leader expressed China's concerns over the Dalai Lama's activities in Nepal. He took a keen interest in Mustang District's development and expressed China's trust in Nepal's version of the Royal Palace Massacre. He said China was ready to provide the RNA with necessary logistics and equipment if needed.

According to a tipoff, the PM has met the Indian Ambassador, Dev Mukherjee; the objective of the meeting remains unknown.

The Ministry of Home Affairs has issued a strong directive to the Maoists for an immediate cancellation of their mass meeting planned in Kathmandu, asking them to also refrain from extortion or intimidation of the people.

On its part, the Army has asked the Ministry of Home Affairs to issue tough directives to the Maoists to immediately stop forcing people to feed them, threaten the families of Army personnel, and extort money from them.

Thursday, September 6, 2001

Journalist Dev Prakash Tripathi has accused the King of trying to take over and concentrate all the executive powers in his hands. A serious instruction has been circulated to interrogate him on what grounds he said it.

Sources say the Maoists are fielding a large number of people from across the country into Kathmandu for their mass meeting. In any eventuality, they are likely to use the ordinary people as human shields.

The Thankot Police Checkpost has been given strict instructions to monitor the vehicles entering and leaving it, after receiving information that the police allows public vehicles to pass through the area without any search or inquiry.

The NC leaders of the Deuba coterie have planted a rumor that the Royal Palace is trying to foil the mass meeting of the Maoists.

GP Koirala (NC President) and Madhav Kumar Nepal (General Secretary, CPN UML) have teamed up to topple down the Deuba Government.

A large number of students of the students' wing of the CPN (UML) have allegedly joined the Maoist Party.

Khum Bahadur Khadka has softened his stance on the monarchy since his recent audience with the King.

The Chinese delegation lauded the King's efforts and his political acumen to strengthen Nepal's relations with its neighbors.

Friday, September 7, 2001

Fondly recalling the late King Birendra as a true friend of China's, the Chinese PM admired his leadership and political acumen.

Following his audience with the King, GP Koirala has softened his stance with a U-turn. He has also disclosed a concept of a greater democratic front in the making in Nepal to accommodate all the parties accepting the core values of democracy. Provided the monarchy supports it, they could alienate the Maoists, he added. He has also recommended a few names to be nominated to the National Assembly.

GP Koirala and Madhav Kumar Nepal have agreed to join hands with the Maoists to embrace a republican setup in Nepal.

India is learned to have prescribed the democratic and Leftist parties to join hands with the Maoists to opt for Nepal as a republic or have a monarchy similar to that of Bhutan's.

A strict instruction has been sent to the CoAS to talk to the PM about former PM Koirala's double talk of a greater democratic alliance and the departure from republicanism. The PM has also been asked to issue a refutation against the reference to Nepal as a republic.

The King commanded me to look for ways to prevent the Maoists' mass meeting from taking place, declaring the important parts of Kathmandu 'Prohibited Areas' for protest gatherings.

Monday, September 10, 2001

The government has taken stringent measures against the planned Maoist mass meeting in Kathmandu, taking into consideration the possibility of any unpleasant eventuality. It has also announced appropriate legal actions against those shouting objectionable slogans against the monarchy.

It's learned that Dalit cadres have objected to Brahmin outfits in the senior ranks of the Maoist Party.

Their Majesties, the PM, the CoAS, and many VVIPs attended the dinner hosted by Her Royal Highness Prekshya at her residence, Jeevan Kunj.

At the dinner, the PM asked me for my help, and I promised to do my best.

I congratulated Jaya Pratap Rana, the newly appointed Royal Nepalese Ambassador to the US. He also congratulated me for having won the [new] King's trust.

Khum Bahadur Khadka reportedly visited Prachanda at Himanshu Pandey's residence in New Delhi.

GP Koirala once again accused the King and the Maoists of intending to sabotage democracy, giving rise to a pressing need for the democratic forces to collectively 'set them right'. It's generally perceived that there's no consistency in Koirala's political position and speeches.

I briefed the King on Koirala's recent public remarks.

Tuesday/Wednesday, September 11/12, 2001

Khum Bahadur Khadka reportedly told the PM that India is willing to supply weapons to the APF, with the offer of 70% of the price as grant and the rest at cost price.

The Chief of the Royal Household called me to receive the sum of Rs 5,000,000 from King Gyanendra for the trust to be named 'Their Majesties King Birendra and Queen Aishwarya Treatment Fund' for the medical care of the Royal Family members and the employees there.

I talked to the King about the present CoAS and his remaining term in office. With only one year left for the CoAS to retire, and as per the system of the Army, it would be a good idea to prepare the next CoAS in advance. CGS Pyar Jung Thapa is one such candidate for the position in every respect, in seniority and for his loyalty. The King told me to talk to him, and when I did so, Thapa sounded very happy and expressed his assurance to work to the best of his ability.

Just when we were discussing some future plans, the breaking news of the terrorist attacks at the Twin Towers of New York flashed on the TV screen. I called the King to inform him, and he said he was watching it, too. He discussed the aftereffects of the terrorist attacks in the world and in Nepal, too. From this attack on the Twin Towers, it's enough indication that no country in the world is safe from terrorism.

The King then told me to issue instructions to the departments concerned to provide security to American offices in Nepal.

He also advised me that, should the state-Maoist peace talk fail, the government must take stringent measures to prevent attacks or violence of any form.

Thursday, September 13, 2001

As planned, the government team headed to Bardia for the second round of the peace talks. Maoist leaders also arrived there. Sources say they announced their so-called central level as 'People's Government'. Such political postures on the part of the Maoists will only sour the atmosphere of trust and the very future of the dialogue.

Monday, September 17, 2001

Today, the former IGP of Nepal Police, Dhruba Bahadur Pradhan, informed me of his plan to write a number of pieces and have them published in order to clear every aspect of the Royal Palace Massacre, then to be followed by an interview with Maj. Gajendra Bohara, the then ADC to the then Crown Prince Dipendra. This will help clarify the mysteries still lingering in the minds of the people about the mass shooting let loose by Dipendra.

In case the peace talks fails, the government must declare the Maoists and their sister organizations terrorists, and ban their activities that curtail the people's individual freedom. The government also should display its wisdom by declaring a general amnesty for those who were forced to join the Maoists under duress. The government also must take actions against Lila Mani Pokharel for his recent critical remarks.

The Maoist politburo meeting in Rolpa claimed the success of the People's War. The meeting singled out the King as the ultimate enemy and prescribed the clarion call of republicanism to bring all the other political parties to join hands with them.

The non-partisan Leftist leader, Padma Ratna Tuladhar, was seen in a vehicle with a diplomatic number plate entering the Indian Embassy.

Girija Prasad Koirala has again ranted the same old rhetoric and criticized the noncooperation of the Army. He and his faction are discussing the formation of an interim government with certain amendments to be made to the Constitution to have the Army under the Parliament, and also raise the issue of the successor to the throne a flexible one in the question of gender.

It's learned that some members of the talks team are leaning to the concept of republicanism, thereby gradually clipping the authority of the monarchy.

Tuesday-Thursday, September 18-20, 2001

Lt.Gen. Sadip Bahadur Shah, in connivance with the former CoAS, Gen. Dharmapal Bar Singh Thapa, is learned to have leaked top confidential information of the Army and the Royal Palace. From his tapped telephone, it's known that he tips off secret information to Kishor Shrestha of the Jana Astha Weekly.

The King has given his Royal Seal of Approval to the APF Bill. He's to present the insignia of 'Banglamukhi' to the newly formed force.

There have been incidences of public disorder, such as vandalism, smashing of furniture, damages to public structures and vehicles in some places of Kathmandu. A vehicle of Radio Nepal has been torched in Patan.

The King today asked me about the actions taken against the Maoists in the Kathmandu Valley.

He also discussed the position of the Royal Palace in the context of the amendments to the Constitution.

Wednesday, September 26, 2001

India has stepped up its security presence along the Nepal-India borders.

China also has similarly expanded its security watch and surveillance in the northern border areas.

Friday, September 28, 2001

Sources say India's West Bengal State Government has strengthened its border security, arresting a handful of Indian Naxalites, thus causing it difficult for the Maoist leaders to enter Siliguri.

The Cabinet has approved of the strength of 51 office staff and the necessary budget for the APF in order for the security force to have its secretariat office at the NSC. It's advisable that the Secretariat be close to the PM's office and staffed by professionals from the RNA, the Nepal Police, the APF, and the NID.

There's information of pamphlets being distributed by one Islamic Front in Nepalgunj, lauding Osama bin Laden as a great leader. Buddhism and Hinduism have also been severely denounced in the pamphlet, with a threat to launch actions against the Shiva Sena.

There's also fresh information of a likely merger of the CPN (ML) and the CPN (UML) anytime while quite a large number of youths from both factions have already joined the Maoist Party.

A team comprising Colonels and Lieutenant Colonels, and headed by Brig. Gen Tika Dhamala, has been commissioned for the efficient security of the Royal Palace.

The King expressed his grave concerns that, with the current situations in Nepal, India has raised its security concerns in the border areas. But when asked about it, the PM claimed to have no knowledge of it.

He said, given the failure of the two rounds of peace talk with the Maoists, the government must develop new measures to take the talk to a logical end.

Monday, October 1, 2001

The Dang District Administration has clamped a curfew following the death of a student in a clash with the police. It's likely that there was Maoist infiltration to incite the mob to their benefits.

A fresh report received here says that the Syangja People's Local Government (the Maoist political structure) Chief has issued a strong order to the people to henceforward have their administrative works done from their government.

The Maoists handed over 17 of the abducted policemen to the ICRC.

Unable to enter Siliguri, Maoists leaders have returned to Kathmandu in a vehicle with the diplomatic number plate 53-CD-1.

Monday, October 8, 2001

The European Union, according to a source of information, is planning to raise funds for the families displaced by the Maoists. The assistance, according to the EU, will be for the reconstruction of the infrastructures destroyed during the insurgency.

The American and British Air Forces have launched massive attacks in Kabul, Jalalabad and Kandahar of Afghanistan. The attacks will have ripple effects in India and Pakistan, and to a lesser degree, in Nepal as well.

It's apt for Nepal to provide additional security to the British and American facilities in the country.

In such an international scenario, the government is likely to take a tough position to deal with the Maoists.

Today, the King commanded me to arrange a helicopter for Swami Kamalanayanaacharya for his religious sojourn to Muktinath, as requested by Her Highness Prekshya. She's the widow of former Prince and the late Dhirendra, brother of the King. Of late, she's been making various demands for resources and services.

The British Military Attaché updated the CoAS on the drone attacks in Afghanistan. Some detachments of the SAS of the British Army have already been moved to some parts of Afghanistan.

Tuesday, October 9, 2001

The IGP of the AFP, Krishna Mohan Shrestha, updated me on Khum Bahadur Khadka's plan to visit the Maoist leaders in Australia.

He also expressed his plan to have the King invited to a dinner after the APF receives its Banglamukhi insignia.

Thursday, October 11, 2001

In the last couple of days, there's been an increase in Maoist activities on the fringes of Simikot District where they are raising funds by trafficking in and selling unauthorized herbs and medicinal plants to purchase essential supplies like food and clothes from China. It calls for surveillance on such activities by keeping the Chinese authorities in contact.

Gyandas Shrestha and Raju Dangol of the Royal Palace have been found leaking confidential information of the Royal Palace to newspapers. It's learned that Gyandas Shrestha, a Royal Palace staffer, is actually a parttimer at the *Nepal Samacharpatra.*

The PM called all the heads of the security forces for a meeting to discuss the current situation of the country.

Monday, October 15, 2001

The King today expressed his wish to announce his son Paras as Crown Prince and confer on him with the honor roll of Prithuladhish on the auspicious occasion of the Dashain, the great annual festivals of the Hindus. I advised him that it would be too early to do so as it would convey an unpleasant message to the people.

I got wind of the CoAS, Gen. Prajwalla SJB Rana, secretly discussing with a group of his trusted officers about replacing me with Lt. Gen Victory SJB Rana as Principal Military Secretary.

Thursday, October 18, 2001

Lately, there's been a perceptible increase in the Maoist peace talk team facilitators' arrivals in and departures from Kathmandu. They were received with bouquets at the Thankot entry pass. Their supporters and sympathizers shouted the same old slogans against the monarchy.

Another report recently received has it that after their lunch in one Moti Krishna Shakya's house at Thankot,they planned to hold their press conference at Orchid Hotel at Tripureshwor.

The King is scheduled to visit the Nawa Durga Temple in Bhaktapur on the auspicious occasion of the full moon day of the Dashain Festivals. His ADC, Colonel Sundar Pratap Rana, already visited the venue earlier for security checks.

Friday, October 19, 2001

His Majesty the King today granted audience to the Colonel Commandant of the Brigade of Gurkhas, General Sir Sam Cowan, who is back in Kathmandu from his visits to a couple of districts. General Cowan briefed the King on various welfare programs operated in the rural areas with British assistance. He also noted that he found the people in those districts furious at the Maoists for their extortion and political intimidation. He described the areas as being beyond the access of the government, and corruption taking deep roots at all levels, and the need for the ISDP program's extension in other areas as well, with local political involvement, and a tactful address to the Maoists' political moves. He cautioned that an alternative of military solutions to the Maoist problem would only result in worst case scenarios.

The King expressed his cordial thanks for the British assistance to the RNA with the hope of similar help in the future, too.

He sounded critical of the unsolicited foreign meddlings in Nepal's internal affairs, even to the extent of enticing many political leaders.

Soon after, a dinner followed which the CoAS, Gen. Prajwalla SJB Rana, the CGS, Lt Gen Pyar Jung Thapa, other Honorary ADCs, and I joined.

Tuesday, October, 30, 2001

Reports have been received about the formation of resistance units against the Maoists in several districts. The King commanded me to direct the Administration to support such forces.

He also asked me to explore ways as to how the Khumbuwan and Madhesi Liberation Fronts could be used.

There's a report of the Chief of the Indian Intelligence Agency, RAW, currently visiting Kathmandu in a holiday sojourn. It's reported that he's scheduled to see the PM and other high officials.

There have been mixed reactions over the King's announcement of Paras as Crown Prince. The likes of Padma Ratna Tuladhar and Krishna Pahadi sounded critical at the move. Tuladhar and Pahadi are human rights activists.

The student wing of the CPN (ML) has also vehemently opposed the decision.

Indian channels, such as Zee TV and Star TV, aired negative comments.

The RNA and the Army Officers' Wives' Association, on the other hand, are set to express their congratulatory felicitations to the Crown Prince.

Similarly, the Kantipur TV has, it's been learnt, approached the Crown Prince for an exclusive interview.

Monday, November 5, 2001

The Indian CoAS's scheduled visit to Nepal has again been postponed.

There's information of the Maoists beating up some locals in Phalate of Nuwakot District. An Army detachment has been mobilized to reach there for action.

The government has scheduled the SAARC Summit in Kathmandu from January 4 to 6. It's the second meet of this kind being held in Nepal.

The Maoists have intensified their opposition to the announcement of the new Crown Prince.

Given the series of retaliation by the locals against the Maoists, their de facto chiefs in these localities have started to resign.

The soldiers deputed at the ISDP have splendidly coordinated with the police for security.

There's been an increase in road accidents and lawlessness in Kathmandu.

Friday, November 9, 2001

Meeting me, the Defense Ministry's Secretary, Padam Kumar Acharya, shared his experiences of his visits in China. He sounded very happy, fondly admiring China's goodwill for Nepal as a good neighbor. He also said the Chinese admired the late King Birendra for his political insightfulness. They advise Nepal to have good relations with India, too.

Brig. Gen Dilip Rayamajhi, who was one of the members in the entourage, admired the Chinese President for his high respect and regards for the late King Birendra. He said the President was deeply affected by the news of the murder of the popular King. The President asked the members of the entourage to convey his best wishes to the new monarch, King Gyanendra, and his eagerness to see him in China. He believes that King Gyanendra's first visit to China will help remove the Chinese people's confusion about the Royal Palace Massacre.

He reiterated China's offer of help to the RNA through proper channels.

The President also expressed his displeasure at the rise and spread of terrorism in the world, and he reiterated that China can no more tolerate American hegemony.

The Koirala faction in the Nepali Congress is discussing the removal of PM Deuba. They think he's working in the interest of the Maoists.

Upendra Yadav and Saraswoti Choudhary of the Madhesi Jana Adhikar Forum (MJF), at a mass meeting in Saptari District, made highly offensive remarks against the monarchy. A report says they garlanded the portrait of the King with shoes and smeared it with soot. The local administration has been directed to proceed with legal actions against these leaders.

The Magar youths of Syangja pleaded with the King for water supply for the people there and issue directives to the Army for the recruitment of Syangja's youths in the RNA.

The King was in Syangja in connection with his worship of the deity Alam Devi. He issued directives to the authorities concerned in the district to start renovation of the temples and shrines across the region.

Monday, November 12, 2001

PM Deuba invited Dr. Narayan Khadka of the Nepali Congress and the Chairman of the State Council Standing Committee, Dr. Keshar Jung Rayamajhi, to a dinner where he asked the Chairman to help him for the state-Maoist peace talk to succeed.

There's bad news of the crash of the helicopter into the Rara Lake and the death of Princess Prekshya along with others onboard.

At around 11 AM, two bodies, identified as those of Dr. Rishi Khatri and Miss Sharada Singh, were found floating on the water. They were rescued.

The body of the Princess was discovered at around 3 PM. At the wish and request of her daughters, the mortal remains of the Princess were brought to Jeevan Kunj where her family, friends and relatives paid their last respects to her. Her last rites were performed at the Pashupati Aryaghat according to Vedic traditions.

Budgetary insufficiency has paralyzed the ISDP for some time now.

It's also been quite some time since soldiers haven't received their allowances.

There's fear of these news sending bad messages.

Complaints from different quarters are also pouring in against civil servants being absent at their offices and workstations.

Tuesday-Monday, November 13-19, 2001

The Government-Maoist talk resumed at the Godawari Resort at around 4 p.m. but it didn't progess satisfactorily as both sides held firmly to their positions.

Offices reopened after the Dashain holidays.

There's been an increase in the incidents of thefts and burglaries in the Kathmandu Valley.

The peace talk ended without reaching a mutually satisfactory stage. Another round of talk is expected to take place by the second week of next month.

The purse of the late Prekshya has been found in the Rara Lake with money, gold coins and cosmetics in it.

During an AFP roll call one evening at Khalanga of Jumla, policeman Yudda Bahadur BC was found in civil clothes and drunk. While being inspected by his superor, Inspector Ishwar Nath Pandey, he fired a shot at the officer and bolted from the camp.

The CoAS shared an interesting story with me about Mukunda Shamsher, a former Secretary at the Ministry of Defense, pleading with him to recommend him to the King for Ambassadorship either to India or Britain.

The Army shall stand behind the Nepal Police during the SAARC Summit for security cordons. The Army, under the command of a Major General, shall operate its round-the-clock security and surveillance of the venue of the Summit.

With a view to studying the modality and feasibility of help to Nepal for its ISDP, the officials of the American, British and Norwegian Embassies have been to Gorkha.

The EU isn't happy with the government for its failure in corruption control. Therefore, it's contemplating its assistance to be carried out through INGOs or the RNA.

The King declined to accept a doctoral degree from a well-known University of the US, a request made to him by Professor Janardan Subedi.

The Maoists, according to a new report, are gearing up for another round of grand mass meetings and a showdown in Kathmandu. They are also lobbying with the leaders of the political parties to win them over to the concept of republicalism in Nepal.

The King told me about the implementation of the Ordinance on Customs wherein the active role of the Army has been delineated.

He also instructed me to concentrate more on the current ISDP rather than expanding it to other districts.

Thursday, November 22, 2001

Prachanda, Chairman of the Maoist Party, has circulated a special decree to the entire party cadres to ready themselves for another round of massive attacks at a time when the CoAS is on a visit to Vienna, in Austria. He's there to inspect the machinery, weapons and explosives to be purchased for the Army.

Due to this new situation, he's immediately returning, cutting his visit short.

Reports recently received say that GP Koirala is gearing up his signature campaign in the NC to have PM Deuba removed. He's trying to secure a minimum of 66 signatures.

Today, the King candidly discussed with me the points of amendments to the Constitution, or the elections to a Constituent Assembly as options to break the political impasse. He instructed me to sort out the points the Royal Palace should attend to and prepare a concept paper in this regard.

Saturday, November 24, 2001

The Maoists launched their massive attacks simultaneously on the Bhagawati Prasad Company, the Police Post, and other government offices of Ghorahi of Dang and Syangja districts.

They robbed the bank and made away with cash and weapons. According to a report, they have taken the District Headquarters under their control.

The attacks on the Army barracks in Dang and the Police Post claimed the lives of 14 soldiers and 22 policemen, including the Company Commander. Subsequently, a large security force has been despatched to the beseiged areas.

When I updated the King on the news of the attacks, he told me to immediately call the CoAS, deploy more security personnel, make a detail report on the losses, and concentrate on the government's efforts to recover the weapons seized by the Maoists.

The scale of the attacks and the resultant losses of lives and property, the absence of warning and the lack of preparedness to avoid the debacles all speak of the laxity of intelligence and of the security forces. Had the security forces adopted vigilance and preparedness, the attacks could've been dealt with most effectively.

Frantic, the PM called an emergency meeting of the heads of the security forces to discuss the need to declare a State of Emergency, put the Maoists on the terrorist watchlist, and launch massive operations against them.

The King consulted with the PM and the heads of the security forces in the wake of the recent attacks.

Monday, November 26, 2001

The Maoists launched an attack in Solu Salleri of Solukhumbu District at 11 p.m. Gunfire was heard throughout the night till dawn. An additional detachment from the No 2 Brigade has been despatched to the site of the clash. The attack has left the CDO, two Inspectors, 27 policemen and four soldiers dead. On the Maoist side, 15 dead bodies have been discovered so far.

Against the backdrop of the fresh attack and the death of the CDO, the PM called an emergency meeting of the Parliament, and the meeting decided to implement the Ordinance for a State of Emergency.

The meeting also decided to mobilize the Army across the country with a timeframe of one year.

For the Army to move against the Maoists, it has proposed to set up a special taskforce in Kathmandu, with Divisional Headquarters in Itahari and Nepalgunj.

In addition, there's also a proposition being made to setting up a Joint Command Center at the Directorate of Military Operations (DMO) at the RNA Headquarters.

Sujata Koirala, a leader of the Nepali Congress and the daughter of Girija Prasad Koirala, is currently in Delhi, and she's scheduled to visit Indian leaders there, bearing the message of her father to them.

Surya Bahadur Thapa, known as pro-India, Arjun Narsingh KC, and some leaders of the CPN (UML) have accused PM Deuba of intending to sabotage Nepal's democracy to serve the interests of the Royal Palace.

Wednesday, November 28, 2001

The Ministry of Home Affairs would do well to circulate a notice asking those who were forced to join the Maoists to surrender themselves with their weapons.

Clear instructions should go to the security forces for the extent they can go to. Their joint operation plans should be clear.

The Army should have efficient communication networking sets.

A working dinner with the King and the heads of the security forces would boost their morale to fight the rebels.

The District Administration Offices of Dang, Khotang and Udaipur have clamped curfew in their respective areas.

Friends of Nepal in India, China and the Western countries have lauded the government's decision to declare the State of Emergency in the greater good and security of the people.

On the spur of the insurgency and Emergency, there's been a trend among the civil servants posted in far-flung districts to seek transfer to Kathmandu.

According to the Military Attaché at the Royal Nepalese Embassy in New Delhi, there's been a telephone conversation between the CoAS of the Nepal Army and the Chief of the Indian Armed Forces.

The PM of India has given his express instructions to supply Nepal with logistics of the required quantity within ten days.

When I briefed the King on India's offer of help, he said there's something up the sleeve behind the offer.

The meeting of the Secretaries of the Ministries reached a decision for the King to address the security personnel of all ranks to work with high morale and confidence.

Thursday, November 29, 2001

The Military Attaché of the Indian Embassy in Kathmandu informed the CoAS's secretary that Indian logistics equipment will reach the borders in five days. He has also asked the Royal Nepal Army Headquarters to send four pilots and two technicians to Ranchi in India to fly the helicopters, provided as India's assistance, to Nepal. India has also provided weapons for the APF.

A large number of Maoists are reported to have been shot dead in a military operation in the Khumel Village of Rolpa District.

In response to the King's inquiry, I said it would be better not to have Nepal Police and APF personnel stationed in the Royal Palace for possible constitutional complications.

Monday, December 3, 2001

The PM's talk of purchasing weapons from India has generated mixed reactions from different quarters in the country.

The construction of tall buildings around the Royal Palace may pose security threats. The PM must talk to the Mayor of Kathmandu Metropolitan City to this effect, to halt or ban the construction of such houses near or around the Royal Palace.

The CPN (UML) has supported the government's decision of declaring the State of Emergency.

Since it's the third month of the imposition of the State of Emergency in the country, it calls for honest evaluations of the situation for guidance to the future course of actions.

It sounds good to form a special committee of experts for the evaluation of the post-Emergency effects.

The King told me to update him and the PM on the progress of the military operations against the Maoists.

Friday, December 7, 2001

The Secretary at the Ministry of Foreign Affairs, Madhu Raman Acharya, and the Secretary of the Defense Ministry, Padam Kumar Acharya, briefed the Chinese Ambassador on the current security situations, and asked him for help.

In response to Nepal's request for help, the Ambassador said China would prefer to provide Nepal with logistics necessary for the security forces.

There's information that Capt. Suresh Hamal's pistol got accidentally discharged by his servant.

Monday, December 10, 2001

Sources have it that, while in training, two soldiers lost their lives in a friendly fire in Surkhet and Nepalgunj.

The Brigade Commander, Brig. Gen Bajra Gurung, has been removed for failing to exercise preparedness prior to the Maoist attack in Dang.

The Ambassador of South Korea in Nepal invited me to attend the Joint Seminar of Army Generals and Military Experts from March 1 to 10 in Seoul. I expressed my thanks for the invitation, but given the situation in the country, there was no question of my going there to attend it.

The King told me to pay attention to the sufficiency of medicines and the quality of food for the soldiers to keep their morale and confidence high.

He also expressed his concerns over the emerging personality clashes and superiority complex among the high-ranking Army officers. He said this was no time to exhibit such petty personal egos in a sensitive period like this.

I assured the King that placements, promotions and transfers in the Army on the basis of merits is the cure-all solution to such problems.

With the King, I discussed the amazing progress of India due to its consistent and efficient diplomacy. India and China are emerging as power centers in the world. India is seeking to obtain a permanent membership at the UN Security Council. The SAARC is of nominal importance to India. Given the regional imbalances and its strategic location, Nepal is walking a tightrope in this aspect. It must observe its immediate neighbors with a policy of equidistance. Any closeness with one may offend the other, and Nepal should minutely observe this without fail.

Monday, December 17, 2001

The Maoists launched another round of attack at Salleri of Solu Khumbu and left Corporal Rudra Raut dead.

In Biratnagar, GP Koirala accused PM Deuba of losing hold on the Ministers of his Cabinet. Sources say he has stepped up his signature campaign to oust the PM. He has, for this purpose, approached the CPN (UML) leaders to form a greater democratic alliance.

The King told me to take special note of the expenses and the budget in the purchase of weapons. Such expenses should reflect transparency because the murmurs of irregularities are coming to the surface. He told me to cautiously watch the activities of the Minister of Home Affairs who is taking unnecessary interest in the purchase of weapons for the APF.

The King expressed his concerns over the Army still being defensive, without causing any noticeable damage to the Maoists even as it's already been three months since the Army's mobilization.

Wednesday, December 19, 2001

There's information about the possible merger of the CPN (ML) with the CPN (UML) soon.

There's also report of some high-ranking police officers tilting to corrupt political leaders.

The Nepal Police IGP, Pradip SJB Rana, seems to be of an easygoing nature, losing his grips on the institution.

Friday, December 21, 2001

As scheduled, the King inspected the proposed sample uniform of the Army at the Royal Guards Brigade.

He also sought information about the long drawnout plan of the Corps Concept in the Army.

On the same occasion, he gave instructions to the CoAS to make an alternative provision for taking oaths with the national flag in hand for female and Muslim soldiers.

The India-Pakistan relations have soured now, and its shadow is likely to reflect on the SAARC Summit in Kathmandu. The Pakistani PM's flight route via China is one such instance to justify such apprehension.

The threat of war between the two countries is heating up. How much can the approaching SAARC Summit contribute to mending the Indo-Pak relations is yet to be seen.

The Indian Ambassador to Nepal hosted a luncheon in reception for the visiting Secretary of the Ministry of External Affairs of India. Diplomats, leaders of the political parties, large or small, writers, and the Secretaries at the Ministries of Home Affairs, the Defense, and the Foreign Affairs, and the heads of the security forces attended the lunch.

Likewise, the Ambassador of Pakistan to Nepal hosted a dinner in honor of the foreign delegates to the SAARC Summit and the Ministers of Nepal. He said the Singapore Model of development best fits in the Asian countries for their progress and advancement.

The Minister of Foreign Affairs, Dr. Ram Sharan Mahat, hosted a reception in honor of the foreign representatives attending the Summit at The Yak & Yeti Hotel.

The water supply systems in some districts have suffered as the Maoists have made their socket bombs out of the metal water pipes.

Monday, December 24, 2001

It's very likely that the Member of the National Assembly, Ramesh Nath Pandey, will join the comprehensive democratic alliance campaign put forward by the former PM and NC President, Girija Prasad Koirala.

On her own part, another NC leader, Shailaja Acharya, has been a vocal critic of PM Deuba. According to her, given the declaration of the State of Emergency and the mobilization of the Army in the country, democracy in Nepal has become something like the search of a black cat in the dark.

I briefed the CoAS on the outpouring of complaints that the general public didn't know about the Army's operations against the Maoists. The Army's thinking of operating a Public Relations Directorate of its own to disseminate information of public importance.

Wednesday, December 26, 2001

Taking into consideration the security situation of the Kathmandu Valley, surveillance of the vehicles passing by and to and from the Singha Durbar has been stepped up.

Some units of the Army have been supplied with the sophisticated M-16 Rifles.

Army intelligence says that some Maoist leaders and cadres are quietly sheltered in Kathmandu.

In this connection, I requested the King to form a team of experts to offer him advice on matters of security.

The Nepal Tech Company needs to be immediately closed and stringent legal actions must be initiated against it for making satellite phones available to the Maoists.

Friday, December 28, 2001

The security agencies have been sent a circular of immediate importance to prepare an inventory of the Maoist combatants and their identities. The rebels have code names, and they use such pseudonyms to prevent easy identification in case of arrest or death in operation.

The preparations for the SAARC Summit have reached its final stage. The Summit, the first in the New Millennium, is going to be very important in a number of ways. Nepal should spare no efforts to make it a grand success, with the inclusion of poverty alleviation, mutual cooperation to fight terrorism, and the establishment of the SAARC Development Fund, and other issues at the Summit.

The CoAS briefed the King on the overall security situation since the declaration of the State of Emergency. He also submitted to the King the latest progress report on the Army's operations against the Maoists, the purchase of essential supplies, and recruitment and trainings.

Monday, December 31, 2001

The King keenly took detailed information on the status of the preparations made at the BICC (Birendra International Convention Center) at New Baneshwor, the venue for the SAARC Summit, and the security arrangements in place for the visiting VVIPs. He then next inspected the Soaltee Hotel and the Himalayan Club to see the management for the stay of the SAARC heads of state during their visits to Kathmandu for the Summit.

The King commanded CoAS Rana to arrange henceforth for Lt. Gen Pyar Jung Thapa to be present at every briefing to be made to the King. Lt Gen Thapa is in line to succeed as the next CoAS in the near future.

The King asked me to look into the process of doctors' appointments at the Military Hospital. The Chairman of the National Assembly, Dr. Mohammad Mohsin, has had appealed to the King for his daughter's position as a doctor there. The CoAS explained the situation being the first of its kind in the Army, having to improvise the traditional criteria for recruitment and for taking the oath of office and secrecy because the candidate is a female and a Muslim to be appointed a doctor in the Army.

A report received says that the Pakistani President, Gen. Pervez Musharraf, has preferred to fly to Nepal via China. This choice of his flight hints at the strained relations between India and Pakistan. The SAARC Summit in Nepal hopes to mend the embittered relations between the two SAARC member states and neighbors.

The Indian Ambassador to Nepal hosted a luncheon in honor of the Indian Secretary of Foreign Affairs. Ministers, Secretaries, heads of the security forces, and key leaders of the political parties joined the luncheon and shared their ideas and experiences ahead of the Summit.

Friday, January 4, 2002

The King separately granted his audience to the heads of state and heads of the government of the SAARC member countries in the Narayanhiti Royal Palace. He also hosted a dinner at the Royal Palace in their honor.

Monday, January 7, 2002

The King attended the dinner hosted by the CoAS at the Tribhuvan Army Officers' Club in honor of the officials at the Secretariat of the NSC.

The King issued his special instructions to the officers to perform their work to the best of their ability and in the interest of the country and the people.

In response to the King's directives, the Joint Coordinator at the NSC Secretariat, Maj. Gen Rookmangud Katawal, assured the King of every effort to be made in the service of the country and monarchy.

The PM and the Secretary at the Ministry of Defense were present on the occasion.

Even as the Army's been moved for operations to check the Maoists, it's still far from achieving success, as it was expected to. For that to be accomplished, the morale and confidence of the soldiers need to be boosted, and suitable weapons must be provided.

It seems that a merger is taking place between the CPN Unity Center and the CPN Mashal.

Some analyses and comments have emerged from the ongoing SAARC Summit in Kathmandu. The representatives at the Summit admired the security arrangements in place. That India didn't accord much importance to the Summit has been perceptible. PM Deuba, representing the host member nation, sounded fickle in his address to the Summit.

The CoAS is retiring soon, on September 8, but lately, he's believed to have planted a rumor of getting an extension of his service by a special command from the King.

There's report of the General Secretary of the CPN (UML), Madhav Kumar Nepal, holding a closed-door meeting with Ramesh Nath Pandey at the latter's residence.

Khum Bahadur Khadka left for Bangkok for his health checkup and treatment.

The unification process of the CPN (ML) and the CPN (UML) has reached the final stage, and the announcement of the merger is expected soon. According to a tipoff, the Maoists have issued a circular to assail pro-Palace people, particularly those close to the PM.

They are also trying to create a fear psychosis by making an excessive use of Improvised Explosive Devices (IED).

They have further adopted a policy of engaging the government in talks while also resorting to individual murders and accumulation of weapons.

The NSC has tipped off Indian authorities regarding the Nepali Maoists for their extradition to Nepal.

Thursday, January 10, 2002

A special investigation team of the Nepal Police has interrogated a diplomatic official of the Embassy of Pakistan on the charge of dealing in counterfeit Indian currency notes.

The government has decided to place an SSP (Senior Superintendent of Police) each from the Nepal Police and the NID, along with a Military Attaché, at the Royal Nepalese Embassy in New Delhi, India.

Friday, January 11, 2002

An Endoscope, costing Rs 27 million then and imported for the Military Hospital, is reportedly lying unused in the store for the last one year. It indicates misappropriation of resources and an abuse of the Army Welfare Fund.

Monday, January 14, 2002

Today the CoAS and I discussed the most appropriate way to receive the American Secretary of State, Gen (Retd) Colin Powell, at the Royal Nepal Army Headquarters. The discussion touched upon the future implications of the precedent of presenting the Guard of Honor to the VVIP visitor at the RNA Headquarters. We decided to take a decision after our consultation with the King. The King, on his part, commanded us to accord the Guard of Honor at our own discretion.

This visit to the RNA Headquarters by a foreign VVIP is the first of its kind.

Thursday, January 17, 2002

There's a report of 30 prisoners bolting out from the Surkhet prison last night. A few of them are Maoist suspects.

Friday, January 18, 2002

As scheduled, the American Secretary of State, Gen Colin Powell, arrived on a visit to Nepal.

While discussing the Maoist problems with the CoAS and senior Generals, he also expressed his extreme pleasure at being presented the Guard of Honor, the first of this kind at the RNA itself – accorded to him at the RNA Headquarters. He expressed his hearty thanks to the host, remarking that this was the first such honor he had received after his retirement from the US Army as its Chief of Staff.

The CoAS briefed the American VVIP on the protracted Maoist insurgency, the State of Emergency, the mobilization of the Army, and the inadequacy of its weapons and logistics. The CoAS also reminded Secretary Powell of RNA's continued tours for the UN Peacekeeping Missions. In the same breath, he also described the difficulty of the transportation of logistics all the way from Nepal to the UN Peacekeeping theaters.

The CoAS thanked the VVIP for his top-priority visit to the RNA Headquarters, and expressed his hope that the visit will contribute to further help and support to the RNA in the future.

In response, the VVIP said he was very happy to hear the briefings made by the CoAS. Once a serving soldier himself, he said he understood the limitations of the army, but still hoped that the RNA would respect human rights while launching its military operations. He pledged support to the RNA with assistance of airplanes and gunship helicopters.

The American Secretary of State, Gen Colin Powell, in his audience with the King, referred to the Royal Palace Massacre, and expressed his deep sorrows on behalf of the American people and himself.

He also lauded the efforts of the RNA to fight the Maoists to bring them into the national mainstream politics.

Present on the occasion were the State Minister of Foreign Affairs, Arjun Jung Bahadur Singh, the American Assistant Secretary for Public Affairs, Richard Boucher, the American Assistant Secretary for South Asia, Christina Rocca, and the US Ambassador to Nepal, Michael E. Malinowski.

The Army has been mobilized in many parts of the country. News of the insufficiency of weapons and logistics is pouring in from those frontlines.

I conveyed the security forces' suggestion of the feasibility of phasewise local elections to the Chief Secretary of Nepal Government.

The Governor of the Rashtra Bank, Dr Tilak Rawal, asked me to arrange an audience with the King for his briefing on the current economic conditions of the country.

Six truckloads of weapons and logistics for the APF have arrived in Kathmandu from India.

GP Koirala, currently on a visit in western Nepal, is said to be busy meeting the local Nepali Congress members, discussing the ways and means to bring an end to the State of Emergency, the commencement of the Parliamentary Session, and the tabling of the motion of no confidence vote against the Deuba Government.

Sunday, January 20, 2002

The former IGP of the Nepal Police, Motilal Bohara, talked to me about his family relations in Burma from where, if he's entrusted with ambassadorship, he would be able to make Burmese investments flow into Nepal.

He also requested me to arrange an audience for him and his son, Maj. Gajendra Bohara, the former ADC to the late Crown Prince Dipendra, with the King. He said he would explain everything to the King in the audience.

The IGP of the Nepal Police, Pradip SJB Rana, updated me on the ongoing frictions between the Police and the Ministry for Home Affairs over the question of the transfer of some DIGs (Deputy Inspector General).

The Ministry is contemplating the transfer of some NID officials to the Nepal Police. It's against the spirit of the law, and it can turn outcomplicated enough to reach the Supreme Court.

Thursday, January 24, 2002

If the proposed commercial joint-venture bank with Chinese banks and the Army Welfare Fund materializes, it would ensure proper utilization of the sums of the Fund to administer to the needs of the Army. The inflow of Chinese investments into Nepal would also open avenues for economic prosperity of the country.

I learned that GP Koirala is desperate to see me. I don't know his reasons for wanting to meet me. Surendra Prasad Choudhary of the Nepali Congress has scheduled the meeting for Sunday.

The Nepali Congress is set to announce the late Krishna Prasad Koirala a martyr of Nepal and found a trust in his name. He's the father of former PMs B.P Koirala and Matrika Prasad Koirala as a well as GP Koirala.

NC leader Shailaja Acharya chaired the meeting, and two former Indian PMs, Chandra Shekhar and VP Singh, were invited as guests.

Monday, January 28, 2002

The General Secretary of the CPN (UML), Madhav Kumar Nepal, while visiting me, said that the State of Emergency has adversely affected the economy of the country, the tourism sector in particular.

He added that it would be wise of us to find a way if the Army could still be mobilized without an extension of the State of Emergency.

He then complained of the NC government for failing to deliver services to the people, and it was high time we sought an alternative to the government. The jumbo Cabinet has only the function of keeping an ineffectual group of people in power, merely to keep the government going.

Misappropriation is soaring in the name of the appointments of advisors, fresh recruitment of temporary employees, populist programs, such as the Ganesh Man Peace Campaign, the Bishweshwor with the Poor Movement, and the like.

I made it clear that the CPN (UML) wasn't in a mood to extend the State of Emergency without some preconditions.

In conclusion, I advised him to elucidate all he told me to the King when he meets him.

According to a communique received from the Royal Nepal Army Headquarters, despite the insufficiency in logistics, the Army's operations have been successful.

The leaders of the political parties have been exerting pressures for the release of the detained Maoist leaders.

The performances of the No 3 Brigade and that of the NID Chief, Devi Ram Sharma, have been praiseworthy.

The Maoists seem to be weakening as they are seeking to extend their contacts with the civil society activists and other organizations.

Public support for the Maoists has indeed drastically waned.

There's been discrimination between policemen and soldiers in terms of daily allowances. It should end forthwith, taking into consideration the shimmering dissatisfaction of the soldiers on the issue.

Many top leaders of the political parties met at the former PM, Surya Bahadur Thapa's residence at Mali Gaon to discuss the current political situations. They reached an agreement to draft a Common Minimum Program to exert pressure on the PM to be serious to downsize his Cabinet, curb corruption, and make amendments to the Constitution at a time when the economic condition of the country has plunged to an all-time low.

In addition, a committee comprising certain leaders has been formed to monitor the Deuba Government's activities. Its members are from Nepali Congress, the CPN (UML), the RPP, and the NSP (Nepal Sadbhavana Party).

As planned earlier, I met Surendra Prasad Choudhary, a leader of the Nepali Congress. Choudhary, close to the Koirala Camp, referred to the signature campaign to remove the Deuba Government as it failed to control corruption and restore normalcy amidst the insurgency. He further disclosed a plan to declare a month's ceasefire, look for an alternative to seek a way out, and steer a political discourse to the extent of holding a referendum as an alternative to the vaunted elections to the purported Constituent Assembly.

Tuesday, January 29, 2002

The RNA has forwarded its suggestion for extending the State of Emergency, taking into consideration the situation of the country.

The Joint Coordinator of the NSC Secretariat, Maj. Gen Rookmangud Katawal, updated me on the progress in selecting the site for the Secretariat of the NSC, and the ongoing preparations to make the office fully operational.

Wednesday, January 30, 2002

Speaking at the inauguration of the Kathmandu Municipal Council, NC leader Shailaja Acharya came down heavily upon the Deuba Government, and making utterly critical remarks against the Royal Palace. She termed the State of Emergency as a followup of the Royal Palace Massacre.

The former Indian PM, Chandra Shekhar, on his way back to India after his Nepal visit, is learned to have stopped in Damak to visit the NC leader, Chakra Prasad Bastola.

The environment at the Nepal Police Headquarters and the Ministry of Home Affairs has heated up following the transfer of some officers from the NID to the Nepal Police.

The Home Minister seems to be in the mood to resign soon.

A rumor has again been spread that the former PM, GP Koirala, is soon going to remove Deuba.

Monday, February 4, 2002

The Secretary at the Ministry of Home Affairs, Shree Kanta Regmi, informed me of his planned visit to India to participate in the Secretary-level bilateral meeting. The meeting is expected to discuss the issues of the immediate and longterm importance of the compulsory requirement for identity cards for air passengers recently introduced by India, the treaty of extradition, terrorist activities, the circulation of counterfeit currency notes, criminal activities on the borders, and to work out solutions to these problems.

The Maoists attacked a Police Post at Bhakundebesi of Kavre District, killing 16 policemen, seriously wounding five, and getting away with cash and weapons.

According to a fresh report received here, the Maoists are preparing to mark the anniversary of their launching of the People's War on February 13. They are also preparing for a showdown to convey their message that they are right on their part and enjoy an overwhelming support of the people. They have used the peace talks as a tactical strategy to engage the government.

Student leaders of the Maoist sister wing, All Nepal Revolutionary Students, are reported to be in the capital but the Nepal Police have made no move to arrest them.

Tuesday, February 5, 2002

A demonstration by 4,000 locals from different parts of Rolpa was organized, shouting slogans for the King and multi-party democracy and against Maoist dictatorship. They demanded actions against the Maoists, an immediate return of their confiscated property, and an end to violence.

There's an intelligence report of the Maoists robbing Rs 20 lakhs from the Rashtriya Banijya Bank in Lukla.

Today, the CPN (ML) and the CPN (UML) officially decided to reunite. But still, some dissidents have decided to retain the CPN (ML) intact.

Friday, February 8, 2002

An informal meeting of the Nepali Congress leaders, held at NC leader Sujata Koirala's residence at Mandikhatar, discussed in detail the agenda to be presented for approval by the Central Committee Meeting of the Party.

Ramesh Nath Pandey has lauded the concept of the Greater Democratic Alliance, a blueprint for entrapping Deuba into a corner and then derailing his government. Pandey's intent behind his support of the Alliance seems more directed to bag for himself the portfolio of foreign minister and have his son appointed ambassador to America than anything else.

PM Deuba reportedly met the CPN (UML) General Secretary, Madhav Kumar Nepal, seeking support to stabilize his government.

Monday, February 11, 2002

A report on the improvement of the Royal Palace security system has been prepared by a taskforce.

Crown Prince Paras is to commence studying under the curriculum of the training, the organization, and the hierarchy of the Army.

The Central Committee Meeting of the Nepali Congress has directed PM Deuba to downsize his Cabinet, which the former PM Koirala thinks will serve his interest in enticing the disgruntled Ministers to his side to have a majority of members to oust Deuba.

Koirala is speculating on the King to ask Deuba to resign as PM for underperformance.

Wednesday/Thursday, February 13/14, 2002

The PM discussed the current security situation with the CoAS. Later, talking to the Secretary at the Ministry of Defense, the PM vented his dissatisfaction over the CoAS's noncooperation and also for ignoring some of his instructions.

To mark the anniversary of their so-called 'People's War' (Jana Yuddha), the Maoists shot dead a retired Sergeant named Tek Bahadur Gurung in the Nagin Village of Panchthar District and Sergeant Thakur Prasad Choudhary. The murders took place on Thursday, February 15.

The Army discovered eight kilograms of gold jewelleries in a house in Pokhara.

Sunday, February 17, 2002

The Maoists simultaneously attacked the police posts at Mangalsen and Sanphe Bagar of Achham, and Lalbandi in Siraha districts. In the clash, contact with the Achham police unit was disconnected. The RNA

Headquarters had alerted the No 5 Brigade in Dipayal of a possible attack there, but it didn't maintain the necessary preparedness.

According to a preliminary report, in Mangalsen alone, 48 soldiers and 34 policemen lost their lives, eight soldiers have gone unaccounted for, and 18 policemen were seriously injured.

The Maoist attack in Lalbandi of Sarlahi has left four policemen dead, another six wounded, the condition of four of them critical. The Maoists took away their weapons.

Monday, February 18, 2002

The operations of the security forces have been merely defensive, and concentrated only at the district headquarters.

The CoAS called me to say that a majority in the Nepali Congress Party is opposed to the ratification of the State of Emergency. The question of whether or not to ratify it has widened the differences within the party. The likes of Govinda Raj Joshi are in favor of ratification rather than extension.

Some key leaders in the Nepali Congress Party are making vocal demand for the PM's resignation.

The leader of the Opposition Party, Madhav Kumar Nepal, accused the Army commander in Mangalsen of overconfidence, without a bit of preparedness.

Wednesday, February 20, 2002

The Maoists are playing dual tactics of talk and attack on the security bases. Emboldened by their successes in Achham and Sarlahi, they are mulling over yet another series of attack in the hilly districts.

They are also approaching Leftist intellectuals to exert pressures on the political parties not to ratify the State of Emergency.

Captain RP Pradhan called me to inform of America's likely provision of gunship helicopters for the Army. He wished to talk to the King about it.

Thursday, February 21, 2002

The Nepali Congress has, after the rough and tumble of arguments and counterarguments, agreed to ratify the State of Emergency for the present, and force the PM to resign by showing him the signatures of most Parliamentarians of the Party to compel him to leave.

The CPN (UML), too, has agreed to ratify the State of Emergency, on the condition that the present government step down.

Scared of possible Maoist attacks, the police and some government employees in Dadeldhura have fled their duty stations, so says a report recently received.

The King, talking to me, recalled the timeframe of one month for the PM and three months to the CoAS he had given them to bring about perceptible changes in their performances and military operations. Any lackluster performance by the PM and the Army would embolden the Maoists, and it in turn will greatly erode the public trust in them.

Monday, February 25, 2002

Today, at a regular briefing, I reminded the King of the government's fresh initiative of economizing on expenses by slashing the number of official parties and glamorous receptions, and asked him if the Army, too, would likely cancel the annual dinner on the New Year's Day to mark the Army Day. He told me to consult the CoAS and do accordingly.

Following a recent briefing to the American Ambassador, Michael E Malinowski, on the mobilization of the Army and its need of logistics, a section of the journalists and intellectuals have raised their eyebrows as though it would constitute interference in Nepal's internal affairs.

Tuesday, February 26, 2002

Today, the Joint Coordinator of the NSC, Maj. Gen Rookmangud Katawal, told me that he had briefed the PM on the elaborate preparations being made for the opening of the Secretariat Office of the NSC on March 11.

He said he also dismissed the PM's suspicions and fears of the militarization of Singha Durbar by the Army, a rumor planted by some people.

He added of the PM complaining against Brig. Gen Dilip Rayamajhi.

He further said the PM sounded determined to fight corruption even at the expense of his own premiership.

I updated the King on all what Maj. Gen. Rookmangud Katawal had told me. In response, the King instructed me to advise the PM to control untoward and unnecessary expenses and instead increase the allotments to security-related matters, to strengthen the law and order system, to set up a fund to provide training to the unemployed youths for foreign employment, and to increase people's participation in development works.

Some police officers have similarly suggested that there should be a Police Service Commission to ensure free and fair selection of police cadets and also to isolate the entire institution from political interferences.

As ploys to have his side of the balance tilt in his favor, GP Koirala is contemplating a reshuffle of the Central Committee of the Nepali Congress Party. The intent behind his move seems to make himself PM again.

The chief of the NID, Devi Ram Sharma, updated me on the fresh information of the British Government's pledge of help to Nepal.

In the first round of the assistance, British experts will provide a 15-day special training to a few police officers of the Intelligence Bureau on British expenses.

It'll also bear the costs of a consultant who'll study the different aspects of the Bureau to suggest reforms therein.

Sharma also recalled a similar visit by the Chief of the Indian Intelligence Agency, RAW, two years earlier.

Now that the RAW has invited him to visit India, he requested me to arrange an audience for him with the King to brief him on it himself.

Wednesday, February 27, 2002

During the regular briefing, the King instructed the CoAS to advise the PM to rethink the feasibility of holding phasewise polls should the House of Representatives be dissolved to resolve the ensuing political deadlock.

He also told the CoAS to assure the PM of every possible help of the Army and the Royal Palace to facilitate his core group to win in the elections.

Next, the King instructed me to provide him a briefing on the role the Royal Palace can play, and the justification the Palace can explain on the dissolution of the Parliament and the immediate needs of the Election Commission for it to carry out the polls.

The King, the CoAS, and I held serious discussions on new psychological operations to be executed by the Army, particularly against

Maoist Chairman, 'Prachanda'. We decided to entrust the responsibility for this to the Joint Coordinator, Maj. Gen Rookmangud Katawal.

Monday, March 4, 2002

The Ministry of Foreign Affairs has completed the schedules of the King's official visits to China and India, sending the details to the respective embassies.

The Chinese Ambassador called me to update on the information of the schedule of the royal visit forwarded to Beijing for confirmation. He asked me to update him on the King's visits to specific places or special programs, if any, so that he could set the programs accordingly.

The Chinese Ambassador termed the political condition in Nepal as fluid, and raised his heightened concerns over the frequent visits of American teams to Nepal. He said even India is unhappy about this.

The editor of the *Sanghu Weekly*, Gopal Budhathoki, has been taken in Army custody for interrogation for his negative comments against the Army. People from different quarters, however, have opined that the Army could've simply interrogated him through correspondence. This particular arrest could send out bad impressions.

The *Time* Magazine has brought out a news story, alleging that the late King Birendra refused to use the Army against the Maoist insurgency, this point leading to speculations that he had reached an informal understanding with the Maoists which in turn caused his death.

Tuesday, March 5, 2002

The Maoists have resorted to hurting the religious sentiments of the people by vandalizing temples, creating obstructions to the performances of pujas, even murdering those mourning the death of their kinsfolk, as well as bombing schools.

The program for the King to observe and inspect the 'Mine Clearing Vehicle' and other equipment in the premises of the Royal Guards Military Police Unit has concluded.

To accept the invitation by his Indian counterpart, the Chief of the NID is scheduled to visit India from March 7 to 12.

The signature campaign to oust PM Deuba is likely to slow down following his assurance to form a corruption control commission.

The PM has asked the Army for additional security for the Prime Minister's Office. The CoAS has taken it positively, and the new security system will go into effect soon.

Friday, March 8, 2002

Some PSOs have complained of the existing system that allows only the CoAS, the DMO and the DMI to have knowledge and information of the military operations against the Maoists. They say that keeping them in the dark may cause confusion for the military operations to be effective.

I convinced them that the Army operates collectively, but still I promised to talk to the CoAS to rectify matters if anything has gone amiss.

As I shared the matter with the King, he instructed me to develop a mechanism for the PSOs to know about the details of the military operations.

Consequently, briefing of the military operations to the PSOs commenced soon thereafter.

Although it's already been over three months since the Army was mobilized, the military operations haven't been able to produce the desired results to the expected level of success. It calls for strategic changes in the operations against the rebels. Such operations should be launched in coordination with all the security forces, with a unified command control the nature and modality of which need to be clear beforehand.

The numbers and standards of the living quarters in many of the Army camps and barracks are reported to be poor and insufficient compared to the strength of the Army. It calls for an immediate construction of more quarters and their maintenance for which more funds need to be allotted. The restrictions imposed on the utilization of the Army Welfare Fund for such infrastructure development need to be liberalized without any further delay.

The King called me to circulate his instructions for all-round security arrangements regarding the general elections to be held in August-September as an environment is building up for the dissolution of the present Parliament.

He also asked me to discuss the exact needs and quality of weapons the Army would require in the context of India exerting its pressures to provide the RNA with the Insas rifles.

He then told me to keep certain places of Parsa District under surveillance as the Maoists are reportedly trafficking in marijuana and generating income from its cultivation.

The CoAS asked me to arrange the transfer of Brig. Gen Dilip Rayamajhi, currently the Commander of the No 6 Brigade, to the Military Operations Department at the RNA Headquarters.

Sunday, March 10, 2002

The National Wildlife Conservation Department is prepared to relocate a pair of rhinos from the Royal Chitwan National Park to the Bardiya National Park in Nepalgunj. The Army has been instructed to facilitate the passage of the wild animals to the designated place.

Today, two Army soldiers lost their lives and 18 others were seriously injured when a landmine set by the Maoists blew the truck that was carrying personnel from Mugling to Narayangarh.

Monday, March 11, 2002

The Honorary Royal ADC, Maj. Gen Bharat Keshar Simha, called me to arrange for him a leave effective from March 24 to April 5. He was seeking the leave for his puja in Jammu and Kashmir and to attend a meeting of the World Hindu Federation in Delhi.

Now the military operations are being conducted under the Unified Command Control. However, the program lacks the inclusion of the NID which would additionally contribute to the further success of such operations.

PM Deuba inaugurated the NSC Secretariat Office at the Putali Bagaincha of Singha Durbar.

Tuesday, March 12, 2002

Sources say a police van blew off in the ambush set by the Maoists at Dolalghat. It claimed the lives of five policemen.

Complaints have been received from the locals in some districts that the Army has occupied their lands to establish their bases.

DFID has reportedly distributed an analytical paper to a couple of European Ambassadors, saying that against the backdrop of the tussle for power between Koirala and Deuba, the King may take advantage of this situation to assume state powers for himself.

Wednesday, March 13, 2002

Devi Ram Sharma, the NID chief, shared with me his experiences of meeting with the Chief and Deputy Chief of the Indian Intelligence Agency, RAW. They said India always wished for peace and stability in Nepal and expressed their readiness to help by providing the NID with training and communication equipment.

They also suggested that Nepal shouldn't privatize the Nepal Telecommunications until the Maoist problem is resolved.

Relating to his Indian counterparts' remarks, Sharma said it's surprising to see the way the leaders of the Nepali Congress and the CPN (UML) regard certain RAW operatives as though they were their gurus.

Friday, March 15, 2002

As preplanned, the RNA's high ranking officers' collective briefing on the post-Army mobilization situation, progress and problems to the King concluded.

The PSOs discussed the activities and future plans of their departments. They pointed out some lapses in the Army intelligence system, and the neglected status of the earlier plan to set up Army Divisions in some hilly districts. They described the remedies to rectify them. They realized that there are mounting complaints from soldiers about inferior weapons, logistics, foods, and clothing for the soldiers deputed to the field, and also figured out ways to solve the problems before it would be too late.

The agenda of the briefing included the timeframe for the Army to remain in field position, providing relief to those long in the field, and mount preemptive operations against the rebels, maintain the logistics, the standards of the arms and ammunition, etc.

The officers also noted the priority of destroying Maoist bases and training centers and to impair and block the supply routes of their essential supplies of food and clothing.

The officers are unanimous that the NSC should suggest to the PM to discuss security issues with his Indian counterpart when he visits India.

The King, in his turn, offered his invaluable advice and instructions to the Army deployed in tactical areas; that the Army see to it that there are no civilian casualties while launching operations, take the political parties into confidence and continue civil programs.

He also gave further instructions to the Army to maintain an effective communication system from the center down to the units.

The Royal Nepalese Ambassador to India, Dr. Bhekh Bahadur Thapa, today shared his bitter experiences with me. He painted a bleak picture of the Nepalese Embassy building in New Delhi, desperately needing repairs.

He also mentioned that Nepali leaders routinely tend to display their fickle and immature sides to their Indian counterparts, unaware of their national identity.

He added that Kathmandu doesn't provide regular and timely guidance and information, and this vacuum often puts him in odd situations to respond to the other Ambassadors' concerns and queries about Nepal.

He shared another bitter experience, this one of meeting the PM the other day. The PM didn't show much enthusiasm about his forthcoming visit to India. Presumably, he was nervous about the possible Vote of No Confidence against him in the Parliament.

Sunday, March 17, 2002

The friction within the Nepali Congress Party has worsened to such an extent that the faction close to GP Koirala is starting a signature campaign of 61 Congressmen against PM Deuba. Sources say the dissident camp is set to table the Motion of No Confidence against the PM when he returns from India.

Contrary to what's speculated, Koirala is making odd statements: such as, the expiry of the State of Emergency on May 25 will determine the future of the next government.

The NC leader, Shailaja Acharya, has opposed Koirala for his proposal of certain amendments to be made to the Constitution.

Other NC leaders, such as Chiranjivi Wagle, Pradip Giri, and Bimalendra Nidhi reportedly wrote their Note of Dissent in the Party against the idea of amendments to the Constitution.

The Home Minister, Khum Bahadur Khadka, also opposed the idea of amendments but didn't go to the extent of writing his own note of dissent.

Troubled by the signature campaign against him, PM Deuba's advisors have counseled that he replace some Ministers with neutral ones to keep the government going.

Koirala wants to meet Shakti Lamsal and Ishwori Dahal, the two confidants of the Maoists, to ask them to open a channel of communication to the Maoists. It's to be noted that earlier, he had sent the NC leader, Chakra Prasad Bastola, to India for the same purpose. Sources say that Koirala is making every effort to contact other Maoists leaders, too.

Monday, March 18, 2002

Sources say that Dr. Baburam Bhattarai, the second most powerful Comrade in the Maoist Party, has asked the embassies not to have their VVIPs visiting Nepal during their scheduled program of the national shutdown and mass showdown.

Today, the King shared his confidential plan with me. He's set to have the PM dissolve the Parliament, then proceed according to the work plan of the Royal Palace, and finally take the entire executive powers in his hands, thus trumping over the personality clashes and the lust for power evident in the Nepali Congress and the repeated failure at good governance of each successive government.

On the eve of his visit to India, the PM consulted with the heads of the security forces. The CoAS handed the PM a copy of the list of weapons and logistics the RNA needs and India has agreed to provide.

The eight-week basic military training for Crown Prince Paras, as prepared by the Royal Guards Brigade, began today. I briefed him on the scope and importance of such training. The chief of the training, Col. Chhabilal Khadka, shed light on the contents of the training.

Tuesday, March 19, 2002

It's been known that a journalist close to the Maoists approached the King's chauffeur, Sudan Thapa, to learn about the King's upcoming visit to Mangalsen. This disclosure of such confidential information related to the Royal Palace is a pointer to the Maoists' infiltration of the most sensitive places, such as the Royal Palace and the RNA.

The PSOs of the RNA expressed their grave concerns over the CoAS's remarks as unbecoming of his rank.

The King expressed his grave concerns over the rapid deforestation and unauthorized transportation of timber, asking me to immediately discuss the problem with the Secretary at the Ministry of Forests and Soil Conservation.

Next, he asked me to discuss with the CoAS if the Army can launch operations even during the PM's visit to India.

Wednesday, March 20, 2002

The factional power politics in the Nepali Congress has reached such a high that fear psychosis is spreading through the party and even forcing PM Deuba to resign in the face of a motion of no confidence vote against him once he's back from India.

GP Koirala has threatened his fellow NC member, Khum Bahadur Khadka, to keep quiet and let things go on their own steam.

Another Nepali Congress leader, Chakra Prasad Bastola, has asked the two key persons, Ishwori Dahal and Shakti Lamsal, close to Maoists, to arrange Koirala's meeting with the Maoist leader, Bamdev Chhetri, in New Delhi.

According to a confidential source of information, the Maoist leader, Dr. Baburam Bhattarai, has asked Koirala to convince the General Secretary of the CPN (UML), Madhav Kumar Nepal, into accepting the agenda of holding elections to their proposed Constituent Assembly through a referendum. Fearing that iron cuts iron, that's to say, the Maoists might enjoy the upper hand, Nepal is learned to have declined the proposal outright.

Maoist sister organizations in India held a protest demonstration against the visiting PM of Nepal by burning his effigy.

Similarly, the Maoists have psychologically pressured the Minister of Education and Sports, Amod Prasad Upadhyaya, to postpone the SLC examinations, which are to begin soon, by three more months.

According to a tipoff, the former PM, Surya Bahadur Thapa, has secretly met the Maoist leader, Dr. Baburam Bhattarai. Thapa is learned to have agreed to the CA elections.

Friday, March 22, 2002

I made arrangements for the King's visits to the western Nepal, including Nepalgunj. It'll be the first visit of its kind after his coronation, and taking place at a time when the Army and the Police are fighting the Maoists. The King meeting people and visiting places will surely convey a good message to the nation and also will boost the morale of the security forces, among other things.

Falling ill, my father has had been admitted to the Military Hospital. His condition was stable, not so bad for us to worry about. He seemed to be recuperating. However, I was shocked to hear the next morning that he had expired. I made arrangements of every official work and deputed people to work in my absence while I rushed to the Pashupati Aryaghat for his final rites and my observance of death rituals.

I lost my parents in the same year, just a couple of months apart. King Birendra had kindly come to express his condolences to my mother's demise. I was expecting the same from King Gyanendra, too, but it didn't happen.

Many high officials and VVIPS visited me at the Arya Ghat to express their sympathies and condolences at the demise of my father.

It was on the ninth day of my observance of my father's death 'pollution' when Shree Ram KC, the deputy head of the Military Secretary, visited me. He told me about the likely restructuring in the Royal Palace Administrative Organization by a series of downscaling of posts and positions.

I was stunned to hear about the planned changes. The Principal Military Secretariat is an office of a most sensitive nature.

With the mourning and the rituals for my father's passing away over, I reached the office to find a discussion going on the proposed restructuring of the Royal Palace Administrative Organization. The Chief of the Royal Palace Household, Sharada Prasad Pradhan, briefed me on the new system, to come into effect from April 14, the New Year's Day of the Bikram Sambat (Era.)

When I asked for the reasons of the new rearrangements, he said it was as per the command of the King.

I said that the Principal Military Secretariat and the Principal Secretariat have been in place at the Royal Palace ever since 1954 when the Army had 15,000 soldiers, and the strength has by now grown to 50,000, so the new organizational restructuring wasn't justified.

I also added that the Principal Military Secretariat can't be placed under the Principal Secretariat by any means.

Friday, April 5, 2002

The Chief of the Royal Palace Household and I reached Nirmal Niwas for our regular briefing. I briefed the King with the facts and figures about the existing structure of the military administration at the Royal Palace, hinting at the legal complications in the future if the new ordered system should go into enforcement.

I also referred to the tacit provision of the Principal Secretary and Principal Military Secretary or Military Secretary to cosign the Ordinances passed by the Parliament and forward them to the Royal Palace for the Royal Seal of Approval.

The new system would surely entail administrative complications and create confusions in the hierarchy. As per the existing system, the Military Secretary is solely responsible for the overall security of the Royal Palace. Placing it under the Principal Secretariat will only complicate the works therein. Further, anomalies as to who should officially correspond with the RNA Headquarters, the Ministry of Defense, and other security forces would certainly arise.

The King listened to me and promptly instructed me to continue with the existing system.

The Crown Prince instructed me to post an energetic official at the Principal Military Secretariat when a new post is created there. He said it would be better to have the posts of two Second Lieutenants (2nd Lt.) and two Captains (Capt.) in the office.

Monday, April 8, 2002

As per the current structure, the Secretary at the Ministry of Defense is the coordinator of the NSC. He has asked the PM to extend his tenure at the NSC for another two years.

According to a reliable source, the CoAS, Gen. Prajwalla SJB Rana, is planning to purchase a house in Eastbourne in East Sussex of Southern England. One Haribibhor Karki is reported to be arranging it for him.

GP Koirala met the Minister and facilitator of the peace talk team, Narayan Singh Pun, to convey to the King his commitment to multiparty democracy and the institution of constitutional monarchy. He also promised to help the King if he, too, reciprocates his gesture of cooperation. He said it was futile to expect good governance and delivery of services and solution to the political chaos from PM Deuba.

Reading between the lines, one could easily sense that he was expecting the King to sack Deuba and make him PM.

When Minister Pun reminded him of his habit of doublespeak, he sounded angry.

The members of the Nepali Congress are going to table their Motion of No Confidence Vote against the PM today.

The heads of the security forces have briefed the PM on the current overall security situations.

Intelligence briefings have revealed that the Maoists' immediate future plan is to target the economic sectors to paralyze the government and the country.

Their next victims are going to be Ministers and top government officials.

Sources say some Christian missions are helping the Maoists with money.

Meanwhile, the Maoists are constantly approaching political party leaders to have them agree with their political agenda.

Wednesday, April 10, 2002

The IGP of the Nepal Police, Pradip SJB Rana, mentioned of GP Koirala, Madhav Kumar Nepal, and Surya Bahadur Thapa joining hands with the Maoists against the Royal Palace.

Krishna Prasad Sitaula, a member of the Nepali Congress, is the emissary of GP Koirala to talk to the Maoists. They have given their assurance to hand over their arms to the government once the date of the CA elections is fixed and an interim government formed. They have also pledged to join the government.

The IGP of the Nepal Police quoted Radha Krishna Mainali and Salim Miya Ansari of the CPN (UML) as advising the King to take over the reins of the government.

The IGP also said that PM Deuba is rapidly losing his hold on the government and the trust of the leaders of his own party.

He further revealed that India has backed the Maoists to serve its own interests. It aims to help the pro-republic forces to obtain two-thirds majority in the CA elections. He added that the Maoists, however, are unaware of how they are being manipulated by India.

I assured him that I would report his findings to the King.

On another occasion, Radha Krishna Mainali shared with me his perception of the current state of affairs. He said India doesn't trust Deuba as he's viewed as pro-America, so it wants GP Koirala and Madhav Kumar Nepal to team up to blunt American influence in Nepal.

Sources say that GP Koirala and Khum Bahadur Khadka had a long meeting, and Khadka advised Koirala to drop his immediate plan to remove PM Deuba because his ouster during the State of Emergency would send out wrong messages of his power-grabbing mindset which can result in displeasure in and disapproval of the security forces.

The government is yet to formulate its policy to deal with the Maoists. Different political parties have varying perceptions, preventing them from forming a unanimous stand.

India tends to soothe Nepal's apprehensions by deporting low-echelon Maoists to Nepal while still providing its sanctuaries to the top leaders.

Friday, April 12, 2002

The Maoists launched simultaneous attacks at the Lamahi and the Satbaria Police Posts of Dang District. They carried away a large number of weapons and killed a couple of policemen and left some critically injured.

The King shared with me the issues he discussed with the American Ambassador, Michael E. Malinowski. He said he asked the envoy to advise GP Koirala to drop his plan to change the government and go for fresh Parliament polls or for the elections to the CA.

Sunday, April 14, 2002

The Nepali New Year of 2059 BS, which happens to today, is marked by political upheavals, inter-and intraparty squabbles, and further escalation of Maoist violence.

The chasms and infighting in the Nepali Congress landed Prime Minister Sher Bahadur Deuba to the point of recommending to the King to have the Parliament dissolved and announce the date for fresh polling.

But the politics of distrust and vengeance had scaled to such a height that the PM was left with no alternative than to recommend to the King for the deferral of the date of the polls by one year.

The very recommendation of the cancellation of the date of the elections cost him very dearly, so much so that it earned him the epithet of being 'incompetent' and summary dismissal from the premiership.

The King then invited petitions from all interested Nepalis for the post of Prime Minister, an exercise neither heard of nor seen in the entire history of Nepal.

Against the backdrop of the monumental failures of successive governments, Lokendra Bahadur Chand was appointed Prime Minister and was entrusted with the responsibility of holding free and fair elections and also proceeding with the peace talks with the Maoists.

Amidst the fears of disturbance to the peace talks in the aftermath of the murder of the IGP of the APF, the people, however, felt a measure of respite.

Monday, April 15, 2002

The ADCs appointed at the Royal Palace for the term of two years are scheduled to be given their farewell on April 29.

The King will grant his collective audience to the soldiers and officers going to the UN Peacekeeping Mission at the UN Peacekeeping Training Center, Panchkhal, on May 9.

I briefed the CoAS on the details of the program.

The leaders attending an all-party meeting criticized the role of the RNA. They said the rough and rude behaviors of some soldiers dented the image of the entire Army. They further elaborated that some soldiers were found drunk and teasing girls while in the field for operations against the Maoists.

The process of change in the Army uniform is progressing.

It'll be better to twin two occasions: The celebration of the King's birthday, and the introduction of the new Army uniform, both on July 7.

Tuesday, April 16, 2002

In an operation launched by the Army's No 3 Brigade of Pokhara, the Maoist Central Committee Members, Krishna Dhoj Khadka and Rekha Sharma, along with a dozen or so district-level cadres, have been arrested.

When briefed on this, the King expressed his pleasure at the success of the Army operation, and also instructed me to take care of the security of the arrested and inform the CoAS, the DMO and the DMI at the Army Headquarters accordingly.

Likewise, attempts are being made to obtain the help of the Indian security forces to apprehend the top Maoist leader, Ram Bahadur Thapa, with the code name of 'Badal.'

GP Koirala is busy visiting the Indian PM, Atal Bihari Vajpayee, the Defense Minister, J. Krishnamurthy, and other leaders at the Hyatt Regency Hotel in New Delhi.

Sources also say that Krishna Prasad Sitaula of Nepali Congress is making tremendous efforts to arrange for Koirala to meet the Maoist leadership in New Delhi.

Koirala's suggestive remarks, that is, taking the help of the Indian security forces to fight the Maoists in Nepal, are viewed as very sensitive and unsolicited. The remarks reflect on the defeatist psyche of the politician, and leaders must be cautioned not to make such irresponsible statements in public which amount to darkening the image of the country itself.

There's information on the leaders close to PM Deuba staying in the guesthouse of the Royal Nepalese Embassy in New Delhi and visiting top Indian leaders. I shared this information with the King.

The Field Marshal, General Nir SJB Rana, telephoned me to express his dissatisfaction with the CoAS. He asked me to brief the King on it. He, however, didn't bother to disclose the reasons for his displeasure. The King remained smiling while listening to my verbatim report.

'I'll talk to the CoAS on this by and by,' was all he said.

Wednesday, April 17, 2002

I've briefed the King on every necessary detail of the state of affairs without fail. I consult and gather details from the RNA, the Nepal Police, the APF, and the NID, and as and when so required, from also the Secretary at the Ministry of Defense, prior to briefing the King.

I reminded the King of his schedule for an unofficial visit to eastern Nepal to worship at the important temples, and he asked me to arrange it soon.

He also instructed me to remind him to advise the PM to implement the Nepal Police Service Commission for fresh recruitment.

Today, I sorted out the areas of immediate importance for the Royal Palace to take into consideration. The areas include the enhancement of the Palace-people relations, narrowing down the differences with the political parties to have them realize the need for a monarchy in Nepal, and taking the media into confidence.

Thursday, April 18, 2002

GP Koirala extended his stay in New Delhi by two more days to visit the Maoist leaders there. Chakra Prasad Bastola approached the Chief of the RAW to arrange Koirala's meeting with the Maoists, as reports received here say. When the latter explained his official inability to oblige Bastola, he, however, advised him to contact the Delhi Police for the purpose.

Koirala met the Maoist leader, Dr. Baburam Bhattarai, at a secret place. They are reported to have reached an understanding to join hands in the CA elections and to phase out the monarchy in a gradual manner.

The draft which was prepared at the meeting was already outlined in Kathmandu prior to Koirala's visit to India.

Monday, April 22, 2002

During his trips in Eastern Nepal, the King is scheduled to visit several temples to worship the deities there. He'll also visit the Myanglung Bazaar of Tehrathum District which was recently gutted by fire.

The Indian CoAS is visiting Nepal on April 14.

Likewise, the British Chief of the Defense Staff is also arriving here on May 24.

Wednesday, April 24, 2002

At a briefing, the King and the CoAS discussed the issues to be raised by him with his visiting Indian counterpart. They finalized the list of weapons and equipment necessary for the RNA, which was agreed earlier to be made available by India.

The CoAS also briefed the King on the preparations to confer on his Indian counterpart the rank of Honorary General of the RNA.

The CoAS sought the King's opinion for taking action against Lt. Cols Yagya Bikram Rana and Ramindra Chhetri for their refusal to take up the posts of Military Attachés. The King gave him a free hand to take action as per the rules.

The Army is seeking the arrest of the Principal of the Kathmandu-based Fluorescent Higher Secondary School, Madhav Marahattha, for his collaborations with the Maoists. The King gave instruction to keep him under surveillance rather than arrest him.

Thursday, April 25, 2002

In the series of visits from friends of Nepal to assess the situation here following the Army's mobilization against the Maoists, teams from the Special Forces of the US Army visited Rolpa to observe an actual operation. They expressed their amazement at the display of professional standard, motivation, morale, and the perseverance of the RNA soldiers in the field of operation. They said the performance of the RNA was praiseworthy. In Nepal, after the 1990 political change, the RNA has proven itself to be the only institution able to retain its professionalism. Impressed by the thorough adherence to professionalism by the RNA, the team assured of increase in American aid for the Army in the near future.

India has raised its eyebrows over the American team's visit and their on-site inspection of military operations against the rebels in the Rolpa District.

Wednesday, May 1, 2002

The King was scheduled to attend a book release and photographic exhibition by former Brig. Gen Khagendra Bahadur Shrestha on Their Late Majesties Birendra and Aishwarya at the Birendra International Conference (BICC).

The King asked me to make arrangements to send a team of technicians to draw a master plan for the renovation of the Deuti Bajai Temple in Surkhet District.

In a series of briefings, the King and I discussed the fresh news of GP Koirala meeting Maoist leaders and the Indian PM in Delhi. From these series of events, it's apparent that India has strong influences on the Maoists. It's strange that India maintains double standards with Nepal, keeping the Maoists on its terrorist watchlist, and at the same time, helping them by providing shelter on its soil.

Anticipating questions from the US and the EU on its double standards, India is fielding the Nepali Congress and the CPN (UML) against the monarchy.

It's known that soldiers deployed in action haven't been provided relief for a long period, and comments and complaints have been made for timely adjustments.

Likewise, there's a talk of fresh recruitment of 5,000 soldiers in the Army for which more weapons will be required.

Now that the local government's tenure of five years has expired, and since some political parties insist on the continuation of the previously elected representatives' incumbency, a decision to this effect needs to be taken without delay.

Since it is imperative to cut costs and provide services to the local people, the King has directed me to advise the PM to rescind the tenures of the past representatives. The Army is planning to have the Public Accounts Committee (PAC) members to visit the No 3 and 4 Brigades soon. A directive has been sent to the Commander-in-Chief to make the necessary arrangements to that effect.

In the observance of a long-held tradition, the King granted his collective audience to the ADCs tenured to the Royal Palace. Lauding their performances, the King expressed his hope of seeing similar praiseworthy works done by them in the days ahead. He also advised them to keep in touch with the Principal Military Secretary Office and the Office of the ADCs, and give them information of importance.

Friday, May 3, 2002

Today, the Joint Coordinator of the NSC, Maj. Gen Rookmangud Katawal, called me to say that he's soon to meet the American Special Forces Team currently in Nepal with a request for an assistance package for Nepal. As asked by him, I advised him to use this opportunity to ask the team to provide the RNA with weapons, logistics, and a helicopter.

I also reminded him of the norms of the RNA to inform the Royal Palace prior to meeting the team, a protocol the RNA always adheres to.

GP Koirala, soon after his return from his visits in India, called on the King. He said he was determined to give the country a way out of the ongoing political deadlock even at the costs of the resignation of PM Deuba once he's back from his visit to the US. He said he sought the King's help in this regard.

According to a source, a BBC team reached Dolakha to meet the Maoists but failed to do so. They returned to Kathmandu, leaving behind a satellite phone for the Maoists to contact them later.

The Principal Secretary, Pashupati Bhakta Maharjan, quoted the Chief Commissioner of the Commission for the Investigation of the Abuse of Authority (CIAA), Surya Nath Upadhyaya, as seeking his advice to look into the corruption as being reported in the Department of Immigration.

The CoAS today shared a fresh piece of information with me, that the Maoists are throwing their Dialogue Card this time around. He said it indicates their weakening position and the state at present shouldn't hurry for a talk without assessing their weapons. The CoAS asked me to consult with GP Koirala and Madhav Kumar Nepal in this regard.

The King instructed me to make arrangements for Crown Princess Himani's delivery in the Trishul Sadan. Her doctors say that the time of the birth of her child is fast approaching.

Tuesday, May 7, 2002

In connection with the military operations against the Maoists in Doti, some equipment made in the Philippines has been discovered. It hints at their association with Filipino rebels.

Following the colossal loss suffered by the Maoists in the military operations in western Nepal, there's been a trend of hundreds of Maoist cadres fleeing to India for rest, treatment and recuperation. They are also approaching Leftist intellectuals to help create a congenial atmosphere for a talk with the state. Simultaneously, India is reportedly exerting its pressures on the Nepali Congress, the CPN (UML), and the Maoists to immediately forge an agreement.

Seven political parties are going to jointly organize their all-party mass meeting and showdown against the government at the Khula Mancha (open theater) grounds tomorrow. To avert unpleasant eventualities, it would be better for the security forces at their checkpoints to control the inflow of unruly supporters to join the mass meeting.

Wednesday, May 8, 2002

As briefed by the CoAS, Gen. Prajwalla SJB Rana, to the King on the purchase of new weapons for the Army, and as recommended by the technical team concerned, the Army has reached a decision to purchase the series of the American M-16 A2 rifles and the Belgian Minimi LMGs.

Lt. Gen Krishna Narayan Singh Thapa, accompanied by the former

CoAS, Gen. Dharmapal Bar Singh Thapa and Maj. Gen Shiva Ram Khatri, is set to fly for Sri Lanka to participate in the UN Peacekeeping Seminar due to commence there from May 10.

The all-party mass meeting organized jointly by seven parties took place with a modest turnout of supporters. The speakers underscored a consensus to be worked out jointly by the seven parties and the Royal Palace. They said it wouldn't be good for the Royal Palace to solely depend upon the Army. They also expressed their doubts over the Maoists' display of their Dialogue Card and spoke of the need to cautiously deal with them.

The Madhesi leader, Badri Prasad Mandal, was met with hooting and catcalls as he spoke in Hindi at the mass gathering.

Two former CoASs, Gens Gadul SJB Rana, and Satchit SJB Rana, together offered their advice to employ retired soldiers below 55 years of age for the protection the National Parks. They also advised an immediate combing operation in Rolpa District to apprehend Maoist rebels.

There's report of a Police Sub Inspector committing suicide in Pokhara. The preliminary report cites a family feud as the reason for the suicide.

The Gorakh Bahadur Battalion has long been deployed in field operations in Rolpa District. There have been calls for their immediate furlough for their morale and physical wellbeing.

Wednesday, May 15, 2002

The Indian TV Channel, Star News, has aired an interview with the former CoAS, Gen Satchit SJB Rana, and his comments seem to reflect on his opinions of the RNA.

The host asked Rana as to how was it that the RNA soldiers have a high level of performance in the UN Peacekeeping Missions but have turned out to be fickle and ineffective when it came to fighting the Maoists at home.

Rana defended the RNA's fights against the Maoists as praiseworthy in causing heavy damages to the rebels and seizing weapons and explosives from them.

He added that India would do well to refrain from providing the rebels with training and weapons.

He also ruled out the need or possibility of the Indian Army's assistance in fighting the Maoists, as the RNA is comfortably capable of doing it on its own.

Thursday, May 16, 2002

The King, as the Supreme Commander in Chief of the Royal Nepal Army (RNA), conferred on the Indian CoAS Padmanavan the rank of Honorary General in the RNA and presented the Sword of Honour to him. The ceremony was held in the Kaski Hall of the Royal Palace.

The Honorary ADC to the King, Maj. Gen. Bharat Keshar Simha, telephoned me to express his reservations about the breach of protocol by presenting the Guard of Honour to the Indian CoAS in the presence of the King.

I told him it would in no way contravene the royal protocol, as the King himself is already present at the venue prior to, say, the presentation of the letters of credentials by ambassadors to him.

In a succession of attacks on NC leaders, the Maoists blew off the house of the Home Minister, Khum Bahadur Khadka, in Dang last night. This isn't the first time the Maoists have destroyed private and public properties.

The Communist Party of China has invited Girija Prasad Koirala to visit China at his convenient time.

Friday, May 17, 2002

The Indian CoAS expressed his extreme pleasure at having visited the Muktinath Temple in Mustang District. He was accompanied by the CoAS of the RNA and his spouse.

The Indian CoAS, however, voiced his displeasure at the tardiness of his Nepali counterpart, as the CoAS, Gen Prajwalla SJB Rana arrived late at the venue of the appointment. He was late by 90 minutes.

Back in Kathmandu, the visiting Indian CoAS expressed his great pleasure at his reception at the Royal Palace.

He said he found the King very concerned and serious about the situation of the country.

On the same occasion, he asked the RNA to forward the list of weapons and logistics India is to provide Nepal soon.

He also mentioned that the relations between India and Pakistan have strained, and there lurks a threat of war any moment.

He spoke of India's offer of supplying the Insas Rifles to the RNA because they are fit for the RNA in every way.

He expected the RNA to pursue professionalism and insulate itself from politics.

Sunday, May 19, 2002

The CPN (UML) has, in an unabashed manner, maintained its double standards on the question of whether or not to ratify the State of Emergency.

The Party's General Secretary, Madhav Kumar Nepal, has consulted other political parties, journalists, and intellectuals over the issue before the Party plans to make its official position public. The experts are reported to have advised him not to opt for ratification.

There's new information on Ramesh Nath Pandey secretly meeting GP Koirala at the residence of his daughter, Sujata Koirala. The topic of their talk has been kept confidential.

The frequent meetings between GP Koirala and Madhav Kumar Nepal are being viewed as indicative of their decision not to ratify the State of Emergency. This means that PM Deuba will have trying times ahead.

A Brigade-level conference with broad discussions began today at the RNA Headquarters. The deliberations dwelt on the successes and flaws during the military operations. The Brigade commanders unanimously concluded that the Maoists were increasingly weakening, with weapons, logistics and other essentials in dwindling supply.

Some commanders voiced their grievances at the lack of allowances and timely relief from field duty, insufficiency in logistics and other essential supplies, especially in the context of the fast-approaching rainy season.

Monday, May 20, 2002

The majority of the Central Committee Members of the Nepali Congress Party has reached a decision not to ratify the State of Emergency. Influential members, like Khum Bahadur Khadka, went to the extreme of writing his Note of Dissent while the former PM, Krishna Prasad Bhattarai, and another member, Pradip Giri, didn't attend the meeting. It's very likely that the rift within the NC will deepen further, leading to the dissolution of the House of Representatives, giving rise to constitutional complications.

The CPN (UML) and other parties have also opposed the ratification of the State of Emergency.

Given the mobilization of the Army and considering the political and economic uncertainties, I humbly asked the King as to what was the Royal Palace contemplating.

He said the PM recommending the dissolution of the Parliament would be the best option, unpleasant though it might sound.

Wednesday, May 22, 2002

Speculations are rife that PM Deuba will be expelled from the Nepali Congress Party, and he will, in turn, call for the dissolution of the Parliament.

A meeting of the Secretaries of the Royal Palace sought the King's role in the creation of an environment for the PM to continue in office until fresh polls take place in the country.

The King and the PM discussed in detail the prevailing situations and reached the point at which the PM would recommend to the King to have the Parliament dissolved, and allowing everything to fall in place in rapid succession.

The news of the dissolution of the Parliament was broadcast by Radio Nepal and Nepal Television (NTV).

Thursday, May 23, 2002

The news of the dissolution of the Parliament has sent shockwaves to the political parties. Madhav Kumar Nepal rushed to consult with GP Koirala at the new turn of events.

Accordingly, a Nepali Congress meeting convened at the party office expelled Deuba from the Party. Even those in the Nepali Congress close to Deuba expressed their dissatisfaction at the sudden dissolution of the Parliament without any consultations being held by Deuba with them.

The PM, in the Cabinet meeting the next morning, explained the reasons behind his move.

The Standing Committee meeting of the CPN (UML) condemned the unexpected turn of events caused by the infighting and rivalry within the

Nepali Congress. The meeting further decided to call an all-party meeting the next day, and prepare itself for free and fair elections.

The RPP termed the turn of events unfortunate, holding the entire Nepali Congress Party responsible for all the political mess at hand, and postponed its regional convention that was due to start.

The frequency of meetings between GP Koirala and Madhav Kumar Nepal has increased. They seem to be seeking the response of the Maoists at this new situation.

Some Congress members are trying their best to bring both factions together for party unity.

Friday, May 24, 2002

The political environment is heating up.

In his audience granted to the British Chief of the Defense Staff, Sir Michael Bois, the King expressed his thanks to the British Government for its assistance to the RNA and asked him to have the ex-British Gurkha soldiers condemn the Maoist activities.

The King also asked him about his views on India's perception of the Maoist problem in Nepal.

The British visitor touched upon the long relations between the two countries and their armies.

He said he found the Defense Minister of India, George Fernandes, committed to disallowing the Nepali Maoists to carry out their terrorist activities from India.

He also said he found India-Pakistan relations embittered to the worst level, threatening an outbreak of war any minute.

He also expressed his hope that the general elections in Nepal would be free and fair and would take place within the declared deadline.

On the occasion, Sir Michael pledged British assistance to the RNA to continue in the days ahead.

Sunday, May 26, 2002

At the lunch hosted at his residence, the Chinese Ambassador requested me to arrange an Army helicopter for his flights to Mustang and Rasuwagadi, the latter presumably to obtain fresh information and the exact status of the construction of the Syabrubesi-Rasuagadi Highway.

He said China would provide the RNA with Armored Personnel Carriers (APCs), useful for Peacekeeping Missions, at cost prices.

Wednesday, May 29, 2002

The IGP of the APF, Krishna Mohan Shrestha, in a regular briefing to the King, asked him to increase the number of awards to be given to all the ranks in the APF on the auspicious occasion of his (the King's) birthday.

He also requested the King to entrust the APF with the role of security escorts in the formal royal autocade.

The King expressed his interest in the modality of payment for the weapons and logistics supplied to the APF by India.

He was also curious to know about the exact status of the communication system of the APF.

In a series of briefing to the King, the Chief of the NID, Devi Ram Sharma, reported that India has been playing the Maoist Card to minimize the American influence in Nepal.

The Maoists have been able to sway the political parties to vote against the extension of the State of Emergency and to seek solutions to the problems politically through dialogue.

The CPN (UML) has promised the Maoists to introduce necessary amendments to the Constitution after the elections, provided the Maoists help the CPN (UML) and its allies gain the two-thirds majority win at the polls.

Friday, May 31, 2002

A helicopter flying for Lukla at 7 a.m. from the Makalu Base Camp has gone missing, according to a report received here.

The Maoists robbed the Agricultural Development Bank and Nepal Bank Limited of Barhathawa of Sarlahi District last night.

Some NC leaders, such as Ram Chandra Poudel, are making tremendous efforts to make peace between GP Koirala and Sher Bahadur Deuba to prevent the party from splitting. They say if the Party disintegrates, it'll reach a point where history will judge the leaders for their personality clashes and personal egos.

Koirala is learned to have said that the Party will take a decision once he's back from his visit to China.

Sources say Chakra Prasad Bastola visited Crown Prince Paras. The latter expressed his displeasure at the dissolution of the Parliament and with the unpleasant turn of events.

Tuesday, June 4, 2002

Narayan Singh Pun called me today to inform that he's leaving the Nepali Congress to form a new party of his own to bring together ex-servicemen, various ethnic groups, and indigenous peoples.

Some Congressmen and members of the dissolved Parliament are preparing to file a case at the Supreme Court, seeking its reinstatement.

There's information of the General Secretary of the Marxist Party of India, AB Bardhan, and another leader, Manjul Ansari, currently in Kathmandu, holding consultations with Madhav Kumar Nepal in his capacity of being the General Secretary of the CPN (UML).

Friday, June 7, 2002

The government has formed a team comprising the Vice Chairman of the National Planning Commission, Dr. Shankar Prasad Sharma; the Joint Coordinator of the NSC, Maj. Gen. Rookmangud Katawal; the Secretary at the Ministry of Foreign Affairs, Gyan Chandra Acharya; Prof. Shridhar Khatri, and Dr. Arzu Rana Deuba, spouse of the PM, to participate in the meeting of the 'Help Nepal Group' to be held in London.

The King granted his audience to the members of the team and advised them to sort out the priorities for Nepal to be raised in the meeting as the agenda for discussion.

India sounds unhappy about the venue of the meeting.

The participation in the meeting is of immense importance for Nepal because it can articulate its problems and convince the friends of Nepal for help with necessary assistance.

The reconstruction of the infrastructures destroyed by the Maoists is one of the priorities certain to gain prominence in the meeting.

The inclusion of Maj. Gen. Rookmangud Katawal from the Army as one of the members of the delegation substantiates the value Nepal wants to attach to the meeting.

Sunday, June 9, 2002

Today I candidly described my assessments to the King: that now the monarchy is gaining a stronghold with its image gradually building up after the Royal Palace Massacre.

I said, therefore, it's time I resign my position as Military Secretary to the King on ethical grounds for the tragedy.

The King advised me not to think of resigning, as it wasn't of my making but a mishap within the Royal family, and there was nothing I could've done to prevent it anyway.

I offered the King my advice on forming a small and unofficial task force to quietly observe the state of affairs of the country and develop strategies and brief him on the same. He then asked me if I could name a couple of individuals who could be entrusted with this responsibility.

The key members' first meeting sorted out the common minimum task of objective analysis of the situation, and the role of the Royal Palace in the upcoming elections in the changed context. We discussed the need to consult with the King about setting up an employment support fund of Rs. 100 million.

The King expressed his keen concerns about the feasibility of elections, the dismissal or continuation of the PM in the government, and the modality of the procedure of the transfers and promotions of the Secretaries.

He told me to work on plans to convey the impression that the monarchy is stronger than is normally thought of.

Next, the King instructed me to figure out the aims and objectives of the agenda to be discussed during his upcoming visit to India.

I said the proposal for the supply of weapons for the security forces, and regulations of the open borders between Nepal and India to curb terrorist and criminal activities and smuggling of goods could be the issues of immediate importance to both India and Nepal.

The King also seemed interested in strengthening the constitutional organizations.

Monday, June 10, 2002

According to confidential information, the Indian Embassy is quietly collecting the personal details of those serving in the Royal Palace; the résumés of some key officials, such as Chiran Shamsher Thapa and Bishwo Bikram Shah, are believed to have already been sent there. I talked harshly to Secretary Sudan Prasad Pokharel on this matter, as he was collecting the CVs of the Royal Palace executives, as asked by the Indian Embassy.

The King sounded very angry at hearing about it.

Sources say that GP Koirala is again in India to get moral support for the stance of the Nepali Congress for the reinstatement of the Parliament. India has exerted its pressures on both factions of the Nepali Congress to patch up for unity so that the Leftist parties can't win a majority of seats in the upcoming elections.

Thursday, June 13, 2002

The planned visit of the Chinese Army delegation to Nepal has been canceled for undisclosed reasons following the return of the Chinese Ambassador and Military Attaché from their visit to the RNA camps in Gorkha District.

GP Koirala, addressing a mass gathering in Biratnagar, underscored the need for a greater democratic alliance at the critical hour of danger looming over democracy in Nepal. He also accused Sher Bahadur Deuba of moving ahead singlehandedly in having the Parliament dissolved on his watch.

There's report of an armed clash between the Maoists and the RNA in Salyan District, causing heavy damage to the Maoist side; 52 dead bodies of the rebels were discovered at the site of the engagement.

Monday, June 17, 2002

As a gesture of showing its high importance to the King's upcoming visit to India, it has deported seven wounded Maoist militants to the security forces.

On the other hand, the Indian Minister of Home Affairs, Lal Krishna Adwani, has accused Nepal of sheltering such terrorist groups as the Lashkar-e-Toiba and Jaish-e-Mohammad by allowing them to set up their camps in the country.

In response to Adwani's remarks, the Pakistani President, Gen Pervez Musharraf, counterattacked India of inciting a regional war.

The King began his first visit to India from June 23 and ended it on the 28th.

Monday, July 1, 2002

Girija Prasad Koirala and Surya Bahadur Thapa quietly visited the former Chief Justice of the Supreme Court, Vishwo Nath Prasad Upadhyaya, who reportedly said that he would've reinstated the House of Representatives were he still in office.

He also added that the sitting Chief Justice, Keshav Prasad Upadhyaya, is very loyal to the King and works to serve and secure his interests.

India and the US are lobbying for the reinstatement of the House of Representatives in Nepal.

A decision has been taken by the Office of the Royal Palace to appoint Retd Lt Gen Fattya Bahadur Limbu as Honorary ADC because he fits all the criteria for the position in every respect.

As part of the proposed structural changes in the Royal Palace Service, the King has given his approval to have Brig. Gen Gajendra Bahadur Limbu, Col. Nepal Bhushan Chand, and Major Himanshu Khadka transferred to the Military Secretariat.

The Acting British Ambassador invited the NID Chief, Devi Ram Sharma, the Joint Coordinator of the NSC, Maj. Gen Rookmangud Katawal, and the Director of the Military Intelligence (DMI), Brig. Gen Dilip Rayamajhi to a dinner where he pledged British help to strengthen the intelligence wing of the NID.

Thursday, July 4, 2002

GP Koirala, in connection with mass meetings in several places, has reached Nepalgunj where he learned about the security situation from the security heads there.

It's learned that no one from the Army attended the briefing.

The security heads described the situation being better with the gradual weakening of the Maoists in the area.

Koirala confessed that the present situation can be attributed to the neglect meted out to the NID in the past 12 years.

Sunday, July 21, 2002

The King's recent visit to China has created good impacts on Nepal-China relations. The visit has been successful to dispel the doubts of the Chinese people about the Royal Palace Massacre.

China expressed its wish to construct a highway in the memory of the late King Birendra. They asked the present King to specify the location for such a highway in Nepal. King Gyanendra expressed his happiness and promised to prepare a detail survey and work plan to this effect soon.

At a reception held at the Royal Nepalese Embassy in Beijing, the former Chinese Ambassador to Nepal, Lee, recalled the yearlong economic blockade imposed by India on Nepal, and the extreme sufferings of the people of Nepal over a weapons deal made between Nepal and China. He added that China was sensitive that nothing Nepal and China say or do should cause troubles to the people of Nepal. He further said that Nepal and China can benefit from tourism, industry, and commerce by encouraging the private sectors of the two countries for investments in these areas.

The Kings official visit to China began from July 9 and ended on the 18th.

Monday, July 22, 2002

The King, in his visit to India, talked to its senior leaders on Nepali and Indian intelligence departments to work together to share information on the hideouts of the Maoists in India and to facilitate the security forces to apprehend and deport them to Nepal.

In my briefing to the King, I mentioned that the Thai monarchy may offer good examples for the monarchy in Nepal: to initiate royal projects in the interest and for the benefits of the common people in the country. The implementation of the royal projects is sure to strengthen monarchy in Nepal.

Listening to my suggestions, the King said, 'Good idea.'

Wednesday, July 31, 2002

GP Koirala, in connection with his health checkup, arrived in his hometown, Biratnagar, whereupon, talking to party activists there, said he had helped the King in every respect, ranging from quietly meeting the Maoist leaders in Kathmandu to convincing them to join the mainstream competitive politics, but the King, in turn, repaid by deceiving him all the time.

He pressed his demand for the old Nepali Congress election symbol of 'rukh' (tree), an image of pride the NC always cherishes.

He also threatened to launch an unprecedented showdown in the capital, in case the establishment doesn't get it.

[It's noted that the NC has splintered into two different parties, each claiming for itself the ownership of the old election symbol.]

He further said that his stance on the reinstatement of the House of Representative is gaining ground and receiving overwhelming support, for it's aimed at the retention of democratic values.

At a meeting today with the leaders of the CPN (UML), Bharat Mohan Adhikari and Dr. Dilli Raj Khanal, I asked them to make their positions and perspectives clear about the monarchy. They said, provided the King have the Army remain neutral during the elections, they would stand by the monarchy as a symbol of unity for once and for all.

I assured them that the Royal Palace would take it positively.

The Chinese Ambassador, back in Kathmandu, called me to ask the government to immediately control the Dalai Lama and the Free Tibet activities in Nepal.

He also asked me to arrange a High Altitude Training course in Mustang for two Chinese Army personnel.

When asked for his permission, the King gave me instructions to have it arranged.

Friday, August 2, 2002

The CoAS, Gen Prajwalla SJB Rana, due to retire on September 9, is seeking personal security provision in place in his retirement. There's rationale behind it because it was during his tenure as the CoAS that the Army was mobilized for action against the Maoists.

There's the common practice for CoASs to take a customary leave one month prior to one's official retirement.

He's seeking an audience with the King on September 4.

It's learned that the Maoists support the stance taken by the Nepali Congress for the reinstatement of the House of Representatives.

They have also spread rumors of massive attacks on the security camps in Rukum, Rolpa, Salyan, and Jajarkot from the helicopter they have captured from the Asian Airlines and now in their control, and painted exactly like an Army chopper.

There's an intelligence report that two Maoists cadres, Dilli Ram Oli and Sudarshan Malla, are leaving for Debghat to meet and talk to Yogi Narahari Nath to mediate a talk between the Maoists and the King.

Sunday, August 4, 2002

Today I asked the King to set the date to confer on the Crown Prince with the title of the Prithuladhish.

I also reminded him of the matter of the utilization of the Army Welfare Fund (AWF) by operating a bank to pay pensions to the retired soldiers of the Indian Army and the British Brigade of Gurkhas. This step would preempt any envisaged security threats at a time when India is also strongly proposing to have more Indian Army pension camps opened across the country. Further, the concept would also augment the funds for investment in Army development projects.

The Full Bench of the Supreme Court gave a verdict in favor of PM Deuba, stating that he had the prerogative to have the Parliament dissolved.

The leader of the opposition party, Madhav Kumar Nepal, is scheduled to visit India from August 10. As a preface to his further talks in India, he reportedly had a closed-door meeting with the Chief of the Nepal Study Wing of the RAW.

It's pertinent to provide the Military Police of the Royal Guards Military Police with allowances at par with that of the Special Forces.

Thursday, August 8, 2002

The Sampati Chhanbin Ayog (Property Investigation Commission) asked Army officers to furnish it with the details of their property without delay.

The law has it that all cases, other than rape and murder, are looked into by a special legal body in the Army.

The joint meeting of the representatives of the Office of the Attorney General and the Defense Ministry agreed to collect the details of the property of army officers and send the details to the Commission.

Lt. Gen. Pyar Jung Thapa will take over the responsibilities as the new CoAS from tomorrow as the outgoing CoAS, Gen Prajwalla SJB Rana, is taking his customary leave of one month prior to retirement.

According to information received from the NID, S. Ramesh, a Councilor at the Indian Embassy, has made remarks to the effect that the RNA is reluctant to be mobilized to Maoist-affected areas, that they can't fight the Maoists, and that the Army isn't under the control and guidance of the government, and so on. Ramesh also allegedly added that although India itself is a developing country, it's providing Nepal with weapons worth Rs. 35 crores (350 million).

He said it was unfortunate for Nepal to question the quality of the weapons provided by India.

He also ruled out the allegations that Nepali Maoists are receiving training in India.

Monday, August 12, 2002

PM Sher Bahadur Deuba has expressed his commitment to holding the general elections at any cost, whereas the Koirala faction of the Nepali Congress, unhappy at the disadvantageous turn of events, is voicing its view on the impossibility of holding elections, given the complexity of the situation, and accusing Deuba of sabotaging democracy.

At the farewell function held for the outgoing CoAS, Gen Prajwalla SJB Rana, the Ambassador of Pakistan to Nepal remarked that his tenure in Nepal wasn't of small challenges, defending Pakistan amidst baseless allegations made against it, although the Nepal-Pakistan relations have always been smooth and good.

The Western Division Commander, Maj. Gen Sadip Bahadur Shah, while talking informally to an officer, allegedly bragged that Maj. Gen Victory SJB Rana would become the Principal Military Secretary, and himself the second man in the Army hierarchy.

The colleagues in the task force appraised and discussed the unsavory activities of the leaders of the political parties in the past 12 years, and concluded that if anything stalls elections, it's the leaders themselves. The colleagues also observed that the CPN (UML) leaders maintain double standards over the question of monarchy and republic.

Today, the Maoists, through a communiqué, announced their overt and covert plans to disrupt the general elections. If no progress is made in the peace talks during September, the communiqué continues, they will take to more stringent measures of attacks to make the government suffer big setbacks.

They are also applying their psychological pressures on those Nepali Congress members close to Deuba to join the Koirala camp.

Friday, August 16, 2002

The Congressmen of the Koirala camp are meeting tomorrow where leaders such as Shailaja Acharya and Devendra Nepali are likely to do their best to stitch together the two splintered factions.

Ajun Narsingh KC of Nepali Congress sounded very critical of the Royal Palace and Deuba.

Prior to the Indian Defense Minister Jaswant Singh's visit to Nepal, the King has offered his advice for the government to prioritize the points of dialogue, as follows:

- Hold discussions on the implementation of the decisions and agreements reached between Nepal and India;
- Prepare answers to the likely issues India would raise;
- Discuss the security concerns of Nepal and the concept of regional security;
- Nepal to convince India about the likely impacts of the Maoist insurgency on both Nepal and India;
- In the event that the upcoming elections can't take place in Nepal, what would be India's response?; and
- How does India look at the double standards maintained by the CPN (UML) with regards to Nepal-India relations?

Monday, August 19, 2002

The Chief and Deputy Chief of the Indian Intelligence Agency, RAW – Bikram Sudh and JK Sinha – arrived in Kathmandu last Friday. By visiting the Pashupatinath Temple on Saturday, they claimed the purpose of their visit to Kathmandu to be purely religious and a respite from the heat in India.

At a meeting with the Chief of the NID, they expressed their concerns about the feasibility of holding the elections, the declaration of ceasefire by the state itself first, the exact status of the state-rebels peace talk, and the future of the State of Emergency.

In response to their queries, the NID Chief replied that the government has already set a precondition for the Maoists to lay down their arms, stop their negative propaganda against the monarchy and the Constitution before coming to the peace talks. He further asked his Indian counterparts about India's own policy when the NC and the UML leaders themselves are meeting the Maoist leaders in India.

The CPN (UML) sounds extremely encouraged by the grand receptions accorded to Madhav Kumar Nepal in India, particularly at being addressed as a prospective Prime Minister of Nepal.

Wednesday, August 21, 2002

The Principal Military Secretariat has now been completely militarized with the withdrawal of non-military or civil servants from there.

The members of the task force met and discussed the fresh developments and also decided to create a situation to freeze the election symbol of 'rukh' (tree) of the Nepali Congress, as both factions are fighting tooth and nail to claim it. The meeting observed that the political parties themselves are vitiating the environment for elections.

An NSC meeting has recommended that the government extend the State of Emergency for another three months, citing the improving operations against the Maoists.

Monday, August 26, 2002

A couple of leaders from the CPN (UML) and the Nepali Congress, in an informal talk, advised me to treat many leaders of their two parties strictly in the context of their open championing of India.

Krishna Prasad Gautam of the Nepali Congress opined that the political situation of the country was getting increasingly intractable, and so the King must take a move in the interest of the country without further delay.

Journalist Vijay Kumar [Pandey] also expressed similar worries over the sorry state of affairs in the country; not even new elections would be able to change the same old characters and attitude persisting in politics.

Today, the King and I candidly discussed the ways to purchase weapons from [South] Korea and Israel.

The proposed visit to Belgium and South Africa by the PM at a time such as this will only arouse public anger.

Sources say that PM Sher Bahadur Deuba has bragged that the new CoAS, Gen Pyar Jung Thapa, is his confederate and he'll do anything at his bidding.

There's information that Tendi Sherpa and Devendra Parajuli, student leaders of the CPN (UML) and the Maoist-affiliated bodies, have labeled allegations against the Royal Palace regarding the death of Pravin Gurung in a vehicle accident.

Thursday, August 29, 2002

Today, the King and I reviewed the security concerns of the Royal Palace in the prevailing precarious situations.

We discussed the measures, among which to ask the Nepal Police to collect details of the people living in the houses near and around the Royal Palace.

The King has asked me to make arrangements for the Crown Prince's two-day sojourn to Mustang.

He also commanded me to make arrangements for the swearing-in and the conferment of the insignia of rank to the new CoAS on September 9.

Wednesday, September 4, 2002

I briefed the King on the precarious situations of the country.

As per the wish of the Crown Prince, a quotation for Pentwall Rifles has been asked from the Bajhang King, Binod Bikram Singh. It calls for sending two Army officers abroad to learn how to operate the rifles. The Crown Prince wishes to have the Pentwall Rifles purchased to have some of them kept at the Military Academy and the Royal Guard Military Police Unit.

To the best of my knowledge, it's neither affordable nor so useful for the Army.

The Crown Prince, however, insisted on buying them from the budget of the Royal Palace.

When I briefed the King on it, he instructed me to arrange it for the Crown Prince's pleasure.

Next, I briefed the King on the understaffing in the technical and administrative departments in the Army.

On the same occasion, I suggested on an autonomous website in the operation room of the Military Secretariat to make it easier for the information cell to prepare reports from the information on the writ petitions, and report them to the King.

The briefing also included the need to find an alternative supplier of weapons as the German Company failed to honor its commitment. Because India wants to sell its Insas Rifles to Nepal, it's been exerting its pressures on Britain not to supply the HK Rifles to Nepal.

The Principal Secretary and I discussed the extremely sensitive write-up by Dipta Prakash Shah, a royal nominee in the National Assembly. We reached a decision to ask him to resign for visiting China without obtaining Royal permission and remaining neutral in the elections of the Vice Chairman to the National Assembly although there was a clear indication from the Royal Palace for him to vote for Ram Preet Paswan.

As instructed by the King, we drafted a resignation letter and called him to the Royal Palace to sign it. He obliged without a second thought.

Friday, September 6, 2002

At the joint meeting of the various agencies held at the Election Commission reviewed the security situation and proposed a phase-wise general elections, whereas the security officers questioned the possibility of elections itself.

The CoAS observed that the campaign of disarming the Maoists by December may slow down if the Army is asked to provide security in the general elections. So he advised the officials of the Election Commission to assess the environment for holding the elections across the country rather than making armchair conclusions.

In continuation of their actions against the security forces, the Maoists blew up the house of the NID Chief in Pyuthan District on September 4.

The National Unity Council today organized a demonstration and a mass meeting at the Khula Manch, the open-air public space in the middle of the City of Kathmandu. It saw a turnout of one thousand or so people. The speakers came heavily upon multiparty democracy and shouted strong slogans in favor of the monarchy.

There's information that the Chairman of the All Nepal Unity

Society, Bam Dev Chhetri, is arrested by the Delhi Police. Talk about his extradition to Nepal is going on.

Monday, September 9, 2002

The shocking news of 49 policemen killed in a Maoist attack in Bhiman of Sindhuli District was received today. After the massive operation, they also carried off all the weapons from the police post. The report spoke of crossfire throughout the night until the dawn. The details of the loss are yet to be confirmed.

The Maoists also attacked Sandhikharka of Arghakhanchi. Details are yet to come.

At a time when the entire machinery of the country is concentrated to fight the Maoists, India has pressurized Nepal to allow it to open an army pension camp at Butwal, a consular office in Birgunj, and to deploy 40 Indian security persons equipped with weapons at the Indian Embassy in Kathmandu.

I briefed the King that the last proposal wouldn't be against the sovereignty of the country.

The Ambassador of Nepal to India, Dr. Bhekh Bahadur Thapa, in an informal talk with me, expressed his unhappiness at the straining Nepal-India relations on the question of Nepal's plan to purchase weapons from a third country, India's proposal to open new Indian pension camps in many places, and have a consular office in Birgunj.

He said India always seeks an excuse to pressurize Nepal to purchase weapons from them. He added that the PM's frequent visits to India and the closed-door meetings with the leaders there all indicate the likelihood of weapons-related treaty to be quietly signed between the two countries.

I made it clear to him that the technical team of the Army had tested several models of weapons from different countries, and the weapons made in India were found to be of low quality.

Weapons, once introduced, can't be replaced at short intervals, as it proves very costly. Therefore, the Army wants to purchase weapons of the best manufacture and quality for the next 20 to 30 years into the future.

I added that the next immediate neighbor, unlike India, never takes undue interest in Nepal's arms deal.

I sensed that India must be exerting its diplomatic pressures on him to convince Nepal to purchase Indian arms and ammunition.

Tuesday, September 10, 2002

The Former Chinese Ambassador to Nepal, Lee, on a visit to Kathmandu with a team of businessmen, talked of the possibilities of Chinese FDI (foreign direct investment) in Nepal. He said it would be a good idea to operate a commercial bank as a joint venture between the Army Welfare Fund and China. He said the development of infrastructure and manufacture of weapons at home would once and for all put an end to the contentious issue of purchasing weapons from another country.

At a meeting with the Governor of the Nepal Rashtra Bank, Dr. Tilak Rawal, Lee told him that the creation of an environment for the free circulation of the Chinese currency is essential if Nepal wishes to promote Chinese tourism in Nepal.

Khum Bahadur, who recently had an audience with the King, has had a change of tone in his statements. He has started saying that this is no time for holding general elections, and the reshuffle of the Cabinet is of immediate importance.

The military advisor to the UN's DPKO (Department of Peacekeeping Operations), Maj. Gen Chitra Bahadur Gurung has recently come back to Nepal to attend to the mourning rites of his mother's death. His present UN tenure is about to expire, but efforts are afoot to have his appointment extended for the next term.

Wednesday, September 11, 2002

IGP Pradip SJB Rana remarked that the Nepal Police isn't as weak as alleged by the Army when it comes to fighting the Maoists.

He also added that there have been baseless allegations made by India of the increase of ISI activities in Nepal.

He also voiced his grievances that the Three-Naught-Three (303) Rifles the Police have had to use are obsolete, very old, and ineffective.

In order to ensure better security arrangements and insulate the Nepal Police from political influences, the IGP should be given a free hand in the transfers of policemen within the districts.

He made a request to the King for scrapping the provision of the

32-year service tenure in Nepal Police and bring it at par with that of the Army.

The King simply smiled while listening to the IGP's proposal that the name of the Nepal Police be changed to the Royal Nepalese Police Force.

The King instructed the IGP to circulate the information of the impossibility of holding the planned general elections in November.

It's learned that India arrests and extradites to Nepal only those Maoist cadres who are opposed to it.

Thursday, September 12, 2002

In a joint meeting of political party officials and all the security heads held in the Office of the Election Commission, IGP Pradip SJB Rana suggested that the elections announced to be held in November be deferred to April or May of the next year.

The leaders of the political parties objected to his remark, alleging that he sounded and was acting more like a spokesman of the Royal Palace than a police officer.

PM Deuba sounded frantic at the impossibility of holding the next elections as pointed out by the leaders of the political parties and the officials concerned. He dropped in at the residence of Prakash Koirala, a son of former PM, the late BP Koirala, and also met the American Ambassador to Nepal, presumably to consult with them about the emergence of the complicated situations.

Monday, September 16, 2002

Judging by the recent speeches made by the leaders of the political parties, it seems to me that they, too, have started using the Maoist phraseology of 'forward-looking change' quite for some time now.

Despite the Maoists' call for a nationwide closedown and strike, the examinations of the Bachelor's Level of the Tribhuvan University (TU) concluded unhindered.

Prakash Koirala, an NC leader, has lately been active to make peace between GP Koirala (his younger paternal uncle) and PM Deuba.

Deuba is leaving for India where he's likely to meet Indian leaders.

Although well aware of the reality, no leader of any party is showing courage to articulate about the impossibility of holding the next elections for the fear of losing face before the public.

Most Nepali media sound critical of the Royal Palace.

At an informal meeting with Ramesh Nath Pandey, GP Koirala requested him to ask the King on his behalf to command the Election Commission to delay its decision pertaining to the case of the election symbol of the Nepali Congress, in the present context of both the split-away groups claiming to obtain it.

The Colonel Commandant of the Brigade of Gurkhas, General Sir Sam Cowan, in a meeting with the King, complained of the Gurkha Army Ex-Servicemen's Organization (GAESO) as creating obstructions to the fresh recruitment of Nepali youths in the British Army.

He also accused the GAESO of extorting money from ex-British Gurkhas to donate it to the Leftists.

The King thanked him for the British assistance to the RNA and Nepal and expected similar help to strengthen the intelligence cell of the Army.

Tuesday, September 17, 2002

Speaking at a confidential meeting of Nepali Congress members in his daughter Sujata Koirala's house at Mandikhatar, GP Koirala said the King was conspiring against democracy by using PM Deuba as his tool. He talked tough, with a challenge to teach a lesson to the Royal Palace if it seeks to sabotage the Congress Party.

Khum Bahadur Khadka has his own list of complaints of betrayal by the King.

PM Deuba himself isn't happy with the King for his noncooperation in helping him to obtain the election symbol of 'rukh' (tree), an iconic election symbol for the Nepali Congress. He also complained that the King promised him help but breached it at the eleventh hour. He believed that the King could've helped him if he really wished to.

The King wished that the outgoing CoAS, Gen Prajwalla SJB Rana, be asked to choose a car from the catalogs as a farewell gift that's customary of the monarch to grant a retiring CoAS.

Govinda Raj Joshi, another NC politician and former Minister, has accused the Royal Palace of manipulating the slated general elections to break the Congress Party.

The CPN (UML) has termed the designated elections 'a sham' and 'an unseasonal rain.'

The Maoist sister organization, the Revolutionary Students Wing, has threatened to go on a nationwide protest with its 13-point demands.

Sunday, September 22, 2002

There's an air of political changes of some kind in the near future: The King has given me a special task as part of the move. I'm working out the strategies to deal with unforeseen constitutional complications, in an extreme manner, not sharing the secret with anyone.

The Chinese Ambassador to Nepal has expressed doubts over the Civil Military Campaign Plan.

I hear that GP Koirala is desperate to meet me; I don't know why.

PM Deuba sounds upset after his visit to the King; it reflects a bickering between the two. He has started firing salvos at the King: that the latter better take over if he so wishes; and they would help him by landing themselves in jail (by courting arrest).

He added that if the King needed his help, then he, too, should extend his help to him to form an all-party government under his leadership so that such a government can deal with the Maoists. He said it's no time for the Palace to risk playing in the ongoing political deadlock.

On the same occasion, the PM added that in case there's no election, the power is sure to concentrate in the King's hands, thereby leading to the state of the 'fight to the finish.'

In an informal talk with the former PM, Lokendra Bahadur Chand, PM Deuba said he'll spill out the secrets of the King if he so wishes.

The King granted his separate audiences to the Chief of the NID, Devi Ram Sharma, and the CoAS, Gen Pyar Jung Thapa. The contents of their discussion have been kept confidential.

In a briefing to the King, I advised him to have a rest for a couple of days, as his heavy workload may entail health complications.

The King instructed all security forces to consult with the Army to make public the information of operations against the Maoists.

It's been agreed that the Army shall inform the Military Attachés at the Royal Nepalese Embassies about the details of its operations.

The British Government has invited the Secretary at the Ministry of Defense and the Joint Coordinator of the NSC and DMI of the Army to visit Britain.

The King has given his consent.

Monday, September 23, 2002

The all-party meeting held in the Office of the Election Commission ended inconclusively as the Minister for Home Affairs, Khum Bahadur Khadka, couldn't assure the political party leaders of providing security during the general elections.

As the General Secretary of the CPN (UML), Madhav Kumar Nepal, presented his political report to the Central Committee at the Party's meeting.

The CPN (UML) is prepared to join the Deuba Government.

Wednesday, September 25, 2002

The Chief of the NID analyzed the political situation: that the opposition parties are flirting with the Maoists' political blueprint.

And on this backdrop, the former PMs, Surya Bahadur Thapa and Lokendra Bahadur Chand, each is seeking to head the government himself.

Even if a government is formed, in pursuance of Article 127 of the Constitution, it should have led by someone from the political parties.

PM Deuba seems worried at the fresh turns of the political events.

Sunday, September 29, 2002

It's known that the Royal Nepalese Ambassador to India, Dr. Bhekh Bahadur Thapa, has arrived in Kathmandu, presumably called by the PM.

The PM has taken his position that Article 115 be enacted prior to Article 127 with the annulment of some of the provisions enshrined in Article 115 of the Constitution.

The Seven-Party meeting held today concluded on the impossibility of holding the general elections and agreed to find an alternative.

The PM has been asked to find a legal and viable way out of the political deadlock, but he's opposed to the restoration of the Parliament.

Monday, September 30, 2002

GP Koirala was received in an audience by the King. The monarch, however, didn't concede to Koirala's proposal to have PM Deuba removed and have him appointed Prime Minister, or somebody from the young generation be appointed to the office.

The Election Commission has allotted 'gamala' (earthen vase) as the new election symbol to the newly-formed Nepali Congress (Democratic) led by Deuba.

Citing the consensus reached by an all-party meeting, the PM proposed the deferral of the date of the general elections.

Some Cabinet Ministers took the PM's step as pushing the elections to uncertainty.

Wednesday and Thursday, October 2 and 3, 2002

Amidst the fluid political situation and looming uncertainty in Nepal, the former Indian Ambassador to Nepal, KV Rajan, has arrived in Kathmandu for fresh consultations with various political leaders.

Although not expressed in so many words, his visit is seen to create an environment for Surya Bahadur Thapa to become the next Prime Minister.

The King granted his audience to the General Secretary of the CPN (UML), Madhav Kumar Nepal.

The King hosted a dinner for the heads of the security forces. There were no specific discussions other than reviewing the current state of affairs.

PM Deuba is mysteriously meeting the Nepali Congress leaders, Pradip Giri and Dr. Prakash Sharan Mahat, perhaps to have the elections dateline extended. Citing the precarious security situation in the country and having no suitable environment of holding the elections on the slated date as the reasons, PM Deuba, in an emergency meeting of the Cabinet, and as per Article 127 of the Constitution of Nepal, formally proposed the deferral of the date of the elections by one year, and rescheduled it for November 19, 2003.

The leaders of the opposition parties and constitutional experts stated that the PM's act of postponing the date of the general elections violated the norms and spirit of the Constitution of Nepal.

Following the fresh turns of events, the King has initiated his fresh consultations with the VIPs, such as the Speaker of the House of Representatives, the Chairman of the National Assembly, the Chairman of the State Council Standing Committee, the President of the Nepali Congress, the General Secretary of the CPN (UML), and other leaders of the political parties.

Given the new turn of events, I directed the heads of the security forces to stay alert and be on a standby position, as the King was set to take a drastic step to find a solution to the political deadlock in the country.

I also informed them of the immediate schedule of the collective audience to be granted by the King to them at 11 a.m. the next day.

The King remained busy in his office till late at night. So we, the Secretaries of the Royal Palace, too, remained in office, speculating on what steps the King is going to take, and what responses it will generate from the people.

As such, we saw two alternatives for the King to take: To announce the rescheduling of the date of the general elections, and/or to opt for himself to take over by removing the present Prime Minister.

The second choice is obviously fraught with risks, as it entails an uncertainty in the time period and the modality of a takeover.

In one sense, the aforethought royal coup is akin to bringing the mainstream political parties and the Maoists closer, which the Royal Palace will never want to happen.

The PM seems to be seeking an alibi to continue in office until the elections, taking recourse to Section 4 of Article 53, and by using Article 127, to recommend the dissolution of the Parliament.

This move will put the King in a difficult situation.

Out of a couple of unpleasant alternatives, the King finally chose one option: to dismiss the PM.

Article 127 has been defined as the ultimate constitutional provision to undo the complications of a constitutional vacuum. It was the need of the hour to use the same Article to break the impasse.

We, the five Secretaries, after our informal meeting on the issue, decided to ask the King to resort to Article 127 to remove the present PM and appoint the Speaker of the House of Representatives as an interim PM to avoid constitutional ramifications.

Friday, October 4, 2002

The King wouldn't afford to watch any further uncertainties continue in Nepali politics as a mere onlooker. I sense that the King was provoked by the Catch-22 situation of the country's politics, and without any possibility of holding the general elections in a near future, and at a time when inter- and intra-party disputes are widening, all these national ailments allowing the King with no other alternative than to spearhead his own move – singlehandedly.

Actually, the King desperately wanted the political parties to bond as an alliance to form an all-party government to isolate the Maoists, and thereby mounting pressures on the rebels to join in the peace talks to solve the national problem.

But Prime Minister Sher Bahadur Deuba was neither able to form a consensus government nor push forward the peace process with the Maoists, nor even hold the constitutionally mandated general elections.

The King had entrusted me with a top-secret assignment at which I had been working most confidentially for over three weeks.

The King granted his group or individual audience to the leaders of the political parties and intellectuals for counsel and discussions. Finally, he issued his instructions to the heads of the security forces at 11 AM, with full information on what was to follow thence.

The astrologers had advised the King that the most auspicious and opportune moment to take his step would be at 10:45 PM.

We, the five Royal Palace Secretaries, after due discussion, prepared the draft of the Royal Address and finally submitted it to the King. We deliberated on the possible charges against the PM in the context of his inability to hold the general elections on the scheduled date. We agreed on the phrase of 'incompetent to hold the elections' as a potent accusation to remove the PM from office.

Thus, the King, through his Royal Address to the country, announced the removal of Prime Minister Sher Bahadur Deuba and taking over the entire executive authority upon himself until another provision would be made to replace it.

He asked the political parties to forward the name of a consensus Prime Minister with a clean image and who has no intention to run for electoral office in the immediate future.

Saturday, October 5, 2002

There have been mixed responses to the royal move. Spontaneous rallies have been held across the country in support of the takeover by the King. The royal putsch has sent shockwaves through the entire political party circles, and they are busy, building strategies in response to the unexpected turn of events.

The heads of the security forces have taken every care to attend to every detail of the security situation. Their joint meeting reviewed the entire security status across the country in the wake of the new turn of events. They observed that the general people took the takeover as a normal course of events. The security forces have been deployed to avert any untoward eventuality. Many people have taken the King's step in a positive light. They say that it in no way curtails their fundamental rights; rather, the coup may contribute to ending the political mess.

The security forces reportedly stopped Rampreet Paswan, the Vice Chairman of the National Assembly, at the Thankot Check Point as he was leaving Kathmandu.

The CPN (UML) called an all-party meeting in its party office at Balkhu. Only the leaders of the NSP and the sacked PM, Deuba, showed up, while the members of the other faction of the Nepali Congress refused to join it simply because the breakaway that Deuba was would be attending the meeting.

We, the Palace Secretaries, are in office, without leave, rest, or relief, today being Saturday notwithstanding. We've been given particular assignments.

The King is busy in his office.

The King has commanded us to start working on corruption control so that people may have the feel of a sense of relief. We've formed cells to listen to people's demands to implement them immediately; the King's command to repair the airport of Dolpa for operations is a case in point. The functioning of the airport is sure to give the people a measure of relief from food scarcity in the region.

Sunday, October 6, 2002

In connection with the formation of a Cabinet with a leader of a clean image to head it, the King today consulted with major political leaders, such as GP Koirala, Madhav Kumar Nepal, Surya Bahadur Thapa, and others.

The outgoing Cabinet Ministers were asked to immediately return the official vehicles they were using and vacate the Ministers' Quarters they were occupying.

The Secretary at the Ministry of Home Affairs arranged security for those Ministers who have threats from the Maoists.

A decision has been taken to cancel all the political appointments of the immediate past government.

A task force has been formed to study and to lowering the prices of essential supplies to give people immediate relief.

An impromptu meeting discussed the ways to bring the political situation under control until the next elected government is in place.

The King granted his audience to the Chief Commissioner of the CIAA, Suryanath Upadhyaya, possibly to command him to look into the political parties' and their leaders' involvement in corruption and prepare cases.

Monday, October 7, 2002

The all-party meeting called by the President of the Nepali Congress took the decision to request the King to find a way out from within the political parties. They have sought a collective audience with the King to apprise him of their side of the situation.

As for the CPN (UML), the Party has taken the decision to go only with the government formed under Article 128 of the Constitution.

Sources say the acting Ambassador of India to Nepal visited the CPN (UML) General Secretary, Madhav Kumar Nepal, perhaps to discuss the fresh political developments in Nepal.

An official of the Ministry of Home Affairs, Raish Khan, has been instructed to look into the cases of stolen vehicles.

The Secretary at the Defense Ministry and the Chief of the NID, on a visit to Britain, didn't display a shred of responsibility to return immediately to Nepal in the fresh turn of political developments. I asked them for clarification for their breach of the code of ethics.

I've made a list of those employed at the NID through political influence and submitted it to the King; political nepotism has had bad effects on the NID's performance.

Tuesday, October 8, 2002

The sister organizations close to the CPN (UML) took out rallies, shouting slogans of 'Down with all forms of autocracy!' and 'No to the seizure of the rights of people!'

Many leaders and social figures have applied for the posts of Ministers in the Cabinet, and others called me personally to recommend their names for the same.

None of the former CoASs has called so far. Nor have they yet commented on the King's October 4 Takeover, though some retired Generals did call me to say they welcome the King's move.

But a couple of IGPs welcomed the King's step and promised to serve the country with whatsoever responsibilities the King entrusts them with.

Sporadic rallies and demonstrations have taken place in the capital and across the country.

Wednesday, October 9, 2002

Five days have gone by since the Royal takeover, yet the process of the formation of a new government is nowhere close to happening.

The British Ambassador to Nepal, Keith Bloomfield, has advised the King to move ahead by taking the political parties in confidence.

The CoAS, Gen Pyar Jung Thapa, has expressed his view that in a situation as at the present, we all must support the King's step, particularly at a time when the security forces are fighting the Maoists.

He said neither Kirti Nidhi Bista nor Lokendra Bahadur Chand is fit enough to deal with the situation as Prime Minister as they are gentlemanly, and so someone shrewd is required to handle the situation now.

The reappointment of Sher Bahadur Deuba as Prime Minister seems to be the best solution to the problem for now because the Army seems comfortable with him.

Former PM Surya Bahadur Thapa seems to be racing up to bag the portfolio of Prime Minister.

The CIAA has issued a warrant against the NC leader and former Minister, Khum Bahadur Khadka, to look into his involvement in corruption and abuse of authority.

Khadka, on his part, angrily said that it was the fault of the PM to be unable to hold the elections. He also accused the King of taking political vendetta on the likes him to send every single of them behind the bars.

Govinda Raj Joshi, another influential NC leader, reportedly rejoiced at the warrant served against Khum Bahadur Khadka, oblivious that it could very soon be his lot, too.

The ex-Lt. Gen, Krishna Narayan Singh Thapa, called me to express his appreciation of the King's step.

Sources say that the former PM, Dr. Tulsi Giri, has arrived in Kathmandu; the timing of his arrival and the objective of his visit are apparent.

The CPN (UML) has expressed its support if the King chooses to appoint the RPP leader, Lokendra Bahadur Chand, as PM. Chand is a polite man of an easygoing nature and with a clean image – ideal credentials to head the government.

The King's consultations with the leaders and intellectuals aren't over yet. The political parties haven't been able to field a consensus candidate for the prime ministership as yet.

The Principal Secretary, Pashupati Bhakta Maharjan, is anxious about the security situation of the country.

A directive has been issued to the intelligence department to maintain their surveillance of GP Koirala, currently in Biratnagar, reportedly for rest, but it's where he's very likely to meet the agents of the Indian Intelligence.

A constitutional void has surfaced, as the country is without a Prime Minister. We've advised the King to appoint Tara Nath Ranabhat, the Speaker of the House of Representatives, to the position for a time being until another proviso would offer itself.

But the King doesn't seem to be convinced that Ranabhat would provide the solutions to the problems at hand.

The Crown Prince, too, flatly rejected the proposal.

It seems that the King prefers Kirti Nidhi Bista for the prime ministership. But we expressed our reservation because the current situation, as it is, demands someone else with political shrewdness and smartness.

Thursday, October 10, 2002

The first round of the King's separate consultations with political leaders is over. Yet an acceptable solution to the political impasse appears nowhere in sight. Political parties haven't as yet been able to find a consensus candidate to head the government. They double talk. Lokendra Bahadur Chand seems to be a suitable candidate for the appointment, as the CPN (UML) has also formally decided to support him. But, overall, each political group is greedy to grab the Prime Ministerial portfolio for their own party. Leaders have repeatedly displayed their trademark unwillingness and their distrust of each other by declining to agree on a common name for the PM; this is particularly true of the President of the

Nepali Congress, Girija Prasad Koirala, and the General Secretary of the CPN (UML), Madhav Kumar Nepal.

Friday, October 11, 2002

Non-Resident Nepali National (NRN) Janak Basnet called me to convey to the King that the paper he had been asked to prepare is ready and asked me to fix a date for him to meet the King.

Sources say that GP Koirala, Madhav Kumar Nepal, and Surya Bahadur Thapa held a secret meeting at Thapa's residence to weave a strategy of forwarding Madhav Kumar Nepal's name as a consensus PM, and if things went against their plan, their parties would hit the streets for huge showdowns once the Dashain Festivals are over.

It's been revealed that the Maoists are running their training camp in the northern parts of Gorkha District.

In a review of the recent turn of events, the Maoists take it that the royal takeover has created an environment in their favor, allowing them to watch the King's action against the corrupt leaders which would ascertain the intensity of their opposition to the King's move.

Sheshadri, the editor of the BJP's (Bharatiya Janata Party) mouthpiece, has written that the BJP has backed the King's move, further expressing his worry that any delay in forming a new government would leave the opportunity for the Leftist forces. He seemed critical of the Leftist parties.

A joint meeting of some 10 small parties has reached a decision to protest if their existence is ignored in the present scenario. Their decision includes their emphasis on their choice of a leader to head the Cabinet who must be a man of clean image and commitment to take stringent actions against the corrupt in the land.

Today, the King formed a Cabinet headed by Lokendra Bahadur Chand as PM, but with the King's own choice of most Ministers.

Apart from the royal list, the Home Minister, Dharma Bahadur Thapa, is a nominee of royalist Sharad Chandra Shah, while many important nominees are the favorites of the Principal Secretary, Pashupati Bhakta Maharjan.

I think it would've been much better if the PM were able to include some major politicians in his Cabinet, as he had indeed wished to do.

In gist, therefore, Chand is merely a cosmetic PM in the present Cabinet. Such a cooked-up setup is hardly expected to deliver the essential services to the people.

Friday, October 18, 2002

Responses in support of the royal takeover are pouring in from across the country, suggesting that the King appoint a Minister each to represent their respective areas.

Journalist Kanak Mani Dixit has applied for the portfolio of the Ministry of Communication and Information, with his assurance to gather ample support for the takeover at home and from abroad.

On behalf of the entire Nepali film industry, film producer Uddhav Poudel has welcomed the King's step taken on October 4.

Similarly, Senior Captain Prachanda Jung Shah is desperate to meet the King to express his thanks for the takeover.

In a good gesture aimed at the people, the PM visited the Birendra Military Hospital to see the wounded in Maoist attacks.

It's expected that the members of the Royal Family should be active in works solely in the people's interest rather than merely showing up at opening ceremonies.

It's also expected that now is the time for the Royal Palace to win the hearts of the people by working for them, just like the Royal Family of Thailand.

Retired Major Mohan Khatri of the Royal Nepal Army was killed by the Maoists in Dolakha last night. He was operating a tourist resort on the Jiri Highway.

There are pressures from the public for the CIAA to step up its anti-corruption drives to take stringent actions against all those tainted with corruption charges.

Highly placed officials, such as Lokman Singh Karki, Dinesh Chandra Pyakurel, and Tika Datta Niroula are the ones, to name only a few, to be axed by the anti-graft body.

In a bid to boost the morale and confidence of the security services, the King hosted a dinner for the heads of the security forces. The Crown Prince was also present at the event.

Those present at the dinner suggested that the Council of Ministers take the big parties on board in the Cabinet, and insulate the

Administration from political interference, and prepare a tacit terms of reference (TOR) for the administrative structure to work for the people in a free, fair and efficient manner to give the public a sense of better changes and just relief.

Similarly, the chiefs of the security forces have expected help from all sectors. They also advised the Council of Ministers to gear up their anti-corruption drives, and bring the Maoists into mainstream competitive politics, and use the good offices of political parties, and eliminate unnecessary expenses.

The Principal Secretary and I reviewed the entire state of affairs after the October 4 move, and also sorted out the priorities of the works to be done henceforth.

Govinda Raj Joshi of Nepali Congress has observed that there was a hand of India behind the King's October 4 move, a truth not known to many people. He said that India's been manipulating the Maoists for a very long time, and now it's started exploiting the Royal Palace, too, to sabotage the political parties.

The CPN (UML)'s Bam Dev Gautam's observation is slightly different. Seemingly supportive of the royal takeover, he, however, advised the Chairman [King Gyanendra's new title] of the Council of Ministers to have the portfolio of the Minister of Home Affairs entrusted to a neutral person for him/her to be able to take the help and support from one and all.

Madhav Kumar Nepal, General Secretary of the CPN (UML), refused to join the government but hoped there would be timely corrections made to the October 4 move by the King.

The Joint Coordinator of the NSC, Maj. Gen Rookmangud Katawal, called me to convey to the King that the representatives of the London School of Economics (LSE) and Radio Netherlands want to talk to him to learn about the future implications of the October 4 takeover. The King granted him his permission to do so.

Rtd. Lt. Gen Krishna Narayan Singh Thapa and Maj. Gen Raghu Chandra Bahadur Singh, backing the King's move, have wished to help him in every way.

A matter of surprise, as said before: Not a single retired CoAS has so far commented on the King's takeover.

Wednesday, October 23, 2002

KP Oli, the Central Committee Member of the CPN (UML), has indicated his willingness to join the government provided he's given the portfolio of either Deputy Prime Minister (DPM) or Minister of Home Affairs.

He also stated his assurance to deal with the Maoists either by dialogue or by use of force.

But the King seems to have already made up his mind on a few names that he can trust with confidence to join the Cabinet. He said to the effect that whether the Nepali Congress or the CPN (UML) joined the government, or declined to do so, will make no difference; his objective is to bring the Maoists to the peace process.

I'm in constant contact with the leaders of the Nepali Congress and the CPN (UML) to help find ways to resolve the Maoist problem.

Thursday, October 24, 2002

One of the 'political contact persons' close to me shared that he had met the former ADC of the Royal Palace, Maj. Gen Tara Bahadur Thapa. He quoted Thapa as saying that he had always kept some of the key persons in the Indian PM's office in good humor which eased works to a great extent, but now the same seems lacking therein. He elaborated that he could still use his level of influence there if the King entrusted him with an opportunity to serve the country.

The former Chief Justice of the Supreme Court and one of the architects of the Constitution of 1990, Laxman Aryal has observed that, in any case, Nepalis are destined to go through a form of autocracy, be it of the King or that of the Maoists.

The Military Attaché of the American Embassy in Nepal has advised the government to talk to the Indian authorities for permission to use its air space for the consignment of arms airfreighted for the RNA from the US.

At a reception and dinner hosted in honor of the British Army General, Sir Sam Cowan, the former member in the dissolved Parliament, Narayan Singh Pun, said that the Ambassadors of the UK and the US backed the King's step.

The Chief of DFID trusts that the Army will respect human rights while launching its operations against the Maoists.

Sunday, October 27, 2002

In our appointed meeting, Khum Bahadur Khadka sounded very angry, and he smelt a conspiracy behind his having to give clarification to the CIAA on corruption charges; he alleged the move to be a purely political vendetta engineered by Madhav Kumar Nepal because he believes that he (Khadka) led to the breakup of the party.

In the same breath, he described GP Koirala as a lunatic and Sher Bahadur Deuba cunning, as the latter kept him in the dark about his move of recommending to the King for the deferral of the date of the elections by one year.

He claimed that Deuba's Cabinet colleagues had requested him to choose the alternative of resignation.

Khadka further disclosed that there was Maoist infiltration of the Nepal Police up to the ranks of SSP. He even alleged IGP Pradip SJB Rana's active involvement in his character assassination because, as Minister of Home Affairs, he hadn't agreed to extend his tenure.

His other grievances were directed against some Nepali Congress and CPN (UML) leaders who had played their active roles to slap corruption charges on him with a view to cripple his political career.

Defending himself, he pleaded with me for help by arranging an audience with the King.

Speaking in whispers, he then disclosed to me that he had spent five crores rupees to split the CPN (UML) Party!

Khadka, for whom I've always had warm feelings, had been trying to meet me for a long time. Today, he came to my place.

Today, the DIG of the Nepal Police, Rajendra Bahadur Singh, in an informal meeting with me, talked at length about the misdeeds of the IGP. For one, it was he (Rana) himself who had previously opposed the tenure of service beyond the 30-year provision in place for police officers. Consequently, many excellent officers faced retirement. Now there's a complete U-turn from his earlier stand when the same criterion has adversely affected his own position: He wants to have his tenure extended.

As if that wasn't enough, he has also incited some police officers to lodge a case, pleading for justice, against the Police Organization, at the Court.

Singh expressed sympathies for those competent officers who had to accept their forced retirement under the 30-year career rules and regulations.

Singh further disclosed that IGP Rana had tried to bribe the then Minister of Home Affairs, Khum Bahadur Khadka, with Rupees two crores [Rs 20 million]. He said it reflected on Rana's ambition and selfish nature.

Kamal Chaulagain, a CPN (UML) member, who visited me at my place, seemed frustrated by the mounting irregularities within a particular party circle, so much so that the clique, without the knowledge of the then Prime Minister [Man Mohan Adhikari], hoarded money when it was in the minority government in 1993/'94.

He then described Bam Dev Gautam as a spendthrift who earns money one day and burns it by the next day, giving it to everybody in the Party. As a result, he's always a pauper. He made so much money for the Party, one way or the other, and deposited every Rupee in the Party's coffer.

Chaulagain does think that Gautam can be useful to the Royal Palace at a time like this.

He also seemed deeply frustrated by his Party's tilt to India. He also complained that in the inner party circle, he's still treated as an 'outsider'.

He further disclosed Ishwor Pokharel as a resource person to collect money for the General Secretary, Madhav Kumar Nepal.

He then hinted at the possibility of several honest UML members quitting the party.

Asked if he would support the King's recent step, he said he would, even at the cost of leaving the Party or being ousted from it. He also said he would form another party with progressive leaders without any Communist ideology. He actually seemed to be expecting an offer of a prestigious portfolio in the King's Cabinet.

I told him that Nepal now needs a party that can accommodate nationalists from other parties with strong mass bases of their own. I said should such a party emerge, many frustrated leaders and cadres from the Nepali Congress and Leftist parties are sure to join it.

Monday, October 28, 2002

A Captain of the RNA, on duty at his post at the Rumjatar Airport in Okhaldhunga, was killed last night in a Maoist attack. Seven or so soldiers have suffered serious injuries in the crossfire.

There's a report of the team dispatched for rescue being ambushed on their way to the site of the clash.

The RNA helicopter flown for Rumjatar for further rescue and operation returned due to bad weather.

At the site of the clash, as many as 16 bodies have been identified as being those of the rebels.

The details of the losses are yet to be established.

The repeatedly unvarying style of Maoist attacks calls for a change in the strategies of the RNA.

The CoAS is scheduled to fly to Rumjatar to have an on-the-site observation and study of the clash.

Wednesday, October 30, 2002

The King granted a separate audience to the newly promoted high-ranking Army officers. On that occasion, speaking to them, the King expressed his belief in a high-level performance from all ranks of the RNA at a time when the entire country was fighting the Maoists.

Speaking on the same occasion, Lt. Gen Sadip Bahadur Shah recommended a new security system for the entire Kathmandu Valley, with the Royal Palace as the core area. He has had a good track record in his performance of duty in the command of a Division in Nepalgunj.

The King commanded him to make the best use of his expertise and experience in his operations by eliminating the flaws in the Army, if any.

Maj. Gen Om Bahadur Pun emphasized on the need for facilitating routine relief for the deployed units in the field areas of operations.

Similarly, Brig. Gen Pawan Jung Thapa put in his observation that because the Maoists were very likely to opt for urban guerrilla warfare, it would be easy for the Army to defend itself from the charge of human rights violations if the government could enforce a code of conduct mandatory for one and all.

Brig. Gen Toran Jung Bahadur Singh forwarded his suggestion for the Army to form a special team to initiate dialogue with the Maoists.

Brig. Gen Chhabilal Khadka explained the current status of the utilization of the Army Welfare Fund.

The King, as Chairman of the Council of Ministers, made it clear that his next move will be to give a final form to the Cabinet and immediately unveil relief packages for the people on priority basis.

He said that PM Chand was actively working to create an environment for both the Nepali Congress or the CPN (UML), or either one of them, to join the government on clear and specific terms and conditions.

Friday, November 1, 2002

Except for a regular meeting and briefing, there was no significant work today.

Kamal Chaulagain, who recently had his audience with the King, shared with me that he forwarded some suggestions to the King:

That he was ready to contribute his share to or make any sacrifice for the country and people; that it was important to take the support of the UK and the US to reduce the ever increasing Indian influence over Nepal; that a disturbing situation in Nepal is seen as rolling over to have a similar fate of Sikkim; that the political parties are far from helping the King; and that the King must take initiatives for a needful move to rectify the situation.

He also said that the King has given him an important task which he, however, didn't specify.

Sunday, November 3, 2002

A person associated with the Indian Intelligence Agency, RAW, and working at the Indian Embassy accused the Royal Palace of spreading a false report of the Maoist leader Dr. Baburam Bhattarai's arrest in India. He claimed that there was a man in the Royal Palace working for Indian interests to the extent of tapping the telephone conversation of the highly placed officials there.

There's a report that says that the US has expressed its good wishes for the government to deliver quick and fair services to the people, control corruption, and assure good governance in Nepal without any further delay.

I briefed the King on the invitation received from the Pakistani Army to the RNA for the inspection of its Army Welfare Trust Scheme and Courses of Training in Pakistan.

It's been reported that the Secretary of the Ministry of Home Affairs had the Undersecretary at the Ministry of Finance arrested, branding him a Maoist. But later, upon scrutiny, it was discovered that the Undersecretary had raised his voice against the irregularities in the Finance Ministry.

When briefed on this, the King instructed me to study the facts and take the necessary action.

Sources say the CPN (UML) is eager to join the government but they won't be given the portfolio of the Ministry of Home Affairs as has been their bargaining chip.

The PM has instructed the RNA to slow down its operations during the Tihar Festivals as a gesture to create conducive environments for peace talks to take place with the Maoists.

At the function of releasing the book, called 'Genealogy of the Ranas' at the Babar Mahal Revisited, some Ranas extolled themselves as belonging to a superior race of people, quite similar to the British in their Raj in India.

The Army patrols detailed for the Customs Department has proved very useful, and therefore, it calls for a further extension and expansion of it.

Thursday, November 7, 2002

PM Chand visited GP Koirala at his daughter's place at Mandikhatar to request him to join the government. Chand also asked Koirala to hold the peace talks with the Maoists since it was Koirala himself who had till recently talked to them on the telephone hotline on different dates.

Saturday, November 9, 2002

The Students' Wing of the newly formed Nepali Congress (Democratic) Nepal Students' Union (NFSU) organized a mass gathering at Basantapur.

Similarly, the CPN (UML)'s sister organization, known as the Democratic National Youth Association (DNYA), also took out a protest march which turned into a corner meeting at the Bhotahiti Square.

Prakash Gautam, a security guard supervisor at the American Embassy, was killed by the Maoists at Buddha Nagar. They also murdered Ramesh Tamrakar working at the Lincoln School. The FBI is looking into the murders.

The former Mayor of the Kathmandu Metropolitan City, Keshav Sthapit, visited me at my residence.

When I asked him to help the King by joining the government, he instead disclosed his schemes to help resolve the Maoist problem and hinted at his wish to be appointed the Mayor of Kathmandu.

Sunday, November 10, 2002

Maj. Dr. Rajiv Shahi, husband of Princess Pooja, and employed at the Military Hospital but currently in London for 'attachment', has overstayed there without informing the RNA Headquarters. Recently back in Nepal, he's again asking for a leave for his MD course. When briefed about it, the King instructed me to arrange it.

Asked to assign a time for Crown Prince Paras to have his Basic Military Training, the King instructed me to consult with himself first and then arrange it accordingly at his own convenience. The training will last eight weeks.

The overall security situation of the country is to be reviewed at the Brigade and Division Commanders' conference scheduled for November 17 and 18 at the Army Headquarters.

GP Koirala is likely to have an audience with the King at a time when it's being speculated that he'll recommend Dr. Ram Sharan Mahat, Amod Prasad Upadhyaya, and Chakra Prasad Bastola for portfolios in the Cabinet.

According to a source, Dipak Bohara of the RPP and Sharad Chandra Shah are arranging for the PM to talk to KP Oli of the CPN (UML) for the latter's party to join the government.

Lately, the Maoists have sent their signal to the government for peace talks but with certain preconditions: the government must remove the tag of terrorists stuck on them; and formally, in writing, invite them to join the peace talks; and make public the whereabouts of the detained Maoist cadres and guerillas.

The King asked me to, first, collect the views of the Army officers prior to proceeding for any peace talks.

The suggestions forwarded by the RNA include:

- There's nothing wrong in calling the Maoists for peace talks
- The Maoists must immediately halt all terrorist activities, such as extortion, forced recruitment, mass meetings, and trafficking of weapons

- They should forthwith cease and desist from any form of negative remarks on the monarchy
- The government must display utmost flexibility
- The status and whereabouts of the detained must be made public from both sides: the government, and the rebels
- The security forces shall continue its regular operations but won't target the Maoists
- The government shall continue its procurement of essentials for the security forces.

I briefed the King on all the above suggestions forwarded by the Army officers. Having heard it, he instructed me to consider that since the Maoists were in contact with the Ministers and leaders of the political parties, it would be wise to adopt a policy of no backtracking, and side by side, launching offensive operations against them.

Likewise, the King issued his firm instructions to the CIAA to quicken its process of investigation into corruption-related cases.

Similarly, he commanded me to look into the facts of the purchase of a Sky Truck M-28 for the Army because there have been serious allegations of irregularities on the purchase deal.

Thursday, November 14, 2002

Prior to visiting the King, GP Koirala has asked some of the Central Committee members for the terms of reference on which to talk with the King.

They are reported to have suggested to Koirala to tell the King that his takeover was against the spirit of the Constitution, and the best option is the revival of the previous Parliament.

KP Oli has shared some facts with the leaders close to him.

The King and Oli candidly discussed the current political situations, some matters being confidential.

If the consensus forged between the six parties goes into effect, the way out isn't far.

He described the easy and viable way out of the ongoing political stalemate.

An explosion targeting the security forces has blown off a public bus in Dolakha which left 16 people severely injured. They were immediately airlifted to Kathmandu for treatment.

There's information of the PM dealing with the Maoists who, if things go right, are very likely to join the government.

Maoist leader Haribol Gajurel said the Maoists have corresponded with the UN and the EU to broker and monitor the peace talks due to start.

According to a grapevine, the Maoists receive large sums of money from some Nepalis in Belgium on a regular basis.

It's also known that the Maoists maintain their central office in Gorakhpur in India. The security forces are rigorously trying to track it.

Sunday, November 17, 2002

Added to the political and economic situations continuing to be precarious, the role of the regional administrators has also been far from effective; so, suggestions have been received from different quarters to form zonal-level mobile administrators with adequate authority and comprising representatives from the army, police, the civil service, and the legal sector.

The King has once again issued his instructions to immediately expand the Cabinet, but maintain the number of Ministries at 18 or so, and minimize expenditures, and make the administrative service delivery to the people effective.

There have been complaints that the Chief Secretary and some other Secretaries of the government aren't working sincerely to cooperate with the King's plans. Therefore, the performances of the Secretaries have been reviewed; that of the Secretary at the Ministry of Defense has been found satisfactory.

The RNA has foiled a Maoist plan to attack the administrative headquarters of the Jumla District where, in a combing sweep, it discovered two sacks full of banknotes and over one hundred dead bodies of the rebels.

The CoAS has exaggeratedly described the operation as a success, as if the entire Maoist problem has been solved by it. The details of the losses on the Maoist side in the Army operation and the details given by the CoAS don't match up. Upon being told that it wasn't such a success as he publicized it, the CoAS sounded angry.

I've had information that a consular officer of the Indian Embassy quietly met GP Koirala and the two talked for an hour behind closed doors.

The Minister of Foreign Affairs, Narendra Bikram Shah, is slated to visit Bhutan with a view to talking to the Bhutanese authorities for the settlement of the Bhutanese refugees in Nepal.

Wednesday, November 20, 2002

Today, the Minister of Physical Planning, Narayan Singh Pun, while briefing the King on the performances of the Ministries and Secretaries, assured him that he would make tremendous efforts to lobby for support for the October 4 takeover from the foreign embassies and donor agencies.

At a dinner at the Chinese Embassy, Ambassador Yu Yong Kong shared his experiences in Nepal. He disclosed the possibility of his transfer to another country before the expiry of his ambassadorial tenure in Nepal. The envoy expressed his wish to see the Maoist problem solved at the earliest.

I briefed him on the developments of the Army operations against the Maoists. Lauding the efforts of the RNA, he expressed his willingness to help with logistics for the Army.

On the same occasion, the Ambassador observed that Nepalis always cherish the monarchy, even helping it at times, but it's sad to note that the political parties now disregard the history. He recalled the King's request for Armored Personnel Carrier (APC) for the Army for its UN Peacekeeping Missions. He said it's wise of the King to ask for help under the cover of the UN Peacekeeping Mission, or else India will find an excuse to put obstruction to this effect.

Turning the conversation to personal matters, the Ambassador said he was going to China for a brief holiday, and by the time he would be back in Kathmandu, the leadership in China will have changed.

Again turning the conversation to political matters, he expressed his concerns over Tibetans pestering the Chinese police in the border areas in various ways. He complained of Tibetans engaged in anti-China activities in Nepal. His displeasure was directed at the UNHCR for giving such Tibetans the status of refugees. He requested Nepal to deport such Tibetans to China.

At a coordinating meeting of the security forces held at the RNA Headquarters, the field commanders reviewed the successes and lapses of their operations in the fields during the past six months. These ranking officers unanimously commented on the inadequacy of logistics and supplies of essentials to those assigned on duty in the field. They took the delay in the supply of essentials as discouraging. They even cited soldiers buying their own clothes and boots to manage in their dire situations.

The lack of space for treatment in the Military Hospital due to the increasing number of casualties was also discussed at length at the meeting. Their suggestion to use some parts of the huge Army Museum for other medical purposes, such as having an OPD and paramedic stations, is a case in point.

The CoAS unveiled the plan of giving relief to those on duty in the fields; especially those from Kathmandu will be sent to replace the soldiers deployed for long stretches out of the barracks.

Thursday, November 21, 2002

Some leaders of the Nepali Congress and the CPN (UML) left for Thailand to participate in a conference held in Bangkok with the major political parties of Asia attending it.

The Nepali Congress (Democratic) Party is preparing to hold an assembly meeting of its youths to work out the Party's strategies for its future courses of action.

It's learned that the Nepali Congress and the CPN (UML) are discussing to join hands with the Maoists to go ahead with a fresh movement.

When briefed on the present political situation, the King said the PM is taking his initiatives for dialogue with the Maoists.

Sunday, November 24, 2002

There's a criterion for Army officers to be promoted to the rank of Brigadier. This requires a one-year course at the NDC or RCDS or NDU.

However, none of the Army officers in the service of the Royal Palace has had this sort of training. Therefore, a policy of making this course mandatory for the officers deputed to the Royal Palace was discussed today.

Brig. Gen Tika Dhamala is required to take this course prior to his promotion to Maj. Gen. I asked the King about appointing him Principal Military Secretary by promoting him to the rank of Major General.

The King merely listened to my briefing without any response.

Finally, he instructed me to ask the CoAS to request a seat for the course. I personally tried my best for it. Despite our efforts, however, the overseas seat for the training wasn't available either in the UK or in the US.

Note:

While the attempt to obtain a seat for the course for Brig. Gen. Tika was underway, I resigned the post I was serving at on Friday, November 7, 2003. The placement was made available by Britain, and the post of Military Secretary remained vacant for a long time even while Brig. Gen Tika went to Britain for the course.

There was an established practice of appointing Military Secretary from among those in long service at the Royal Palace. Tika was promoted to Major General on Monday, July 7, 2004, on the occasion of the King's birthday. Later, Gajendra Limbu, Inspector General at the RNA Headquarters, was promoted to Maj. General and subsequently posted as Acting Military Secretary to the King. Perhaps the King had good intentions behind appointing him in that post. I heard, however, that no good comments came from the RNA Headquarters.

Maj. Gen Dhamala served at the post of ADC even after his completion of the special course. Only later, he was appointed Military Secretary when Maj. Gen Gajendra Limbu had his posting back to the RNA Headquarters.

By that time, the Military Secretariat had remained defunct. As the government didn't extend two more years of terms to their tenures, Maj. Gens Limbu and Dhamala were retired under the current proviso of no terms of extension in place for them.

With regards to the mistreatment of a Chinese Buddhist Monk by the security forces at the Butwal Checkpoint, I advised the King that it would be a good idea for the CoAS to talk to the Chinese Ambassador.

Complaints of brazen breaches in the process of the transfer of some DIGs have surfaced. I briefed the King that there was a long-observed provision of transfer of the DIGs made only by the Cabinet, not by one individual Minister alone.

The King sounded angry at my counsel.

Normally, for one to have the posting as the DIG in the Kathmandu Valley, s/he ought to have the track record of having taken a command position outside the Valley. But the posting of the current DIG is contrary to the established norm, and it has created enough room for discontentment within the Nepal Police.

I had had enough of negative comments hurled at me for pointing out the breach of what is an established practice anyway.

I met the Deputy Chief of Mission, Ashok Kumar, of the Indian Embassy, at a dinner at the Shangri-La Hotel. He described the Indian investment in hydroelectricity in Bhutan which has boosted that country's economy to a better-off state. He said the same model in Nepal would help improve the country's economic conditions to a great deal.

I said many Nepalis feel that Nepal was duped in the Koshi, Gandaki and Mahakali projects which are operated as joint ventures between Nepal and India. In principle, in joint venture projects, both the investors enjoy equal benefits-sharing. Therefore, there's a general view that Nepal should give serious thought when it comes to the generation of hydroelectricity, particularly in collaboration with India.

Presenting technical facts and figures, Kumar claimed there was no loss for Nepal in those projects as is being said.

Turning the conversation to politics, he said that India has always been generous to Nepal, wishing for its continued progress, development, and stability. But in Nepal, there are frequent voices of anti-India sentiments. India is a large country, ever displaying the bigness of its heart to deal with its neighbors, most especially Nepal.

Listening to all this, it seemed to me that India still has its Big Brother hubristic psyche. Normally, when Nepal raises an issue of national interest, it's misinterpreted as being anti-Indian in nature. This is far from showing understanding to a neighbor.

Wednesday, November 27, 2002

GP Koirala prods Madhav Kumar Nepal and Amik Sherchan to prepare grassroots-level cadres for an uprising. It's learned that they are planning strategies to forge an alliance to assert moral pressures on the King.

The CoAS has asked me to arrange a briefing on the Civil Military Campaign (CMC) and the ISDP for the NSC and the Cabinet.

Now that the Military Secretariat has developed its own website, the King instructed me to go ahead to coordinate with the Communications Secretariat and the media.

Thursday, November 28, 2002

The Minister of Labor and Transportation, Kamal Chaulagain, has suggested to make arrangements for appointing representatives at the local levels for people to have their works done in quick and fair manner.

Minister Pun also presented similar suggestions.

The CoAS asked me to arrange an appointment for him to brief the King on the RNA's immediate plan to set up a Unified Security Base at Rajapur of Bardia District where the Army Officers' Wives' Association is scheduled to distribute relief packages to the displaced families and the people of the Tharu community.

The King commanded me to make security arrangements for his unofficial trip to eastern Nepal.

Friday, November 29, 2002

There's an established convention of the King issuing his special instructions to those contingents of the RNA going abroad for the UN Peace Keeping Missions after their preparatory training at the Panchkhal Training Center.

Upon being conveyed the CoAS's invitation to the Training Center to give his words of a sendoff to the Army troop for UN Peace Keeping Mission, the King preferred to address rather than just give advice.

I had prepared a draft of such an address, and Press Secretary Mohan Bahadur Pandey had also prepared one. By combining the two texts into a final form, we offered the draft to the King.

As preplanned, the King reached Panchkhal to hear the briefing on the progress and preparations of the training and the troop's departure. He also saw a training demonstration.

The response to the King's address was impressive.

The previous Indian ambassador to Nepal, KV Rajan, has arrived in Kathmandu. He has dubbed it as a purely personal visit to meet old friends and politicians.

The King commanded me to have Maoist cadres, such as Om Sharma,

Govinda Acharya, and Chandra Man Shrestha, apprehended by the Army released provided it's appropriate to do so. The Principal Secretary, Pashupati Bhakta Maharjan, had pleaded the King for releasing them.

However, while communicating with the Army on the subject, it's been learned that since investigations are still continuing with Om Sharma, his detention will continue, while the other two will be released in two or three days.

Sunday, December 1, 2002

The Minister of Physical Planning and Construction, Narayan Singh Pun, informed me of his party's plan to expand its base at the grassroots levels by organizing a large mass meeting in Kathmandu.

The former Mayor of Birgunj Municipality, Bimal Sriwastav, while discussing his idea with me said it would create a good environment if relations and contacts were cultivated with the political leaders, MPs, intellectuals, and journalists of Uttar Pradesh and Bihar of India by, among other things, arranging their visits to Pashupatinath.

He added that it would be even better if they were encouraged to publish their articles and write-ups in the newspapers in India in support of the King's move. This would specially create ideal moods in the border areas.

The former Mayor also complained that the Army details assigned to the Customs Patrol duties are found involved in irregularities.

He also believes that only the provision of the original copies of the LC (Letter of Credit) will control the unauthorized consignments of goods at the Customs.

He also shared with me his plan to contest the next elections as an independent candidate at the upcoming national polls.

Monday, December 2, 2002

Despite his tremendous efforts to extend his tenure of service by the annulment of the provision in place, or even pleading to the King for his intervention, the IGP of the Nepal Police, Pradip SJB Rana, is soon taking a month of the customary leave prior to his automatic retirement.

To the best of my knowledge, the King is in the mood to appoint him Ambassador to Myanmar.

As the IGP couldn't furnish convincing documents to prove the sources of his wealth and property, he has come under the scanner of the CIAA, and it has swept off his chance of appointment as ambassador.

Negative comments about the Royal Palace Budget Ordinance have emerged. People question the ballooning increments in the budget for the Royal Palace at a time when the economic growth in the country is at its lowest ebb. It only erodes the image of the Royal Family. The King sounded serious when I briefed him on the comments provoked by the increment in the royal budget. 'Nothing passes without a comment,' he replied.

It's learned that Madhav Kumar Nepal has returned to Birgunj from Patna where he had met the Maoist leaders.

Govinda Raj Joshi of the NC is learned to have telephoned Hari Babu Choudhary, the former Chief of the NID, informing him that the environment for the reinstatement of the Parliament was in the making.

Tuesday, December 3, 2002

The NSC Coordinator, Maj. Gen Rookmangud Katawal, in response to the invitation by the British Defense Ministry, has left for Britain.

His Majesty the King has given him a special instruction to specify the reasons behind his October 4 move, and lobby for support from Britain for the takeover, by underscoring the following points:

1. The Maoists' perpetration of violence, and the precarious condition of the country
2. Failure to deliver services and provide a sense of security by the successive governments
3. Ever increasing corruption in the political and administrative sectors
4. Money amassed and creation of factions and cliques by the leaders of the political parties
5. The entire Police Organization paralyzed by sheer politicization
6. Successive governments' apathy and neglect of the entire RNA
7. Inter- and intra-party rifts, and the abuse of politics as a means to amass money
8. Sheer misuse of foreign aid funds received from donor agencies
9. Ever increasing trend of dissolving the Parliament and the misuse of prime ministerial prerogatives at the whims of the Prime Ministers

10. The King could not simply watch all this as a mere onlooker
11. To make peace efforts by bringing the Maoists to dialogue
12. To make efforts to equip and enhance human resources in the Army
13. To make it easier for the Army to launch massive operations to pressurize the Maoists to join the peace process
14. To remove the fear psychosis instilled by the Maoists in the minds of the people; and
15. To convince the Western powers and neighbors of Nepal to help Nepal resolve the Bhutanese refugee problems, and solicit India's help for Nepal to resolve the Maoist problems.

Maj. Gen Katawal has been assigned to prepare a paper and talk to British officials on these points to defend the October 4 takeover.

The King granted an audience to Dr. Sanduk Ruit for his fame and invaluable contributions in cost-effective eye treatment and retina transplantation at home and abroad.

Dr. Ruit talked of Maoists with eye ailments visiting the hospital and his telling them that the choice of their way of serving the people was wrong.

With only a fortnight remaining for the CoAS to visit India, I advised him to consult with the King to ascertain the terms of reference for his talks with his Indian counterpart during his visit.

Today, Radha Krishna Mainali discussed with me his plan of dissociating himself from the CPN (UML), pointing out the need for an alternative political force to the Nepali Congress and the CPN (UML) in the changed context of Nepal, as the two parties have strayed from their core ideologies.

Wednesday, December 4, 2002

It's most shocking that ADC Maj. Mahesh Chand lost his automatic and couldn't find it.

I reminded the King of the Court of Inquiry and Court Martial as per the Army Act for such unprofessional negligence.

As punishment, the King ordered the Major to pay for the lost weapon and withhold his promotion for three years and restrict his duty only inside the Palace complex for one year.

To my mind, the King's lighthearted decrees will inspire only wilfulness in the performances of officers at the Palace.

Later, I heard that Major Chand resigned and moved to America.

Thursday, December 5, 2002

Radha Krishna Mainali requested me to arrange an audience with the King so that he could offer him advice in person. He shared his ideas that at a time such as this, the King needed to go ahead with short- and long-term goals. He added that the King should in no way offend India or overlook the political parties, lest the parties launched their mass movement. He further said that the best course of action for now would be to work with the political parties, bring the Maoists to the peace talks, create an environment for general elections, and collectively go for polls.

In his audience granted to Mainali, the King explained that the political parties always threatened to mobilize the masses, but that would by no means contribute to solving the present problems. The King added, should there be another mass movement, it would only sweep the political parties away. He accused them of persistently flirting with the idea of mass movement, sorely lacking any accountability, and the will to solve the problems by, say, convincing the Maoists to enter competitive politics. Sounding serious, the King further said that mass movements in a situation as of now will most likely prove to be counterproductive to the parties themselves. If, by any means, there were elections to the Constituent Assembly, what if the result wasn't in their favor? Can they quietly accept it? What if other political parties, too, like the Maoists, resorted to weapons?

The CoAS, over the telephone, updated me on the recent visit of the Indian Ambassador to the RNA Headquarters. He said the Ambassador lauded the high-level performances and efforts of the Army, promising an increase in the volume and with the continuity of Indian assistance to the RNA in the days ahead.

The CoAS said the Ambassador's remarks hinted at a change of Indian perspectives on Nepal.

Dr. Duman Thapa told me that Padma Ratna Tuladhar, the independent Leftist leader, was seeking to see me to discuss the peace talks with the Maoists.

Friday, December 6, 2002

JK Sinha, the second in command of the Indian Intelligence Agency, RAW, is in town. Although ostensibly on a personal visit, he's busy consulting with the leaders of the political parties.

It's also learned that he expressed his concerns about the future of the current mass movement, the alliance between the political parties, a consensus Prime Ministerial candidate, and the current speculations on KP Oli's rumored exit from the CPN (UML).

He sounded equally curious about the possibility of the mainstream political parties joining hands with the Maoists, and their tactical unity.

I heard that Sinha also visited the Chief of the NID and asked him about the personality traits and idiosyncrasies of the IGP and the CoAS.

He also reportedly wondered about the King-Maoists alliance, their likely agenda, and the distance or proximity between the political parties and the Maoists.

To make matters even more intriguing, Sinha also asked for the cellphone numbers of the Maoist leaders. Surprisingly, he suggested that the government be led by Surya Bahadur Thapa or Ravindra Nath Sharma for a time being.

Saturday, December 7, 2002

As planned, I met Padma Ratna Tuladhar at the Shangri-La Hotel. Present at the meeting were Dr. Duman Thapa, the Acting Ambassador of Denmark, and Sharad Bhattarai of the Danish Embassy.

Tuladhar explained to me that the Swiss Center for Human Dialogue (HDC) has offered to broker peace between the government and the Maoists, the HDC having experienced peace experts with good track records of brokering such accords in other parts of the world.

He then detailed the technical, confidential, security, and financial aspects of the entire peace initiative, including the charter of a plane and travel documents, such as passports, for the Maoist representatives, to name a few of the procedures. He also added that the entire costs would be borne by the HDC.

Tuladhar asked me to relay all these details to the King and treat everything about it as most confidential.

Sunday, December 8, 2002

Kamal Chaulagain expressed both his surprise and anger at the leakage of what he considered confidential. It so happened that he had just returned from a meeting over tea at the Royal Palace when a journalist of *The Himalayan Times* asked him for the contents of the discussion and conversation thereof.

Chaulagain also disclosed the present status and psyche of Rajendra Shrestha, Jitendra Dev, Sita Nandan Roy, Siddhi Lal Singh and others in the CPN (UML).

Bam Dev Gautam of the same Party asked me to arrange some financial help for him.

Monday, December 9, 2002

The King sounded very concerned when I spoke to him about the idea of the peace talks to be mediated by the HDC. It seemed that he was expecting the assurance of the Maoists' seriousness about such a dialogue, and the future of such a peace initiative. He added that if past records are anything to go by, the preaching and practice of the Maoists don't match; they say they are a nationalist force but they cause inconveniences to the people. The government and the talk facilitators must consider these facts.

The King also seemed extra wary at the possibility of external factors that are likely to affect the results of the CA elections.

However, the King expressed his satisfaction that it's welcome news to hear that the HDC is willing to take the initiative to broker peace in Nepal, but how India will look at it and react must also be taken into consideration.

I advised the King that it would be a good idea to host a dinner for the PSOs of the Army Headquarters to exchange and share ideas to boost their morale.

Turning the conversation to another matter, I told him that the behavior of Sharad Chandra Shah has been increasingly audacious for some time now; there have been mounting complaints against him.

Tuesday, December 10, 2002

The all-party meeting held at the Central Office of the Nepali Congress seems to concentrate on the modalities of the mass movement. Top leaders of the political parties were present at the meeting.

The former Chinese Ambassador to Nepal, Lee, has arrived in Kathmandu with a team of investors and businessmen. They are discussing the idea of defense production for the Army in Nepal itself, with the collaboration of the industrial and commercial sectors of both countries.

Friday, December 13, 2002

In the audience granted to GP Koirala, the King criticized that the Nepali Congress at present doesn't seem to have a clear vision. He made it clear that in a situation like it is at the present, the monarchy can't compromise on its three common minimum goals: peace and security, general elections, and nationality. The King told Koirala, amidst the speculations that the LTTE is training the Maoists, the best course of action is to provide the security forces with the essentials they need to boost their courage and morale.

Shortly after visiting the King, Koirala told his Party members that he found the King in a confused state of mind. He said he told the King to hand the leadership of the government over to the Nepali Congress for the way out.

Madhav Kumar Nepal, as General Secretary of the CPN (UML), has made the Party's position clear that they aren't inclined to side with the Nepali Congress; rather it's the Nepali Congress itself which is so eager to team up with the CPN (UML) because of their huge mass base and support. He claimed the leadership of the government for his Party to hold the elections as per Article 128.

Speaking to the leaders and cadres of his Party at the general convention held in Pokhara, leader of the RPP Nepal, Kamal Thapa, charged that corruption has become rampant in the last 12 years of governance by the big parties, which is much beyond the level seen during the entire 30 years of the Panchayat rule. Referring to statistics,

Thapa blamed the big political parties for imposing the debt of Rs. 10,000 on each single Nepali. He said it indicates the incompetence and failure of the big parties, and therefore, they should publicly beg for forgiveness for their misrule during their administrations.

Tuesday, December 17, 2002

GP Koirala has been to Biratnagar, accompanied by Ramchandra Poudel. It seems that the entire Nepali Congress Party takes the King's visit to eastern Nepal as a challenge. Asking the party members to arrange mass meetings for him to address the people, Koirala has said perhaps the King is daydreaming that huge masses are in his favor simply because a bunch of people showed up during his sojourns here.

Poudel reiterated the Party's stand on the reinstatement of the Parliament to enable the government to enter into talks with the Maoists and also to have the Constitution amended for the said purpose.

The CPN (UML) also similarly organized a mass meeting in Kathmandu, but it's heard that they trucked in hired people from outside to show a big gathering.

The IGP of the APF, Krishna Mohan Shrestha, has returned from India. His visit to India on the invitation of the Indian Embassy and his Indian counterpart has breached the norms and practice of senior security officers observed so far. However, he defended his visit by citing his inspection of weapons India is to supply to the security forces. Later, he realized his error, admitting that he 'overlooked.'

Wednesday, December 18, 2002

Dipak Bohara of the RPP has requested Madhav Kumar Nepal to join the government to find a solution to the political quagmire and put in motion a process to appoint representatives at the local levels to prepare an environment for holding the general elections.

I advised the King to address the leaders of the political parties, asking them to rise above the petty and partisan interests in order to resolve the problems and create an environment for holding elections.

Documentary proofs have been found of the Maoist leader Dr. Baburam Bhattarai's informal talks with the officials of the South Block of the Ministry of Foreign Affairs of India. In fact, the Maoists are seeking Indian support to introduce republicanism in Nepal. On that occasion, the Indian officialdom asked the Maoist leader why they destroyed the infrastructures built for Nepal by India. Bhattarai reportedly defended himself by saying that they targeted only those installations built by those countries identified as enemies by them in their analysis.

This is enough indication of how the Maoists kowtow to India for its support for republicanism in Nepal and have the mainstream political parties agree to their agenda.

Narayan Singh Pun asked me to arrange an appointment with the King to brief him on his talks with the Maoists.

Addressing a mass gathering in Jhapa, GP Koirala made some critical remarks against the RNA.

Thursday, December 19, 2002

Prime Minister Lokendra Bahadur Chand has appointed Gita Bhakta Joshi as his principal advisor.

It's heard that GP Koirala is to visit India. Sources close to him say he's going there to attend a wedding ceremony of one of his relatives. He would do well to let the nation know about such visits that he often makes, considering his national political position and his track records of being Prime Minister so many times.

The programs for Their Majesties' three-day visit to the Eastern Development Region and their civic receptions there have been set. I notified and issued directives to the security departments concerned for foolproof security arrangements during Their Majesties' visits.

Monday, December 23, 2002

As India disallowed the flight over its skies, the airplane freighting the consignment of weapons and light machine guns (LMGs) purchased from Belgium for the RNA took an emergency landing in Azerbaijan.

The Belgian Embassy and the company exporting the weapons are constantly in touch with India seeking permission for their overflight through India to Nepal.

A vehicle of the Indian Ambassador with diplomatic number plate was seen at the residence of Madhav Kumar Nepal at Koteshwor. Sources say they held a closed-door one-on-one talk. The contents of the talk have been kept top confidential.

Minister Narayan Singh Pun has asked for a leave in order to deal with the Maoists outside the Kathmandu Valley.

Tuesday, December 24, 2002

Sources in New Delhi say GP Koirala held consultations with Indian leaders, such as IK Gujral, George Fernandes, and Sheela Dixit. It's also known that he met the Maoist leader Dr. Baburam Bhattarai there.

The Maoists shot dead Pradip Thapa, an Inspector at the NID, while he was on his way home from a walk.

The Paintball Rifles arrived today from Hong Kong for firing practice for the security forces now in the service of the Royal Palace.

With a view to disrupting the King's visits to the Eastern Development Region and his Civic Receptions there, the Maoists hastily called a two-day strike and shutdown of shops and educational institutions on January 3 and 4.

Thursday, December 26, 2002

At a collective briefing by us, the Secretaries of the Royal Palace, to the King, we discussed several aspects, the complications, and dimensions of local elections. The King gave us his instructions to strategically conclude the local elections before the general convention of the CPN (UML).

It's also set that the King will address the mass gathering on the occasion of his Civic Reception.

Similarly, another program for him to visit the Western Development Region and grace a Civic Reception has also been scheduled.

The King said that India doesn't want a third country to broker peace in Nepal.

Friday, December 27, 2002

In response to the Army's inquiry about its plan to operate a bank in a joint venture with an investment of the Army Welfare Fund, the Ministry of Finance has sent the RNA Headquarters the code of rules. According to the rules, a bank can have a foreign investment of 67% or

50% or and 20% percent. I suggested the proposed army bank have 20% foreign investment, 30% investment from for the general public, 30% investment from the Army Welfare Fund, 10% investment from the army officer - retired or still working – and 5% investment each from the Nepal Police and APF.

I shared my concept of the modality and pattern of investment with the King and he was positive about it. The King said, 'How about giving some shares to government employees listed in the Provident Fund, too?'

Note:

The operating of such a bank didn't materialize due to the obstructions put to it by bankers at home and abroad by exerting their influence on Nepal Rashtra Bank. I think several financial institutions feared the resources of the AWF wouldn't be available to them, as before. The Army's long cherished desire to operate such a bank that would contribute to the medical, educational and financial needs of the serving men and women in the security sectors was thus aborted which would otherwise institutionalize the AWF.

According to a reliable source, Nepali Congress and CPN (UML) activists in the guise of Maoist cadres have threatened people not to go to the King's Civic Reception on January 3.

Tuesday, December 31, 2002

The security forces have brought the Maoist Central Committee Member Chaturman Singh Rajbanshi to Jhapa; he was arrested in Siliguri of India.

As requested, I arranged an audience with the King for Minister Narayan Singh Pun who sounded very happy and hopeful following his talks with the Maoists.

He said he found the Maoists flexible enough to accept the institution of constitutional monarchy and respect the Preamble to the Constitution of 1990.

He added that they were also ready to reconsider their demand for holding elections to a Constituent Assembly.

He narrated their version of their recent decision to opt for competitive politics.

Minister Pun also mentioned that the Maoists eagerly awaited a formal letter from the Royal Palace; if that was inconceivable for technical reasons, a letter from the PM would equally do.

He also reported that they respected the King's wish for holding peace dialogue, and so expected the fitting gestures from him to create an environment for the same.

The King looked serious to hear the Maoists' newfound love for a peace dialogue and gave instructions to Minister Pun to ensure they are true to their words and action.

He added that it doesn't sound proper for the Royal Palace to write them a formal letter; so it would be a good idea to send a letter signed by the PM or a Minister, as peace talk is more important than who signs and sends the letter.

I offered to talk to the PM on this matter, assuring him to bring an indicator to proceed for peace talks.

When it came to the arrangement for holding the peace talks, legal complications arose as to how to maintain secrecy and the way to steer it at a time when the Maoists have been placed on the state's terrorist watchlist.

It occurred to us then that the Principal Secretary, Pashupati Bhakta Maharjan, too, should be made privy to the agenda as it concerned holding dialogues with the rebels. Therefore, the details of the matters at hand were shared with him as well.

Accordingly, I called on PM Chand to explain the procedure of the dialogues to be held with the Maoists.

Since he sounded so encouraged about the peace endeavor, I called the facilitator, Minister Pun himself, to collectively draft a letter.

With everything set ready, I updated the King on the new developments.

But unbeknownst to me, it turned out that at that very moment Kamal Thapa and Sharad Chandra Shah were also waiting to meet the PM in another room at the Prime Minister's Residence. I also didn't know that in connection with the government-Maoist dialogues at the HDC Center in Geneva, the two were asked to go there for the purpose.

The rigorously planned and the most secret government-Maoist rounds of talk, however, had come to the knowledge of Sharad Chandra Shah. It occurs to me that he felt sidelined because, perhaps, he had also been asked to contribute to the peace endeavor, but through another channel.

Since that day, he turned against me, and would go to any extreme to wish me ill. I do think that this man had his full hand in the creation of the environment in which I had to resign. I have every reason to speculate on it because the news story published with Babita Basnet's byline in the *Ghatana Ra Bichar* weekly on Wednesday, November 12, 2003 was in fact ghostwritten by Sharad Chandra Shah, for the wording and the angle of criticism in the publication neatly fits his style.

Thursday, January 2, 2003

Addressing a mass meeting at Malangwa of Sarlahi District, GP Koirala reiterated his Party's stand on the return to the previous Parliament.

He said he was scared of ghosts on Fridays because shocking events, ranging from the Royal Palace Massacre to the October 4 takeover happened on Fridays.

He further said that the King had no right to ignore the people; what if the populace turned against him?

He added that the Army and the Royal Palace should act under the purview of the Parliament.

Koirala continued that he always defended the monarchy even at those times when foreigners asked him for reasons for his differences with the King.

Speaking in a challenging mood, Koirala declared that it was by no means good for the monarchy to help revisionist forces.

Friday, January 3, 2003

The people of the Eastern Development Region accorded the King a Civic Reception where he addressed them.

It generated mixed responses from different quarters, one section terming it as balanced and positive.

Quite others took the speech as an indication of an exit from the present political stalemate.

There are, of course, those who find fault with anything, no matter what's said or done. These people alleged that the King's Civic Reception was filled with all hired men ferried from other parts of the country.

Many people felt a sort of respite because they said they found clear indications of the solutions to the Maoist problem through dialogue.

I briefed the King on all the reported responses from the public when he granted separate audiences to different individuals serving the country in their respective fields.

The Civic Reception was followed by a dinner at the Divisional Headquarters of the Army, and it was attended by the PM, the DPM, Ministers and high officials.

The King keenly observed the cultural programs presented by the Army.

Thursday, January 9, 2003

In connection with the auspicious rice-feeding day observed for the new Royal, Prince Hridayendra, the puja procession to the Ganeshsthan, Lord Ganesh's Temple, concluded peacefully.

People watched the procession with joy and enthusiasm that reflected on their unflinching faith in their monarchy.

The security forces deserve admiration for their security arrangements.

Friday, January 10, 2003

Minister Pun sounded very angry on the telephone while he reported on his utter surprise at the unexpected turn of events.

Referring to Ramesh Nath Pandey's purported letter to the Maoists, Pun quoted Pandey as asking the Maoists to close all the previous channels of talk and also drop the idea of any peace talks as he (Pun) wasn't capable of handling it.

Pun further quoted Pandey as saying that it was he who was the first person approached to proceed with the peace talks with the Maoists; he also insisted that they commence their peace talk with the right person for the talk to reach a logical end.

It was really surprising how such a classified matter was leaked to Ramesh Nath Pandey.

It didn't take me much time, however, to learn that it must've been the Principal Secretary, Pashupati Bhakta Maharjan, who could've leaked the confidential matter.

I assured facilitator and interlocutor Pun to proceed as he was the authorized person with the formal letter of the PM for that purpose.

The King, too, was stunned too to hear about the fresh developments.

[It occurs to me now that the likes of Ramesh Nath Pandey, and the then Principal Secretary, Pashupati Bhakta Maharjan, along with Sharad Chandra Shah, pushed Nepal's monarchy to this state where it's reduced now.]

The *Janadharana* weekly published quite another news story about yet another fresh peace initiative by another channel of communication, this time by Prabhakar SJB Rana.

The CoAS quoted the PM as asking him to arrange a meeting with the Maoist leaders, now in Army detention, with the human rights activist, Sudip Pathak.

Apprehensive of likely pressures for the release of the detainees, he said the arrest of the rebel leaders hasn't yet been made public.

He added that it would be better to arrange the meeting after the Maoist leaders state their views about the monarchy.

Accordingly, I briefed the King on it.

The King accepted my advice on the postponement of the Fun Fair to be organized by the Army Officers' Wife's Association, a long-held tradition in the Army, taking into consideration the present security situation.

Saturday, January 11, 2003

Human rights activists, such as Malla K Sundar, Sudip Pathak, and Sindhunath Pyakurel have hyped much on the Maoists' six-point bottom line for talks. They include:

1. The entire executive authority must reside in the political parties, and the King should be institutionalized as a ceremonial figurehead;
2. Nepal's monarchy should be similar to that of Japan or Britain;
3. Elections to the Constituent Assembly must be held, with the provision of constitutional monarchy;
4. The Army must be placed under the control of the Parliament;
5. The Maoist PLA (People's Liberation Army, or the Jana Mukti Sena) must be integrated with the RNA or merged with Nepal Police, as was done in 1950/'51 when the revolutionary forces were merged with the state's Army; and
6. The Maoists are liable to launch a massive attack should there be

no sincere efforts to solve the state-Maoist stalemate within one week of this communiqué being made public.

Tuesday, January 14, 2003

For two days, the Royal Danish Ambassador to Nepal has been seeking to meet the King to learn about the present status of the peace talks with the Maoists.

The Central Committee Meeting of the Nepali Congress, describing the King's recent visits to the Eastern Development Region as a rehearsal for his 'Divide and Rule' stratagem, reached a decision to opt for public agitation by forging its alliance with all the democratic forces. Consequently, the leaders of the Nepali Congress, the CPN (UML), and the CPN (United People's Front), with the expectation of covert Maoist support, are planning to call for a mass movement soon at a future date.

Minister and peace talk facilitator Narayan Singh Pun called to update me that perhaps the PM was too busy to find time to send his response to the Maoists letter. He added they (Maoists) were in a 'wait and see' mood till February 26 for the government's response, and then go for a massive attack in the districts, such as Baglung, Pyuthan, and Gulmi.

When I briefed the King on Pun's input, he instructed me to immediately talk to the PM in this regard.

Thursday, January 16, 2003

Sources say Khum Bahadur Khadka is headed to Delhi for what's thought to be his health checkup where he's expected to visit some leaders. Shortly after his release on bail, he said he would join the parent group of the Nepali Congress.

Leaders of the Nepali Congress, such as Chakra Prasad Bastola and Narahari Acharya, are making their tremendous efforts to reunite the splintered factions of the Nepali Congress.

Talking to the party people, Bastola said that GP Koirala himself is stalling the Party from unification by unaccountably trying to pass the Party's leadership mantle to his daughter, Sujata Koirala.

He also added that India wouldn't support the planned mass movement with Koirala in the leadership.

It's been reported that some Germans, posted at the German Embassy in Kathmandu, are secretly in the Terai in search of the Nag Mani, a kind of 'serpent gem' supposedly 'worn' by a certain species of cobras on their hoods, and believed to be endemic of the region. I had heard of a row over the price of one such gemstone there but I know nothing about the existence of such a jewel.

The Lokendra Bahadur Chand Government is completing its honeymoon period of 100 days in office. Comments on its performance are expected to pour in from all quarters.

Friday, January 17, 2003

Yogi Narahari Nath has reached Gorkha from where he called me to convey to the King to visit the religious gatherings of people; perhaps he thinks such participation will help build the King-people rapport.

The King told me to convey his thanks to him for the advice. A Central Committee Meeting of the Nepali Congress Party has taken important decisions on the monarchy: that only the King, the Queen, and the Crown Prince should have the royal title of the *Shree Panch* ('Majesty').

The meeting also termed the ballyhoos raised for the local elections a total sham.

Monday, January 20, 2003

I was baffled to hear from PM Chand on the telephone yesterday when he told me that the HDC wanted to further the process of the government-Maoist peace talk. I also came to learn that he had already written a letter, on December 31, to the HDC to that effect. I told the PM that I would first like to talk to facilitator Pun and let him know soon.

When I asked Minister Pun about the developments of the process of talk, he described the process being at a standstill at present, with the PM cooling his heels in replying to the Maoists.

When I briefed the King on the delay of the process, he curtly said such indecision on the part of the Prime Minister may have adverse effects on the future of the talk.

The PM, talk facilitator Pun, and I quietly met at the Prime Ministerial residence at Baluwatar to consider the Maoist proposal positively, putting on hold the Swiss proposal for the time being.

Pun said that the Ambassadors of the UK, the US, and India are eager to learn about the fresh developments in the process of the peace talks. He quoted the Indian Ambassador as saying that he's ready to contribute to the peace talks to have it reach its logical conclusion.

The Chinese seem to be concerned about the arrival of an American team in the pretext of training the RNA and the increase in the Tibetan activities in Nepal.

It's no secret that both China and India have one thing in common: that they can't stand the presence and influence of the American Army in Nepal.

Khum Bahadur Khadka has expressed his grievances at the sheer injustice meted out to him by being held under the scanner of the CIAA in a vindictive manner. He complained against the King of repaying his help by having him arrested and placed in jail on corruption charges.

Sunday, January 26, 2003

I was shocked to hear from the Chief of the NID who broke the unpleasant news of the IGP of the APF, Krishna Mohan Shrestha, being shot dead along with his wife, Nudup Shrestha, and his bodyguard, Surya Regmi, at 7:30 o'clock this morning while the three were on the way home from their morning walk.

The search for the late IGP's equal to fill his place surged on my mind.

It'll be wise for the King to express his condolences at the death of the late IGP,

An emergency meeting of the Cabinet was called following the fatal shootout.

The final rites for the deceased will be performed tomorrow once their daughters arrive in Nepal from abroad.

There was another rumor of the Maoists shooting the Minister of Education and Sports, Devi Prasad Ojha, today. But nothing untoward has been reported.

Following the murder of the IGP of the APF, the PM called me to his office at Baluwatar where Narayan Singh Pun was already present.The PM looked shocked and frantic at the incident.

Reviewing the situation, we decided to condemn the cowardly and heinous act of the Maoists to murder the IGP of the APF at a time when the process of the government-Maoist peace talk was gradually gaining momentum. Now it seemed that the murder of the police officer of the highest rank shed doubts on the very future of the peace talks. But exercising our best possible fortitude and foresight, we, on our part, decided to proceed with the peace talks by first having the terrorist tag pinned on the Maoists removed, and then suspend the Interpol's Red Corner Notice on them, and canceling the bounties set on their heads.

We hoped our decisions would address the key demands of the Maoists to create an environment of trust for a talk.

Shortly after the conclusion of the meeting, I briefed the King on it, and he expressed his interest to see the draft of the letter to be sent to the Maoists on the government's commitments.

Monday, January 27, 2003

Narayan Singh Pun said to me today that he was surprised to find Ramesh Nath Pandey already there at the PM's residence.

He said the PM described the developments of the peace efforts, including the exchange of letters, and the creation of an environment to both Pun and Pandey.

Then the three men held a formal meeting and agreed to have the agenda on the peace efforts passed by the Cabinet, and the PM to consult with the heads of the security forces.

I telephoned the security heads to inform them about the above developments and asked them to help for the talk to succeed.

The Chairman of the National Assembly, Dr. Mohammed Mohsin, the Chairman of the RPP, Pashupati Shumsher Rana, and former PM and the President of the Nepali Congress (Democratic), Sher Bahadur Deuba, separately called on the General Secretary of the CPN (UML), Madhav Kumar Nepal, at his residence.

The Joint Coordinator of the NSC, Maj. Gen Rookmangud Katawal, quoted the Military Attaché at the British Embassy as showing his curiosity with him about the facts, the progress, and the modalities of the peace talks.

Tuesday, January 28, 2003

I heard that the PM this morning told the CoAS about the imminent peace talks with the Maoists through Minister Pun, and instructed him to cooperate by declaring a ceasefire during the talk.

In response, the CoAS made it clear that the Army was ready to do so but without disrupting its regular course of activities, such as training, purchase and transport of weapons and logistics.

The PM, on his part, instantly instructed Minister Pun to ask the Maoists to send to the government their letter with the clear statement of their position, agenda, modalities, and the duration of talk with the government.

He said that he told the PM that the process had already gone ahead.

Because Ramesh Nath Pandey was at the scene to claim his credits for the dialogue, Pun she felt redundant. That was the reason why he seemed reluctant to take further initiative.

Wednesday, January 29, 2003

To the King, I described the fresh complications, owing to two resource persons separately trying to proceed with the peace talks, which was very likely to create confusion, rendering the entire process to a failure.

The King instructed me to convey to the PM and Pun to continue the peace process from his (Pun's) channel of communication.

The Chief of the NID said that the Chief and the Deputy Chief of the Indian Intelligence Agency, RAW, have expressed their wish to come to Nepal on a ten-day visit from February 15 to March 2, and on that occasion to go on a pilgrimage to Muktinath. He said they have requested him to send his formal confirmation.

The King gave his consent to issue them the required permission for their purpose.

With only a few days left for the long-awaited peace talk to commence, the CoAS consulted with the King and decided to reciprocate with his announcement of a ceasefire; the Maoists' own ceasefire starts from today.

Thursday, January 30, 2003

A newspaper today carried a news story about the minister and talk facilitator, Narayan Singh Pun, flying the Maoist leaders in a helicopter to meet the King at the Royal Palace.

I am amazed at the extents of fabrications one resorts to serve his petty interests. It's learned that one particular Minister had the news story ghostwritten. I wondered if it Ramesh Nath Pandey himself who planted such a news item!

I conveyed the King's instructions to Minister Pun to avoid the media and media persons until the talk turns to be fruitful, because revealing something classified of this nature may jeopardize the peace talk itself.

Former IGP Achyut Krishna Kharel now bluntly talks about having some journalists on the Nepal Police payroll while he was in office. He has divulged some details, too. It's particularly been learned that he had one Tika Ram Rai, a journalist, to write news stories against the monarchy.

Friday, January 31, 2003

Following the reciprocal ceasefire declared by the state and the Maoists, there have emerged mixed responses in the public.

Journalists are on their 'wait and watch' mode whereas many others take it as a positive turn.

GP Koirala has termed it a sham, part of a 'Grand Design', a phrase he has so often used in recent times.

The common populace has taken it very positively, as something 'solely in the interests of the people.' They want the government to prepare its groundwork for the peace talks to reach its logical end.

Intellectuals and security chiefs opine that the government must be serious enough not to release the Maoist detainees without the rebels' assurances that things won't turn otherwise.

It's been revealed that an Undersecretary at the Press and Communication Secretariat of the Royal Palace works as a part-timer at the Nepal Samacharpatra daily, something unethical and incompatible with the status of his post and responsibilities; a case of flagrant breach of the code of conduct.

On this, the King instructed me to consult first with the Chief of the Royal Household, and do the necessary as rules and regulations warrant.

While the PM is seeking quietly to form a team of experts from the legal, diplomatic, security, and political realms to facilitate the peace talk, the King has advised the government to publicly form a task force comprising the former CoAS, Gen Dharmapalwar Singh Thapa; the former Chief Secretary, Karnadhoj Adhikari; the former IGP, Dhruba Bahadur Pradhan; Professor Dr. Mohan Prasad Lohani, and Advocate

Bal Krishna Neupane, and have it function under the aegis of the NSC Secretariat.

Sunday, February 2, 2003

Without informing the Peace Talk facilitator, Narayan Singh Pun, as protocol would warrant it, the Maoists publicized the names of their representatives for the roundtable. They are Dr. Baburam Bhattarai, as coordinator; and Ram Bahadur Thapa, Dev Gurung, Krishna Bahadur Mahara, and Matrika Yadav as members.

The PM and Pun felt perturbed at the deliberate breach of the understanding.

The PM directed the state-owned media to ignore the news.

Sources close to the PM said he was going to form another talk team, led by DPM Badri Prasad Mandal.

The PM is set to address the nation with regards to the government's preparation for the long-awaited peace process. He's also likely to make public the names of the government team of facilitators.

Several INGOs have wished their best for the success of the peace process.

Monday, February 3, 2003

According to fresh information, the CPN (UML) has decided, in advance, to approve of its agenda on the general elections to the Constituent Assembly (CA) at its 7th General Convention, soon to be held in Janakpur.

It's advisable that the Royal Palace explore every possible means to avert it from happening.

Perhaps, as part of that strategy, the King granted his audience to the Party's General Secretary, Madhav Kumar Nepal who, after his audience with the King, did a total about-face by dropping the phrase of 'elections to the CA' from his speeches, and started singing an altogether a new tune of the Party's possible entry into the present government.

That's to say, the CPN (UML) will most likely join the bandwagon of the current government, provided they get their choice Ministerial berths.

Speaking at an orientation program of his Party, Chairman of the Nepal Peasants and Workers Party, Narayan Man Bijhukchhe, is reported to have tersely criticized the peace talks. He remarked that the upcoming peace talk was taking place at the initiative of the King himself who is the Supreme Commander of the Royal Nepalese Army. He also added that it was very likely that Prachanda and Baburam will either be killed or extradited to Nepal, at the pleasure of India.

Dhani Ram Lamichhane, who facilitated the peace talks, has mysteriously headed to Janakpur to meet Madhav Kumar Nepal, The reason for the meeting was kept confidential.

Tuesday, February 4, 2003

At a joint audience with the King, the British Ambassador to Nepal, Keith Bloomfield, and the Chief of DFID, David Wood, reiterated British concerns over the Maoist insurgency, the deteriorating security situation, the violation of human rights, and the chronic assaults on democracy in Nepal.

They added that the recent commencement of the peace process was a matter of pleasure, and they, therefore, promised to provide Nepal with experts to facilitate the initiative, if needed and asked for.

They wished both sides to exercise restraint, bury the past discords, and refrain from trading accusations but work to contribute to peace.

The British Ambassador promised a continuity of British assistance to the RNA in the days ahead.

Recalling the age-old Nepal-Britain relations, he said the two countries have always remained friends ever since, and therefore, Britain is always concerned about the problems of Nepal.

In response, the King expressed his thanks to Britain for her continued support and assistance to the RNA. He said Nepal always rated highly the British assistance to the RNA.

Speaking about the ongoing government-Maoist peace talk, the King explained that the spirit of the endeavor was first to settle the humanitarian and other fundamental issues, and then to address the political agenda, and finally work out a lasting and durable peace and security.

The King further related that looking at Maoist talk team, they seemed to be honest and serious about the peace process.

He concluded that Nepal expected cooperation and goodwill from its friends for the peace process to reach a logical conclusion.

Wednesday, February 5, 2003

Speaking with Western diplomats, British Ambassador to Nepal, Keith Bloomfield, admired the King as being serious to resolve the crisis with a road map with the future course of actions after the conclusion of the peace process.

It's learned that independent Leftist leader, Padma Ratna Tuladhar, is displeased that he wasn't entrusted to facilitate the peace talk between the government and the Maoists for two reasons: one, because the peace talk was taking place at the King's initiative; and two, because he didn't enjoy access to the Royal Palace.

The IGP of the Nepal Police, Shyam Bhakta Thapa, speaking on the 48th year of the organization, said that since the neo-democratic year of 1990, the service has seen individuals gaining the upper hand in its administration rather than the institutionalization of the institution. Though a domestic constabulary force, the Nepal Police were deployed across the country to fight Maoist insurgents, in addition to general policing. Consequently, the insurgency has so far claimed the lives of 966 policemen, leaving 339 critically injured, and the total of 1,600 police posts shrinking to just 700, the rest being demolished by the Maoists.

The handover of the Kakani Training Center and the one at Danda Pakhar to the APF has caused inconveniences in training for the Nepal Police, he said.

He disclosed the Nepal Police's aim to contest in the elections of the Interpol of the Asia region.

When briefed, the King gave me instructions to work out a mechanism to do away with political influences in the police organization. He specifically asked me to keep track of the police officers who serve the interests of the political parties.

The King, the PM, the peace talk team facilitator Pun, and the heads of the security forces, over a private dinner, discussed the progress of the peace talk and the likely agenda thereof.

Sunday, February 9, 2003

Even when the government-Maoist peace talk is progressing, the Maoists, quite out of the blue, issued a strongly-worded press release.

In a quick succession of events, the Indian Ambassador to Nepal quietly visited Madhav Kumar Nepal at his residence in Koteshwor.

It occurs to me that such clandestine meetings of personalities of national statures are by no means respectable.

It's speculated that the Nepali Congress and the CPN (UML) are going to oppose the ongoing peace talk.

Monday, February 10, 2003

The Joint Coordinator of the NSC, Maj. Gen Rookmangud Katawal, called me to inform that the PM has asked him to recommend a representative to participate in the program on 'Conflict Management' organized at the British Embassy. I advised him to select the representatives from the NSC.

The Chinese Ambassador to Nepal has invited Princess Prerana, along with her husband, Raj Bahadur Singh, for dinner.

The King gave his consent to the social visit.

Tuesday, February 11, 2003

The Nepal Peasants and Workers' Party is organizing an all-party meet tomorrow.

The CPN (UML) is trying to forge a Leftist tactical unity.

The Chairman of the CPN (Maoist), 'Prachanda', is likely to propose to the Nepali Congress President, Girija Prasad Koirala and Madhav Kumar Nepal, the General Secretary of the CPN (UML), and the Chairman of the Peasants and Workers' Party, Narayan Man Bijhukchhe, to form an alliance against the King to secure their common objectives.

Wednesday, February 12, 2003

With only two months remaining for my military rank of Major General to expire, I advised the King to decide on my promotion or retirement so I could either continue in office or take the customary leave of one month prior to my automatic retirement.

The King didn't give his immediate reply.

I requested the King for his permission to publish the book, *Rajya Arohan Ko Ek Barsha* ('One Year of Ascent to the Throne') in the English version, with the addition of some important episodes, and coinciding with his birthday this year.

The CoAS updated me on the program of the inauguration of the Bailey bridge in Gorkha District, constructed by the Army under the ISDP.

Thursday, February 13, 2003

Narayan Singh Pun updated me with the information of his informal consultations on the vital issues of the dialogue on peace with the Maoist leader, Dr. Baburam Bhattarai, who was, according to him, very positive about the peace talks.

Pun again quoted the leader as saying that since the political parties are operated by a third force, they will obviously be unable to forward solutions to the problem. He added that he believes that the Maoists seek to use this as a rare opportunity to take the peace talks to its logical conclusion.

Friday, February 14, 2003

The CoAS informed me of the program finalized to summon all the Brigade and Division Commanders to Kathmandu two days before the Army Day, coinciding with the Maha Shiva Ratri, in order to study their periodic reports and to issue necessary policy directives to them.

He said he has sent his invitation to the King for a dinner at his official residence at Shashi Bhawan on February 28.

I was instructed by the King to make available for him the list of the invitees at the dinner prior to his attendance, a routine the King always observes when it comes to attending a party.

At a joint briefing held for us, the five Secretaries of the Royal Palace, the King referred to the ongoing peace talk in detail, and in the meantime, instructed us to work in perfect coordination for all possible humanitarian accountability.

He gave his special command to the Principal Secretary, Pashupati Bhakta Maharjan, to maintain the personal files on corrupt officials ready for his reference, and for the investigations into their property details to be conducted by the CIAA.

Monday, February 17, 2003

Peace talk facilitator and Minister Pun drove the Maoist peace talk team from Bhairahawa to Kathmandu. It comprised Dr. Baburam Bhattarai, Krishna Bahadur Mahara, Dinanath Sharma, and others. Sources say that it was agreed not to make public Dr. Baburam Bhattarai immediately.

Thursday, February 20, 2003

The marital engagement of Her Royal Highness Sitashma concluded today at Jeevan Kunj in the gracious presence of Their Majesties.

The eminent persons' group to facilitate the peace talks met the government facilitator Pun to take fresh reports and documents on the progress of the peace talks. The team agreed to operate from its office in the NSC complex, with Col. Baldev Raj Mahat as the contact person.

Kamal Chaulagain said it's crucial that the CPN (UML) will split again, which would be like beating the iron while it's red hot, as its senior leaders, such as Radha Krishna Mainali and Bamdev Gautam, feel extremely dissatisfied and seeking just an alibi to quit the Party.

Even the Maoists are making efforts to court Bamdev on their side, Chaulagain added. He commented that donor agencies are hesitant to extend help as the local bodies are void of representatives.

He further disclosed that the Minister of Education and Sports, Devi Prasad Ojha, is leaking every decision of the Cabinet to a top CPN (UML) leader.

There's information that Maoist leaders, Dr. Baburam Bhattarai, Krishna Bahadur Mahara, and Dinanath Sharma are visiting the former PM, Surya Bahadur Thapa.

In a series of political leaders visiting other leaders, the Chairman of the RPP, Pashupati Shumsher Rana, visited the General Secretary of the CPN (UML), Madhav Kumar Nepal, at his residence in Koteshwor to discuss the current political affairs.

The Maoists made public their feelings, going even to the extent of including the UN representatives as facilitators, if the need arose, in the imminent peace deal to succeed.

Likewise, it seems to the Maoists, that in the changed context in the country, as of now, it wouldn't be wise of them to write off the monarchy.

In response to the question as to what if the King didn't accept the proposal of holding the general elections to the CA, they hinted at their willingness to go along by overhauling the current Constitution. This can be in the form of introducing a provision to have the Army under the Parliament which will address the issue of removing the absolute power of the King. The Maoists seem confident of resolving the problem through an all-party roundtable conference.

There's every reason for one to smell a rat in the Maoists' newfound love for the peace talks to reach a logical end. One shouldn't forget that India has a game plan behind this newfound flexibility of the Maoists because, only recently, SD Muni, Professor at the Jawaharlal Nehru University (JNU) of New Delhi, had visited Kathmandu to consult with the Maoist leaders, such as Dr. Baburam Bhattarai, at The Everest Hotel.

Friday, February 21, 2003

The King and the Secretaries (us) of the Royal Palace thoroughly discussed the current turn of political events. On the occasion, the King issued his directives to prepare a draft of the code of conduct, a guideline for the peace talks, by the middle of March, and have a copy of it sent to the Royal Palace.

The King made it clear that the regular activities of the Army shall continue even during the peace talks.

He gave further instructions to prepare a 'white paper' on the progress of the peace initiative for his reference during his visit to India.

I got reports of Jagat Gauchan speaking ill of me and publishing his rude comments against me in various publications. His main grouse is that Narayan Singh Pun was made Minister in his place because I took a huge bribe for the appointment.

My fellow Secretaries guffawed at my report to the King in our joint meeting, while His Majesty remained quiet.

Monday, February 24, 2003

Narayan Singh Pun shared with me his experiences of the first round of his informal talks with the Maoists.He said the preliminary round progressed satisfactorily with good rapport and confidence-building niceties between the two sides. He said the Maoists tend to appear sound in their reasoning, steadfast in their position and well-prepared in their presentation.

Pun, however, complained of the political parties for their noncooperation in the peace talks.

He also sensed the Maoist leaders' wish to meet the King.

I asked him to keep updating me on every bit of information from his side.

The King asked me to first confirm the agenda with which the Maoist leaders wanted to talk to him before deciding on the time and place.

On being briefed, the King gave me his permission to meet the Deputy Chief of the Embassy of India.

The CoAS informed me that an Army delegation, led by the Defense Secretary, Madan Kumar Aryal, was visiting China, and he wanted to include Brig. Gen Dambar Singh Gurung in the team.

According to a tipoff, GP Koirala and the Maoist supremo 'Prachanda' met secretly at an unknown place.

Sources say that nine Leftist parties met at the CPN (UML) headquarters at Balkhu where the Maoists requested all the parties to agree to a roundtable conference and the reinstatement of the Parliament.

The General Secretary of the CPN (UML), Madhav Kumar Nepal, wanted to know about the objective, modality, and end results of such a roundtable conference.

Wednesday, February 26, 2003

Narayan Singh Pun phoned to inform me that the Indian Ambassador to Nepal has invited him for lunch. He promised to update me on all the important matters, if any.

The Lamsal Commission, formed to look into the income sources and property details of the politicians and top government officials, has shown its soft spots for the Army and its officers. It suggests that the Army itself compile the property details of the top-level Army officers. The Commission then will look into it and present the facts and figures to the King before making it public.

The CoAS today proposed fresh reforms in the scale of ration (supplies of grains and other essentials) for the Army.

The King instructed me to tell the Principal Secretary to consult with the PM and put the process into action accordingly.

The CoAS proposed the replacement of the old helicopters by advanced ones (the AB139 or HUEY-2).

He said he was soon to start his inspection tour of the barracks in eastern Nepal.

Friday, February 28, 2003

Speaking at the joint meeting organized at the Tribhuvan Army Officers' Club, the Brigade and Division Commanders unanimously commented on the same chronic problems.

They claimed that the Maoists have joined the process of dialogue due to the Army's operations.

The ADC to the King, Col. Binoj Basnyat, called me to convey the King's instruction to collect and send the dossiers of the Maj. Gens and Lt. Gens to the Royal Palace for study. I, in turn, communicated it to the CoAS, and the process went into motion instantly.

Sunday, March 2, 2003

Narayan Singh Pun rang me up to inform about the complications developing in the Maoist team for the peace talks. He disclosed a surprising piece of information, that the Minister of Communication, Ramesh Nath Pandey, has had a news story carried in the *Kantipur Daily*. It's very likely that the published piece can vitiate the environment of the talk.

Pun said there are two sorts of speculations doing the rounds on my promotion and the consequent continuity of service at the Principal Military Secretariat during the ongoing peace talk for it to reach a logical conclusion, or my retirement and the appointment of another Military Secretary in my place.

Now, the King's asking for the dossiers on the Maj. Gens and Lt. Gens begin to make sense to me.

Monday, March 3, 2003

Narayan Singh Pun has cautioned the facilitators of the Maoist talk team about the likelihood of media persons exaggerating and misinterpreting the political agenda they have raised during the peace talks. He added that the Maoist supremo, 'Prachanda', himself was likely to talk to him any day regarding this.

The PM called the heads of the security forces, and collectively discussing the current state of affairs, requested them to be positive and hopeful about the future of the peace talks. He also asked the political parties to join the government to contribute to peace.

The members of the Maoist talk team, Krishna Bahadur Mahara and Dinanath Sharma, are known to have visited and held consultations with the leaders of the Nepali Congress, Chakra Prasad Bastola, and Krishna Prasad Sitaula, on political issues and the current state of affairs.

From the consultations, it's been revealed that the Nepali Congress doesn't want the peace talk to be successful.

Friday, March 7, 2003

Pun updated me on the fresh news of the positive turn of the peace talks, and the finalization on the code of conduct for it.

The King granted his permission to the CoAS to visit eastern Nepal for the inspection of the Army barracks there.

The Maoists expressed their unease over the concept revealed by the US Government that was made public yesterday. The US has included the Maoists on its watchlist of terrorist organizations that calls for responsive operations against them.

Sunday, March 9, 2003

I had invited the Deputy Chief of Mission of the Indian Embassy, Ashok Kumar, for dinner at the Shangri-La Hotel yesterday. He was keen to learn about the developments of the ongoing peace talk.

He also expressed his displeasure that although India was always positive about Nepal, it always declined to sign the Extradition Treaty, which eased criminals' infiltration into the country.

The DCM added that the Koshi and Tanakpur projects haven't incurred Nepal any loss, as is often ranted in Nepal. Referring to statistics, he said that India's purchase rate of electricity from Bhutan is far lower than that of Nepal.

He also tried to convince me that there was no need for Nepal to have other countries involved in installing large hydroelectricity plants.

It seemed that India wants to have the sole influence over Nepal's hydroelectricity projects.

In response, I said proper regulations of the Nepal-India borders is a priority for Nepal.

Monday, March 10, 2003

GP Koirala, Madhav Kumar Nepal, Narayan Man Bijhukchhe and other political leaders are seeking their collective audience with the King. When I briefed him on the request, the King showed no reaction.

A vehicle carrying Army soldiers on a rotation had an accident in Bardia, killing four soldiers and leaving 23 injured. Some 13 of the critically injured have been airlifted to Kathmandu for treatment.

Wednesday, March 12, 2003

According to diverse sources of information collected by the Chief of the NID, the people have felt a sense of respite after the ceasefire and the commencement of the government-Maoist peace talk. As a result, it has caused the political parties to lose supporters.

Lately, the Maoists have adopted a tactical strategy of the continuance of the war with another phase – training their cadres, obtaining weapons, and extortion of money. They seem far from putting their full faith in the peace talks with the government. They have also aired their new stand to talk to the King himself for the full and final resolution.

The King commanded me to tell the talk facilitator and Minister Pun to be serious about the emergence of new situations.

The Chief of the NID repeated the same old problems: lack of communication equipment, cut in its budget, and the suspension of allowances for the employees in the office, all causing to hamper the efficient delivery of services from the office.

The King gave me instructions to consult with the Principal Secretary to instantly rectify this condition.

Col. Vijaya Bikram Khadka from the ADC Office rang me about the audience for the CoAS, Gen Pyar Jung Thapa, with the King, made for 11 a.m. tomorrow. I passed the detailed message to the CoAS and told him to instantly return to Kathmandu.

I had only just put down the receiver of the phone when it rang again. It was Brig. Gen Tika Dhamala who broke the news of the King's mood to retire me. He said he had advised the King to hear from the CoAS also before making such a major decision.

The reference to my retirement wouldn't adversely affect me much, as I'm used to hearing it often. There was already a rumor about it earlier. My retirement or retention is up to the King.

Thursday, March 13, 2003

As required by the norms and standards, I received the CoAS at 10.45 a.m, a good 15 minutes earlier than the appointed time, for him to brief the King. Brig. Gen Tika Dhamala conveyed to the King of the CoAS waiting for his audience. He later told me that the King sounded angry at my presence.

Normally, the King consults the Military Secretary prior to granting audience to the CoAS. But today, no such norm was observed. Just when the CoAS was entering the chamber, the King instructed the ADC to stop the CoAS and send me in. I sensed something unusual. The King asked me to wait until he finished his consultations with the CoAS about his upcoming visit to the Western Development Region.

Later, the King summoned me in to say that he had decided to promote me to Lieutenant General and assured me to consider the post of the Principal Military Secretary later.

In response, I thanked the King for his generosity and expressed my commitment to serve from whatever post the King wished fit for me. I resumed, as usual, my duty of briefing the King.

Referring to the information received from the security forces, I said that the rapport between the government and the Maoist facilitators' team hasn't been cordial to the required degree. The leaders of the major political parties seem more influenced by the South Block and the Indian Intelligence Agency, RAW, to act collectively against the monarchy. India isn't happy about the ongoing peace talk without its role therein.

I advised the King that the best course of action for the government now is to convince the friends of Nepal to support Nepal's fresh efforts at peace.

The Cabinet has prepared its Code of Conduct mandatory for the facilitators and participants alike to observe during the peace talks, and it's to be signed by Minister and peace talk facilitator Narayan Singh Pun for the government, and by Dr. Baburam Bhattarai on behalf of the Maoists, respectively, before it's made public today.

All the leaders of the Communist parties seem to be scared by the fresh turn of event over the US including Nepal's Maoists on the terrorist watchlist last Friday.

The Indian Ambassador to Nepal, Shyam Sharan, visited PM Chand yesterday morning at 9 o'clock.

I briefed the King on the breach of norms in the transfer of Joint Secretary Baman Prasad Neupane from the Secretariat of the NSC.

I couldn't help imagining about my promotion to Lt. Gen, and the promise of reconsidering me for the position of Principal Military Secretary.

There are already speculations of posting either Lt. Gen Victory SJB Rana or Lt. Gen Sadip Bahadur Shah as Principal Military Secretary.

If the existing norms and standards are any guidance, the officer who is to be posted as Principal Military Secretary should have at least five years of continued service as Military Secretary. So far as the question of my posting as Principal Military Secretary is concerned, I still have to cover more time to complete the full five years of service.

And, by nature, I'm not the type to race for post and promotion.

I had frankly advised the King to have an officer transferred to the Principal Military Secretariat one year in advance so that I could hand the responsibilities of the office over to him before I take my retirement.

The King's remark, 'I have my own people' hurt me, but I was committed to my duty and had no reason to fret and fear.

I need not publicly claim my share of the hard work I put in for bringing the monarchy to normalcy from that period of turbulence created by the Royal Palace Massacre. I was prepared to have my retirement without cherishing any dream of further extension of my term or promotion. As Military Secretary, I had served the King for one year and eight months during which I briefed him on everything on every issue of national importance. One needn't beat around the bush.

I have good knowledge of the King asking the CoAS to find my alternative in the office. I merely wished there would be an experienced Military Secretary to replace me. What I still can't make sense of is why a sudden change occurred in the King's heart.

Monday, March 17, 2003

I walked to my office, pleased that the King had promised me the promotion to Lt.Gen, the highest rank an Army officer serving at the Royal Palace is promoted to. The CoAS from the RNA Headquarters congratulated me on my advancement. We discussed ways to improve the Army's logistics, the inadequacies being an old malady in the Army barracks.

The CoAS also showed me a handwritten letter addressed to him. The anonymous letter had bitterly denounced the CoAS. We guessed it must've been written by one of the PSOs (Principal Staff Officer), dissatisfied with the CoAS and his mannerisms.

I discussed the new structure of the Military Secretariat, its fiscal plan, and expected help from the Army for my staff in the office. Our discussions included immediate reforms in the Royal Guards Military Police Unit and the complications in the promotion of the Royal Guards Military Transport (MT).

Disclosing to me the secrets of the Party, a leader of the CPN (UML), representing the Rapti Zone, expressed his sheer dissatisfaction that the Party has dropped its core ideology and deviated from it, and as a result, many of its leaders are mired in self-seeking. He said that Radha Krishna Mainali criticized Kamal Chaulagain and Devi Prasad Ojha of incompetence in their performance as Ministers.

The leader stressed the need for a third force to counter the unaccountable political parties to steer ideal politics ahead.

Wednesday, March 19, 2003

In my briefing today, I pointed out to the King on the need for facilitating a vibrant media to rise and the creation of an alternative political force with nationalistic leaders from the mainstream political parties.

Turning the meeting to the King's upcoming visit to India, I advised him on certain points to be prepared, with answers, prior to the visit.

Referring to the context of India's persistent pressure that Nepal purchases the Insas rifles from India, the King should have answers ready in his mind, as the issue is very likely to be raised during his visit.

I added that India may raise the issue of the Treaty of Commerce and Transit this time again, as it tends to reiterate its security concerns each time.

India always accuses Nepal of turning a blind eye to the ISI's active presence in the country. So I advised the King to put in succinctly, that it would be very easy for Nepal to look into the matters of security and take immediate actions if India provided Nepal with its Detail Intelligence Report.

The US Army's attack on Iraq has created a condition of no functioning government there. The US move on Iraq lends ample grounds to justify Nepal's fight against terrorists in this part of the world.

I told the King that we should think about the likelihood of India attacking its neighbors on the pretext of its security concerns. It's no surprise that India, with its arrogance of being a predominant power or with its big-brotherly attitude to South Asia, is very likely to do so.

As instructed by the King, I asked the CoAS to prepare an analytical paper as to what extent 'America's Attack on Iraq' affects Nepal, and that such an analysis may be useful for the government in developing its strategy.

According to a tipoff, the Maoists have assembled weapons and guerillas to launch their simultaneous attacks on the administrative headquarters of the Districts of Jajarkot, Rukum, Rolpa, Kalikot and Dailekh.

It would be a good idea to arrange a dinner for Prime Minister Lokendra Bahadur Chand, the talk team coordinator/Minister, Narayan Singh Pun, and the heads of the security forces, to hear about their opinions and suggestions about the peace process, a question India is very likely to raise, and the King is required to prepare his answers before his India visit.

It's advisable that the King should visit places across the country for bonding with the people, especially with the youths, and also by forming of Youth Associations.

The IGP of the Nepal Police, Shyam Bhakta Thapa, quoted the Chinese Ambassador to Nepal as emphasizing on the need for better rapport and exchange of information between Nepal Police, the Chinese Police, and the Chinese Embassy.

A program has been slated to hand over the MI-17 helicopter to the

No 11 Brigade at the Tribhuvan International Airport. The helicopters provided by Britain to the RNA will be received by the Quarter Master General of the RNA.

I briefed the King on the dubious roles of the Minister of Education, Devi Prasad Ojha, and the Minister of Transportation, Kamal Prasad Chaulagain. The King told me to inform the Principal Secretary about it and ask him to keep an eye on their activities for action.

In the context of Madhav Kumar Nepal and Radha Krishna Mainali both wishing to meet the King individually, he instructed me to first find out about the purpose and the agenda for the talk.

As planned, the King is scheduled to leave Kathmandu for Delhi on March 20 by a NAC flight and return on the 31st on an Avro airplane of the RNA that shall pick him up in Bangalore.

Monday, March 24, 2003

Narayan Singh Pun shared with me the Maoists' 'strange and incompatible' demands, which they made perhaps on the prodding of political leaders and civil society activists.

The costs borne by the Military Hospital of the free treatment for the Royal Family, their relatives, and the employees serving at the Royal Palace are burgeoning steadily. So the idea of operating a fund in the names of the late King and Queen would contribute to minimizing such costs incurred by the Military Hospital. I'm thinking of advising the King to establish the King Birendra and Queen Aishwarya Memorial Trust once the King returns from his trip to India.

Speaking at the four-party meeting convened at the party office of the Nepal Peasants and Workers' Party in Bhaktapur, Chairman of the Party, Narayan Man Bijhukchhe, said that the King's takeover came about in a response to the serious flaws and weaknesses on the part of the political parties.

Speaking on the same occasion, the General Secretary of the CPN (UML), Madhav Kumar Nepal, remarked that the political parties should collectively mobilize the people in the streets once and for all, particularly because the King and the Maoists have made their pact to weaken and sideline the mainstream political parties.

In his turn of delivering his speech, the President of the Nepali Congress Party and former PM, Girija Prasad Koirala, said he had read the King's mind, and if he ever thought of weakening the parties, it would create a question mark on the very existence of the monarchy.

A protest march gathered at Ratna Park shouted slogans against the US attack in Iraq. It's learned that the march was organized by Shyam Shrestha, editor of the monthly magazine Mulyankan. Present in the march were Leftist students and leaders of the likes of Lila Mani Pokharel, Shakti Lamsal, and some Maoist leaders.

Tuesday, March 25, 2003

The unofficial meeting of the RPP, held in its party office, decided to forge an alliance with other political parties and mount pressures on PM Chand to resign, although the PM himself is a founding leader of the same Party !

The China visit scheduled for the Secretary at the Ministry of Defense, Padam Kumar Acharya, has been rescheduled for April 2.

Wednesday, March 26, 2003

Narayan Singh Pun shared his fresh information with me. He said the international media is giving wide coverage to the government-Maoist peace talk.

I attended the dinner hosted by the Chinese Ambassador to Nepal. He congratulated me on my promotion. In response, I expressed my thanks.

From the exchange of pleasantries, the conversation turned to politics. He expressed his keen concerns over the ongoing government–Maoist peace talk and wished for its success and sustainability.

He said both China and India have common concerns on the increased US activities in Nepal.

Next, the conversation turned to the change of leadership in China. The Ambassador said that leadership change in China goes in a normal way. The National Assembly Chairman and the Premier have taken their retirement to make way for the new leadership to come forward and show their talents to steer China to a more prosperous future.

The NSP has suffered a vertical split which may affect Badri Prasad Mandal's political career. Mandal is the present DPM.

Similarly, the Chairman of the RPP, Pashupati Shumsher Rana, alleged the PM of incompetence and demanded his resignation.

Thursday, March 27, 2003

Narayan Singh Pun updated me with the latest news of Maoist leader Krishna Bahadur Mahara and the independent Leftist leader, Padma Ratna Tuladhar, heading to Butwal where the Maoists' Regional Bureau and the Central Committee meeting is to finalize the names of the Maoist talk-team's facilitators/interlocutors. According to Pun, as the facilitator of the government talk team, everything was going as planned. He added that the Maoist talk team was to arrive in Kathmandu to make public the names of the facilitators at a press conference once they get to Kathmandu. He said that an agreement was reached to start the first round of goodwill talk on April 1.

According to information received from the NID, complications are likely to arise between the PM and the facilitator of the government talk team, Narayan Singh Pun, regarding the question of the expansion of the talk teams.

As arranged by Dr. Duman Thapa, I met Madhav Kumar Nepal at the residence of Dr. Dilli Raj Khanal. Speaking on the occasion, and putting his observations of the current political situation, Nepal said that although they assert that the Nepali Congress utterly failed to steer the country, they can't support the current government continuing in office because it lacks the popular mandate. He stressed on the need of an immediate change of government and making of another one led by his party.

When I assured him of finding time for him to meet the King when he himself can present his opinions, he asked me to talk to the King first and let him know about it.

Monday, March 31, 2003

I hosted a lunch at my residence for the visiting Chinese delegation.

After the King's return from his visit in India, I briefed him on the fresh developments of the overall peace talk process, the political state of affairs, and the security situations in the country. I also referred to the discourse with the General Secretary of the CPN (UML), Madhav Kumar Nepal, and the Chinese Ambassador to Nepal.

I also told the King about the political parties' expectation of the announcement of something new from the King before, during, or after his departure for and stay in Dhangadhi for the civic receptions.

The press termed the King's visit to India only partially successful, and that he had no alternative to taking the political parties in confidence.

Narayan Singh Pun sounded very happy and hopeful about the direction and future of the peace talks. He said he wouldn't mind whatever role is entrusted him with, but the success of the peace talk is the prime concern of the general populace.

He disclosed that the Maoist second-in-command comrade, Dr. Baburam Bhattarai, is staying in the house of retired Major Man Bahadur Shah and is seeking a viable channel to meet the King.

Reports continue to reveal that Maoist political party organs keep holding occasional discourses against the government. It's unsavory on the part of the Maoists to apply such double standards at a time when the peace talks is hoped to progress further.

Now that the NSP has split, the same malady plagues the RPP, too, threatening it to a splinter, a common malady plaguing almost all the political parties in the country for quite a long time now.

The former PM and the President of the Nepali Congress (Democratic), Sher Bahadur Deuba, is currently on a visit to the US and the UK. He has at times voiced his dissatisfaction against the government.

Every security arrangement for the King's upcoming visit to Dhangadhi for the Civic Reception is now complete.

On the same occasion, the CoAS is scheduled to brief the King on the military operations, the success stories thereof, and the experiences gained, and the security situation of the country over the last six months.

There was another protest demonstration against the US attack on Iraq and the rise in the price of petroleum products here at home.

Tuesday, April 1, 2003

To everyone's surprise, a program hosted at the Birendra International Conference Center (BICC) by the Reporters' Club blatantly breached protocol by inviting PM Lokendra Bahadur Chand as its chief guest and the Maoist leader, Dr. Baburam Bhattarai, as the special guest, barely a week after the commencement of the peace talks.

Sources say that Gunaraj Luintel, news coordinator at the *Kantipur Daily*, and journalist Rishi Dhamala of the Reporters' Club, visited Maoist leader Dr. Baburam Bhattarai but the contents of their discussion are kept confidential.

The ANNFSU (Revolutionary) has announced a reception for the members of the Maoist talk team tomorrow at the Tundikhel Khula Manch. Citing security reasons, the Central Security Committee has asked the government to disallow the event.

The Indian Chief of the Army Staff, Gen. Nirmal Chander Vij, is slated to visit Nepal from April 22 to 25. The CoAS has asked me to arrange a time for the King to confer on the guest with the rank of Honorary General of the RNA on April 23.

The King said it's important that the code of conduct for the peace talks be strictly observed and monitored by both sides.

He gave me instructions to implement a plan to infiltrate the political parties and start imploding them from within.

Thursday, April 3, 2003

Due to technical complications, the MI-17 helicopter flying from Kathmandu to Hetauda made an emergency landing today, leaving the members on board slightly injured.

Monday, April 7, 2003

The Maoist talk team yesterday visited the Chairman of the RPP, Pashupati Shumsher JB Rana, and soon after the former PM, GP Koirala.

While crossing the River Arun in Tumlingtar of Sankhuwashabha, four policemen of the Nepal Police and the APF have gone missing and eight rifles drowned as the boat carrying them capsized midstream.

Lt. Dr. Ushav Mohsin, daughter of the Chairman of the National Assembly, Dr. Mohammad Mohsin, is the first ever woman Army officer to be appointed to the Medical Service from the Muslim community. There arose some confusion as to how she should take her oath of service and secrecy of the Army; so, it was advised that she do so by touching the Army flag.

Assistant Minister Rabi Bhakta Shrestha hosted a dinner at his residence for the Ambassadors of India, America, Japan, and Israel, the Secretary General of the SAARC, the chief of DFID, and other high diplomats of the Embassies.

Also present at the dinner were the Minister of Information and Communication, Ramesh Nath Pandey, the Minister of Finance, Dr. Badri Prasad Shrestha, the Minister of Physical Planning and Construction, Narayan Singh Pun, former CoASs, the Royal Palace's Principal Secretary, Palace ADCs, and others, including myself.

The others present at the dinner were the industrialist Prabhakar SJB Rana, royalist Prabhu SJB Rana, tycoon Basant Chaudhary, and publisher/editor Kanak Mani Dixit.

In the series of their visits and consultations with distinguished individuals and personalities, the Maoists also are seeking to visit and consult with the diplomats. The US Ambassador to Nepal has made it clear that he was discussing it with his British and Indian counterparts on whether or not to meet them. If agreed, then what should be the points included for discussion in the meeting.

The Japanese Ambassador to Nepal tacitly observed that there isn't any question of the Envoys meeting the Maoists as their countries fully support the Nepal Government and not the Maoists.

Tuesday, April 8, 2003

Peace talk facilitator and Minister Narayan Singh Pun told me that he found a Maoist leader, Dr. Baburam Bhattarai, quite different from his image in the public and the press: he's rather unpolished in his talk and crude in mannerism.

Wednesday, April 9, 2003

The King is scheduled to visit the Western Development Region where, among other programs, he'll also inspect the Divisional Headquarters. He'll stay in Nepalgunj.

I made arrangements for the dinner to be given by the King for the Army officers at the Nepalgunj barracks, the royal dinner being a tradition long held in the Army.

Today, I presented the file to the King for the promotion of Brig. Gen Tika Dhamala for the second time, as he's due for promotion to Major General. If he's promoted, Dhamala is to be sent for a course at the National Defense College in the UK, the US, or India.The King remained silent, this time again, without bothering to respond to my queries.

I've received information that Bijay Kumar Gachhadar of the NC (Democratic) and Radha Krishna Mainali of the CPN (UML) are wishing to meet me, perhaps to discuss the current political situations in the country.

Princess Prerana's vehicle hit a taxi at Kalimati. It's advisable that its driver/owner be given due compensation for the damage, and prevent the news of the accident from reaching the media that overly sensationalize such things when they get one.

There have been sporadic protests and demonstrations in the capital and across the country against the government's recent decision to raise the price of fuels. The situation in Butwal grew tense and went out of control for some time as a student protesting the price hike was killed in a clash between the police and the demonstrators, and in order to bring the situation under control, a curfew was immediately declared by the District Administration. The students held an employee of the NID, and the police had a hard time to set him free from the clutches of the students.

In response to my advice for the King to visit and inspect the Headquarters of the Nepal Police, the APF and the NID, he said that was a good idea but the government lately appeared weak in maintaining security.

Expressing his worries at the rampant poaching of the rhino and other wildlife, the King gave me express instructions to mobilize, if need be, even the Companies of the Bhairavnath Battalion to control it instantly. He told me to take tough measures to identify the individuals involved in the smuggling of leopard skin and bring them to book.

Next, he asked me to quicken the process of investigation of the murder of the IGP of the APF, Krishna Mohan Shrestha, and proceed to take legal actions against the culprits.

His next concern for an urgent control was the thuggish activities of the Maoists' sister organizations.

The King congratulated the newly appointed IGP of the APF, Sahabir Thapa, with the hope that he will work with foresight and fortitude as his appointment has come at a time when the post was lying vacant, following the murder of IGP Krishna Mohan Shrestha.

On the occasion, the new IGP outlined the plans and strategies of the APF. The plan includes an approach to the JICA for the construction of a separate hospital for the AFP's wounded, and the establishment of separate schools for the children of the police personnel serving in the APF.

He also advised the King that it would be better for the APF to remain in the NSC, to prevent it from the same fate of the Nepal Police – chronic politicization.

Thursday, April 10, 2003

The CoAS, General Pyar Jung Thapa, invited a couple of Ambassadors to a dinner at his residence. On the occasion, recounting the unprecedented economic growth, stability, and development in Pakistan following the Army's takeover, the Pakistani Ambassador said the same opportunity has presented itself in Nepal, and which the country should utilize by any means. The National Accountable Bureau in key places in Pakistan made it possible for the government to effect quick, free, and fair delivery of services; he elaborated and narrated the instances of the success stories in his country.

I briefed the King on it.

Coinciding with the Mass Movement Day, the students' organizations of the major political parties took out demonstrations. The leader of the Nepal Free Students Union, Kishor Singh Rathod, led the demonstration. It's known that the demonstration is funded by former Ministers Khum Bahadur Khadka and Govinda Raj Joshi.

According to a source from the NID, the old and senior republican leader, Ram Raja Prasad Singh, who has lived in exile in India for a long time, has quietly come to Nepal, and has held consultations with the leaders of the political parties to work for a common agenda, such as holding the proposed general elections to the CA and the establishment of a federal democratic republic state in Nepal at a time when the government-Maoist peace talk has just started.

It seems that the old leader hasn't come to Nepal for nothing but as an emissary of India, as it seems. It's apparent that India isn't happy about the peace talks going on without its involvement in it, and it wants to derail it under any pretext.

Friday, April 11, 2003

In the wake of the intense protests and demonstrations against the price rise of fuels, the King rang me to consult on ways and means to diffuse the protest by having the three Cabinet Ministers concerned fired. He also wanted me to consult with the heads of the security forces regarding the matter. I promised to inform him once I consulted with the Secretaries of the Royal Palace, and accordingly, described the newfangled situation to them. In an impromptu meeting, we Secretaries discussed all the alternative ways, reaching a decision to have the Minister of Education, the Minister of Home Affairs, and the Minister of Supply resign.

The emergency Cabinet meeting has taken the decision to release Maoists Krishna Dhoj Khadka and Rekha Sharma, a demand the Maoists have been persistently making since the beginning as a precondition to the peace talks. Their release also looks right on humanitarian grounds at a time when Khadka's father has expired in his home district, Pyuthan.

Narayan Singh Pun telephoned me to say that the rise in the price of petroleum goods is an 'invitation to complications' at a time when the Cabinet has remained ineffective throughout its tenure.

I told him to request the Maoist Party to instruct its students' wing to immediately stop their demonstration at this hard earned and opportune environment of dialogue and negotiation.

Saturday, April 12, 2003

In view of the above reasons, the Maoist leader, Krishna Dhoj Khadka, was duly released as a positive gesture displayed by the government, hoping it to be contributory to the creation of an environment of trust for dialogue.

I told Pun that now that the government has released their two leaders, it was the Maoist Party's turn to display similar flexibility and positive gestures to ask their students' wing to dissociate themselves from any form of destructive activities, such as protests and demonstrations.

~~~~~

Amidst the second round of the peace talks, and citing the noncooperation of the political parties as the main reason, Lokendra
~~~~~

Bahadur Chand resigned as Prime Minister, leaving the space as though made vacant for the convenience of Surya Bahadur Thapa to succeed as PM.

Thapa took office with his vow to bring either the Nepali Congress or the CPN (UML), or both, to the government.

Soon after, however, the table turned against him, as the hard-earned peace initiative failed, let alone the parties joining the government.

I was appalled to hear what an officer of the APF told me: That Nepal's Maoist guerrillas were trained in Chakrata of India.

But I left the Royal Palace before I could learn further details of it.

Monday, April 14, 2003

To celebrate the New Year's Eve of the Bikram Sambat (calendar), I hosted a dinner at my place where only my relatives and close friends were invited.

In an elated and jubilant mood of my promotion to Lt. Gen, I wore the insignia of the rank from the hands of my wife, Geeta, and Brig. Gen Gajendra Limbu, as there's no tradition of holding any official program to offer insignias to the newly promoted officers of the ADC and the Military Office of the Royal Palace at the RNA Headquarters.

I vividly remember donning the insignia of Major General from the hands of Crown Prince Dipendra, and that was a long time ago.

To mark the New Year, a dinner was hosted, followed by a cultural show, at the Tribhuvan Army Officers' Club of the RNA Headquarters.

Their Majesties graced the occasion.

I sensed that the Royal Palace had regained its strength because many VIPs seemed desperate to have an audience with the King.

Only a few RPP leaders were present at the celebrations while the leaders of the other political parties were absent.

It all means something significant.

Wednesday, April 16, 2003

At my next regular briefing to the King, I thanked him for having promoted me and informed him of Khum Bahadur Khadka visit to the PM with his earnest promise of cooperation on the condition of the PM's reconsideration of the corruption charges slapped against him.

Sources say that the Secretary at the Ministry of Home Affairs, Tika Datta Niroula, clandestinely met the representative of the Dalai Lama. The PM was angry when he heard about it. He's in the mood to consign the Secretary to the Reserve Pool when the time comes for his transfer.

There's also speculation of Khum Bahadur Khadka secretly meeting a representative of the 'Free Tibet' Campaign.

Gopi Hamal, a businessman from Dhangadhi, rang me to say that the King's visit to the region has left good impressions on the people from all walks of life.

A group of intellectuals has reportedly advised that Nepal adopt the Pakistani model of politics and development.

There's information of Ram Raja Prasad Singh meeting and holding confidential talks with GP Koirala.

Narayan Singh Pun is requesting me to arrange an audience with the King.

It's important that leaders like Bam Dev Gautam, Radha Krishna Mainali, Bijay Kumar Gachhadar, and Shailaja Acharya be brought together to form a third alternative political force, as they drastically differ from the other leaders of their respective political parties in their conviction.

It seems that no moment is as opportune as now to make efforts to break away for some CPN (UML) leaders, as the Party suffers from an economic crisis.

A serious thought needs to be given to soothe the occasional noise that comes on the surface regarding the irregularities seen in the procurement process of the RNA.

The King gave me instruction to find ways on how the irregularities in the Army's procurement procedures can be curbed and controlled for once and all.

He also wanted me to unmask and bring to public notice activities of some of the political leaders by expediting the process of the CIAA's investigations into the corruption cases of the political leaders.

Thursday, April 17, 2003

The appointment of the former Chief Secretary of the Nepal Government, Tirtha Man Shakya, as Ambassador to Japan will soften the complaints of preferences given only to the Ranas and Thakuris when it comes to ambassadorial appointments.

Active as a facilitator in the government-Maoist peace talk and well known to the DMI of the RNA under Brig. Gen Dilip Rayamajhi, Dadhiram Lamichhane), revealed the fresh Maoist gesture in the form of their willingness to lay down their arms once the King grants them audience and talks in person to the Maoist Supremo, 'Prachanda' or leaders, Mohan Baidhya, or Ram Bahadur Thapa.

Terming the other Maoist leaders, Dr. Baburam Bhattarai and Matrika Yadav, as unreliable, facilitator Lamichhane reassured that the Maoist Students' Wing, ANNFSU (Revolutionary), will withdraw itself from the demonstrations once the government makes public the whereabouts of its four disappeared student leaders.

He further disclosed that the Maoists were looking forward to meeting the CoAS and the Military Secretary (me). Lamichhane added that the Maoists were looking for a face-saving way to lay down their arms, drop their demand for elections to be held to lead to the CA, and accept monarchy as a guardian figurehead once the political agenda are settled.

Analysts say that the political parties are increasingly losing their shine and hold on the people; they are carrying out their protests and demonstrations merely to engage their cadres and prevent them from quitting the parties while causing inconveniences to the lives of the common people with their routine disturbances. Analysts further believe that the success of the peace talk is sure to weaken the parties' bases.

Narayan Singh Pun said he's saddened by the negative comments now being made about the status and the future of the peace talks. He defends the ongoing peace talk when asked about its future by foreign friends. Appreciating his role, the King wants him to continue with the mission.

If the fresh piece of information is any indication and guidance, it'll be no surprise if the veteran leaders of the CPN (UML), Radha Krishna Mainali and Bam Dev Gautam, fed up with the party line, quit it whereas another leader, KP Oli, is intent and active to weaken other leaders in the Party.

Sharat Singh Bhandari, the Madhesh-based leader of the Nepali Congress, met me and disclosed that Madhav Kumar Nepal, who is behind the ongoing demonstrations by students, wishes the environment turned to their favor with bullets fired by the police so that the situation will grow edgier for the movement to regain its momentum.

Speaking at a reception to mark the birthday celebrations of the Danish Queen, the US Ambassador to Nepal, Michael E Malinowski, reportedly asked GP Koirala what made him oppose the ongoing peace talk. The source says Koirala remained silent when the Ambassador asked him about the immediate priority of his Party and the country.

The Crown Prince reportedly visited a couple of nightclubs last night where he behaved in ways which are unbecoming of a royal. Such unsocial conduct doesn't at all convey good messages about the monarchy at a most sensitive time such as this.

The Minister of Home Affairs and the IGP of the Nepal Police are going to express condolences and provide relief to the family of the student who was killed during the Police-students clash in Butwal.

In an interview given to the *Kantipur Daily*, Professor Dr. Surya Subedi of the University of Leeds, England, admired the King's step as a necessary and timely move, taken in the interest of the people and the country.

The King hoped for the ongoing peace talk to reach its logical end. The best course of action, for now, would be to have some media houses unmask the leaders of the political parties' intent on foiling the peace process. The Maoists, on their part, would do well to contribute to peace rather than develop links with the political parties and flirt with unattainable goals.

It's important that leaders with mass base join the government and concentrate on rehabilitation and reconstruction in all the five development regions, the King observed.

Saturday, April 19, 2003

Worried over the students' protracted demonstrations, the King gave me instructions to find ways to end the disturbances even at the cost of closing down the students' hostels and campuses, and accordingly issue directives to the Registrar and Deans of Tribhuvan University to this effect.

Sunday, April 20, 2003

The Minister of Home Affairs accused former Ministers Khum Bahadur Khadka and Devendra Raj Kandel, and a former Secretary, all three of amassing huge wealth by unlawfully issuing citizenship to Tibetan nationals. More facts should come to light for everything to be clear. It's being increasingly felt that while the people are fast losing faith in the government, the South Block and the Indian Intelligence Agency, RAW, are pushing the political parties to agree to the agenda of holding elections to CA.

Charging that the CIAA is harassing him unnecessarily, industrialist Piyush Bahadur Amatya is inciting the entrepreneurs and businessmen of the Newar community to rise against the monarchy.

Should the peace talks fail, the cause of the failure must be directed at the Maoists, or the government must also be liable to blame. If necessary, the PM must resign. In such a scenario, the Royal Palace must engineer its timely strategies with due consultations with the security forces. Another option would be to have a government formed by a political party with the grassroots base. The exploration of other alternatives must also continue.

By now, it's no surprise that the US and India have identical security concerns regarding Nepal, the former looking at Nepal through the latter's lens.

These were some of the points the Cell Committee meeting discussed. We decided to submit them to the King.

With uncertainty looming large on the peace talks, the King commanded talk team facilitator Narayan Singh Pun to report on the progress made and the complications seen in the peace talks.

Pun described the role of the Minister of Communication and Information, Ramesh Nath Pandey, as negative, giving rise to an environment of distrust in the continuance of the peace talks.

Responding to the King's idea of an all-party government, Pun said he advised the King that it would be a bad option.

Regarding the Maoists impatiently wishing to meet the King, Pun made it clear that unless and until the Maoists accept the monarchy, there would be no question for the King granting them an audience.

Tuesday, April 22, 2003

Today, Radha Krishna Mainali, a senior leader of the CPN (UML), came to my house to speak his mind. He said he was ready to help the King with any role he found fit for him. He added that he was waiting for the King's response to his earlier proposal, that of forming a third political force as an alternative to the political parties which now are held in contempt by the mass for the fibers they have displayed. According to him, the expansion of the Cabinet and the induction of the likes of Rajeshwor Devkota, Shailaja Acharya, Tara Nath Ranabhat, Marich Man Singh Shrestha, and Kirti Nidhi Bista would convey good messages to the people. He also made it clear that the present government won't last beyond its first month.

To let go of the available opportunity would be tantamount to weakening the monarchy and strengthening the Maoists, according to Mainali: As the ongoing agitation called by the political parties can't gain public support, the best course of action, for now, would be to expand the Cabinet with leaders of clean image.

Peace talk team facilitator Pun rang me up early this morning. He said he had invited the facilitators and interlocutors of the Maoist talk team, Krishna Bahadur Mahara and Ram Bahadur Thapa, to dinner at his place yesterday where, according to him, they discussed at length the progress made and experiences gained at the talk meetings. Pun quoted the Maoists as saying that since the King and the Maoists have more or less identical vision and mission, why not sit and discuss the plan for the future course of action? They also said that the present political model and polity already prove a failure and is defunct now; The Singapore model of politics would best suit the soil of Nepal.

It seems that the Maoists are speaking their mind. I described to the King what Radha Krishna Mainali and the Maoists said. He said he would consider giving them an audience later once the peace talks reached an irreversible point of success.

Wednesday, April 23, 2003

According to a news received from the RNA Headquarters, in an effort to remove the landmine set by the Maoists on the repeater tower of the Mahadev hillock, Capt. Rupendra Raj Pandey sustained serious injuries as the landmine blew up.

Sources say that around one thousand people attended the cultural program hosted by the Maoists in Dhangadhi.

The King today gave his approval to send Brig. Gen Tika Dhamala to pursue a course at the National Defense College in India.

There's information that the dialogue between the TU Administration and the Students' Organization ended inconclusively as the students didn't budge an inch from their position.

As is suggested earlier, students' organizations are demonstrating against the government in protest to the price rise, apparently an excuse to turn it into a mass movement.

Speaking to the Nepali Congress Central Committee Members, District Committee Chairmen, and student leaders, GP Koirala spelled out his newfound conviction that the monarchy is the source of the political instability in Nepal.

According to Pun, the peace talk, on hold at present, will resume when the Maoist talk team member, Dr. Baburam Bhattarai, arrives in Kathmandu.

The Students' organizations have called for the closedown of the Kathmandu Valley today. There's news of coercion by students to shut down shops, and they vandalized vehicles. Students found pelting stones at the police were arrested from the Pashupati Campus at Chabahil.

According to a reliable source, Maoist leader Rabindra Shrestha quietly visited GP Koirala. The purpose of the meeting and the contents of the dialogue have been kept top secret.

The Indian Chief of the Army Staff, Gen. Nirmal Chander Vij, reportedly advised CoAS Pyar Jung Thapa to increase the current strength of the RNA and modernize it accordingly.

The General also asked the CoAS to cooperate with India to allow it to operate a pension camp in Butwal to ease the hassles of pension distribution to the Indian Army's Gurkha retirees in that area.

The CoAS, in turn, said that the RNA has planned to operate a bank in a joint venture to utilize the money of the Army Welfare Fund which, when it comes into operation, would address the problems, if any.

In his visit to PM Chand, the Indian CoAS expressed his grave concerns over the recent increase of the ISI activities in Nepal and asked for immediate stringent measures to control it.

He also disclosed to PM Chand India's plan to deploy its paramilitary border security force, the SSB (Seema Surakchha Bal), along the Nepal-India borderlines.

Thursday, April 24, 2003

The merger of the Scholars Council and the Ex-Army Organization is sure to strengthen the World Hindu Federation which, in turn, helps Nepal retain its identity as the only Hindu nation in the world.

According to Pun, Ramesh Nath Pandey wants to steer the peace talk singlehandedly as though he were the sole champion of dialogue and negotiation with the Maoists.

Friday, April 25, 2003

The British Military Attaché, Col. Sutherland, who came to see me in my office, expressed his hope that Nepal will make every effort to steer the peace process to its success. I assured him of every effort to make it happen.

Turning the conversation to another subject, the Military Attaché disclosed the plan of British Crown Prince Charles, along with his eldest son, Prince William, to visit Nepal in or around March. Prince William, soon to finish his graduation from Sandhurst and join the Welsh Guards, will have an additional advantage of learning about Nepal and the Gorkhalis once he's here.

The Prince of Wales is the Colonel in Chief of the British Gurkha Rifles.

The schedule of the royal visit is yet to be confirmed mutually through the diplomatic channels.

The CoAS of the RNA hosted a dinner in honor of the Indian CoAS where the General expressed his assurance of every possible Indian help to the RNA.

The Indian CoAS admired the valor of the RNA soldiers in fighting the insurgency despite the inadequacy of arms and resources.

The PM reiterated the need for help from the Congress Party for the peace talks to succeed.

In a series of calls on the top leaders, the PM visited Bam Dev Gautam who assured to have the CPN (UML) join the government provided they are offered important portfolios.

The facilitators and interlocutors of the Maoist talk team, Dr. Baburam Bhattarai and Krishna Bahadur Mahara, were met with an icy reception at the entrance to the US Embassy where they were subjected to minute search.

On receiving the Maoist leaders, the Second Secretary of the Embassy asked them to display their honesty by making public the culprits involved in the killing of the IGP of the APF.

It's clear to all that the US adopts tough measures against terrorism of any form in any part of the world.

The US official's speech was enough indication that the US wasn't happy about India's role in Nepal. The US corresponded with India for the apprehension and extradition of 'Prachanda' to Nepal but India professed to have no knowledge of that man being in India, the American Embassy officer complained.

Monday, April 28, 2003

The leaders of the political parties have constantly been vocal critics regarding the role of the RNA. It isn't the first time they said that the Army must be under the elected government.

Bharat Mohan Adhikari, a leader of the CPN (UML), believes as though an all-party government led by the CPN (UML) will be the cure for all ills. He said if the other parties can't agree on Madhav Kumar Nepal as PM, they were open to other options, such as the House Speaker, Tara Nath Ranabhat, or even Surya Bahadur Thapa, or the leader of the Nepali Congress Shailaja Acharya.

Adhikari also claimed his ability to negotiate with the other parties for certain rights reserved for the King if he entrusts him as PM.

On the CoAS's recommendation of conferring the Army Medal upon the US Assistant Military Attaché, Peter Fowler, the King has given instruction to forward the process to the Awards Committee of the Army.

The Chinese Ambassador to Nepal asked me if I could arrange an Army helicopter for his trips to Jomsom and Muktinath this time again. The same had been arranged for him earlier too. When briefed on it, the King gave his consent.

Tuesday, April 29, 2003

A couple of days ago, His Majesty the King had asked me to look into the possibility of finding employment for two engineers, Keshav Prasad Khatiwada and Manoj Kumar Acharya, with the RNA. That seems impossible given the age limit for recruitment in the RNA. So, I advised the King that perhaps the two candidates could fit into the vacancies at Surya Nepal. He asked me to approach Prabhakar Shumsher to that effect.

The Indian Ambassador to Nepal this morning went to see the NC President, Girija Prasad Koirala, who subsequently met former PM Surya Bahadur Thapa. The contents of the discussions have been kept confidential.

Wednesday, April 30, 2003

As per the King's consent and instruction, I went to meet the Maoist leaders, who were also the facilitators and interlocutors in the talk team, in the office of the Karnali Air, the airline owned by the facilitator of the talk team, Narayan Singh Pun, in Sinamangal, close to Tribhuvan International Airport.

The Maoist leaders, Dr. Baburam Bhattarai, Krishna Bahadur Mahara, and Dinanath Sharma arrived at the appointed time and soon our discussion began.

I explained to them that their wish to meet the King in the present context won't send out any good message, but still I assured them to convey to the King whatsoever their anxieties, confusions or concerns were.

Of course, the contents of this meeting would remain classified except for briefing the King on the points discussed.

I said that since the King and the Maoists had identical visions and missions in the interest of the people and the nation, a merger of their plans and visions was still possible and these national sentiments could be combined to preserve Nepal's sovereignty and maintain its integrity from the influences of external elements.

I emphasized that the best course of action, for now, would be to let the peace talk achieve its success and form an all-party government, including the Maoists.

I also reminded the Maoists that the recent arrival in Kathmandu of Prof. SD Muni and the Republican leader Ram Raja Prasad Singh from India simultaneously can't have been a mere coincidence. Such happenings will by no means help create an environment of trust, I explained.

Relating the series of events since their launching of the 'People's

War', Ram Bahadur Thapa, alias 'Badal', said they had come to the peace talks with utmost seriousness, honesty and flexibility by keeping the people and nation above all else.

He also didn't miss to remind the government not to misinterpret their flexibility as weakness.

However, 'Badal' expressed his team's dissatisfaction and confusion over the composition of the government talk team: Of the two Ministers, he asked, who was the real authority to represent the government at the peace talks – Narayan Singh Pun, or Ramesh Nath Pandey? Though apprehensive about Pun in the beginning, Badal seemed to trust him now with more confidence.

When Pun and I briefed the King on every detail of the fresh consultation with the Maoists, he instructed us to remain in contact and engage them in the talk to convince them into accepting the indispensability of monarchy for the country.

Expressing his ascertained futility at the dragging of their feet by the political parties, the General Secretary of the CPN (UML), Madhav Kumar Nepal, today disclosed that there would be no objection on the part of the CPN (UML) for the King to head the government himself.

I heard a rumor from within a CPN (UML) circle that the Military Secretary (that's me) was quietly working to cause a splinter in the party. When I asked about it, the General Secretary laughed it away, saying the Party wasn't so weak as to be split just because someone wished or dreamt it to happen.

The PM's visit to Karna Dhoj Adhikari and Shailendra Kumar Upadhyaya – two principal functionaries of the erstwhile Panchayat System – was followed by another meeting between the former Speaker of the House, Daman Nath Dhungana, and Dr. Baburam Bhattarai, Padma Ratna Tuladhar as well as the duo of Adhikari and Upadhyaya.

The CoAS expressed his conviction that it's important for the Army to continue its regular functions, and concurrently keep a close watch on and surveillance of the activities of the Maoists.

Friday, May 2, 2003

In an informal meeting in the house of Dr. Dilli Raj Khanal of the CPN (UML), and my question to the Party's General Secretary, Madhav Kumar Nepal, to clarify their position on the monarchy, Dr.

Khanal, on behalf of the entire Party, asked me not to doubt their commitments to monarchy. To this effect, he even offered to make the manifesto of the Party available for me to study it and see for myself.

I asked the General Secretary if his recent inflammatory speech delivered against the monarchy didn't contradict his wish to see the King heading the government, he took my question easily, openly saying that one's party line and speechmaking at mass meetings are different things.

Sunday, May 4, 2003

Sources say that the Maoist talk team visited GP Koirala yesterday at Shriharsha Koirala's house, and shortly after the consultation with him, visited the General Secretary of the CPN (UML) at his residence in Koteshwor. It's learned that they held consultations on arranging a joint mass meeting.

Asked for what purpose were the Maoist leaders visiting the leaders of the political parties, talk facilitator Narayan Singh Pun quoted them as saying that there was no possibility of their going together with the political parties until they agreed on their (Maoists) three-point proposals – 1) a roundtable conference, 2) an all-party government, and 3) holding general elections to the Constituent Assembly (CA).

The mass meeting attended by some 10,000 people concluded with incendiary and objectionable speeches made against the monarchy by the leaders of the political parties.

It's advisable and important that the government introduce strong strategies to unmask the leaders' double standards by having the intellectuals and authors write articles on their misrule and misdeeds for years.

It's also equally timely to reshuffle the Cabinet with a clear roadmap for holding general elections within one and a half years, conclude the peace talk successfully, and bringing the Maoists into the mainstream of competitive politics.

The reshuffled Cabinet should include individuals from the political and intellectual spheres to have the people feel the fairness in the delivery of services to them.

There have been complications in installing the telephone monitoring system equipment provided by India to the NID. Technicians have been brought all the way from India and it's hoped that the system will come into operation soon. Once the system starts functioning, the NID will have its efficiency substantially enhanced.

Wednesday, May 7, 2003

The Royal Palace's 'Situation Center' has been set up and which has already started its operation in the office of the Principal Military Secretariat with the tacit objectives of preparing matter of factly analyses based on the pieces of information received at the Center from the intelligence office about the activities of the political parties.

The objectives of the Center also include the retrieval of information about the activities of the leaders, analysis of the current state of affairs prepared by Nepali and foreign analysts, all with a view to briefing it in a concrete manner to the King.

The Situation Center has its own strong belief that the majority of the broadsheet dailies (newspapers) are operated too commercially, and many of them are critical of the monarchy due to the publications being chiefly guided by cheap popularity rather than inclined to the dissemination of decent information to the people.

The Center is to attract investors to operate its own media to counter the other broadsheets with fresh tastes and feels of professionalism to disseminate the right information to the common populace.

Facilitator and interlocutor of the Maoist talk team Ram Bahadur Thapa's press statement came seemingly in support of the political parties.

Given the demand for manpower in South Korea, Nepal would do well to send Maoist victims from the government quota for foreign employment.

Simultaneously, there have been complaints for action against Moon Drops Overseas that supplies manpower to South Korea.

It's better for the Army to employ ex-army personnel in its Central Mess System and place its trained soldiers on active duty instead of deploying them in the mess halls of the Army units. It's a proper utilization of an important resource of the Army: its serving soldiers posted at their appropriate work stations.

The introduction of a central mess system for the soldiers serving at the Royal Palace is a glowing example of this kind, and it has received great reviews both from soldiers and civilians being catered to in the dining areas.

Asked as to why he gave his contrarian press statements, Ram Bahadur Thapa 'Badal' said his intentions were misinterpreted by the press, a typical safe escape many leaders resort to nowadays.

The political parties are making a mountain out of a molehill over the slight increment in the budget allocated for the Royal Palace.

Leaders of the CPN (UML) made very critical remarks against the monarchy.

Dr. Prakash Chandra Lohani of the RPP shared with me his subtle analysis of the current political situation. The analysis sounded correct as he didn't just say the PM in office was incapable to deal with the problems but also offered alternatives: have Madhav Kumar Nepal or Surya Bahadur Thapa to head the government.

He hinted for himself as a claimant to that portfolio although he didn't say so in so many words.

Khum Bahadur Khadka begged the PM to have the action taken by the CIAA against him on corruption charges reconsidered.

He also sounded very critical of GP Koirala.

According to information received, irate and agitating students burnt the effigies of the King at several places in the Kathmandu Valley.

Thursday, May 8, 2003

It's heard that the former Chief Justice, Keshav Prasad Upadhyaya, who looked into the Royal Palace Massacre as the Chief Commissioner of the Panel of Investigation, has remained dissatisfied and upset for some time now for not having his contributions evaluated.

Although there were speculations on his possible Ambassadorship to France, he didn't bag such an appointment.

Asked to find if the Royal Palace possessed land in his command areas, the Commander of the No 4 Brigade discovered that there were 34 bighas of land in Latikoili and Jhupra registered in the name of the Royal Palace.

The RNA also found that the Royal Palace owned large plots of land in Tarahara of Sunsari, Dhankuta and Chisapani as well.

In his response to my suggestion that transfer of those lands be made to the Army to set up its camps, and the royal decision would look commendable to the people, the King instructed me to produce the documents of ownership so that he could consider doing so later.

Members of the Maoist talk team, such as Dr. Baburam Bhattarai and Matrika Yadav, quietly held their consultations with the Deputy Chief of the Mission at the Indian Embassy, Ashok Kumar, where the latter expressed his grave concerns, whether the Maoists had links with the terrorist groups of India and those of Pakistan.

Dr. Bhattarai made every effort to clarify that the Maoists were a political force, and therefore there arose no question of any nexus of theirs with any terrorist groups.

He also took the opportunity to be critical of America for its intention to sabotage their revolution.

He also expressed his displeasure at India for arresting some of their leaders and deporting them to Nepal.

Sources say that the Indian Ambassador to Nepal expressed his displeasure at GP Koirala for backbiting him with the US Ambassador to Nepal.

Dr Ram Sharan Mahat of the Nepali Congress has come heavily upon the government for its utter failure to maintain law and order for peace in the country.

The former Mayor of the Birgunj Sub Metropolitan City, Bimal Prasad Shrivastav, asked me to brief the King on the accomplishment of his share of work he was assigned.

As referred to earlier, he was asked to keep in touch with the leaders and intellectuals of the Nepal-India border areas.

He elaborated that he had established a channel of communication with Dr. Madan Jaiswal, an MP of Betiya in Bihar from the BJP (Bharatiya Janata Party). The Indian politician wants to visit Kathmandu and worship at the Pashupatinath Temple. Shrivastav advised that an audience with the King for the Indian politician would be a capital idea.

As a gesture of reciprocity to the Maoists' demand, and with a view to building mutual trust and confidence, the government is about to release Maoist leaders Muma Ram Khanal, Rabindra Shrestha, and Bamdev Chhetri from Army detention.

But the Maoist demand that the Army limit its activities within the perimeters of 5 km from the nearest barracks is absolutely illogical and unacceptable.

I asked Pun to convey this position to the Maoists.

Asked about conferring on Prem Kishor Singh, a Nepali national working as defense specialist in the office of the Military Attaché at the American Embassy, with a medal, the King asked me to consult the Medal Award Committee and do as per its norms.

Singh is taking his retirement soon after 35 years of service at the US Embassy.

Monday, May 12, 2003

Taking into consideration the likely torch rally to be held today by the political parties, and in order to avert any untoward incidents, the government had circulated a firm notice to the general people to stay away from any such disruptive activities.

The police exercised their restraint to a great extent but took to light baton charges on the demonstrators as they pelted stones at the security personnel.

There's news of some leaders receiving minor injuries in the scuffle.

Likewise, as sources say, some demonstrators were taken under control in Salyan and Nepalgunj and released soon after.

The Honorary ADC, Maj. Gen Bharat Keshar Simha, today asked me to arrange an audience for the former PM Surya Bahadur Thapa with the King, as the latter so wishes for an important briefing.

Likewise, Shailaja Acharya of the Nepali Congress also has sent me a message to arrange her audience with the King.

The *Ghatana Ra Bichar* and *Tarun* weeklies have published fake and misleading news regarding relations between Sharad Chandra Shah and the Military Secretary (me!). The publications describe my involvement in the confidential meetings held at the residence of Damodar SJB Rana.

According to Narayan Singh Pun, he met the CoAS at the No 3 Brigade in Pokhara where they discussed the future of the ongoing talk and the necessary preparedness to make it successful and to contribute to peace.

With the political parties agitating and demonstrating, leader of the Nepal Peasants and Workers' Party, Buddhi Goshain, remarked that if this is like what's going to happen, the situation is likely to slip off our hands to a dreadful uncertainty: there's every risk of the very mass movement going out of track and control.

Wednesday, May 14, 2003

Narayan Singh Pun mentioned his sheer dissatisfaction at the Cabinet's noncooperation for the peace talk. He said he was thinking of resigning the responsibilities given him.

The process of operating an offset press and bringing out a professional and balanced newspaper with the investment of the Royal Palace has gone well ahead.

At a time when the demonstrations held by the political parties have become fickle and ineffective with low turnouts of participants, Govinda Raj Joshi of the Nepali Congress has conveyed his regrets for flirting with the Communists to steer the demonstrations jointly with them. He added it was unwise of the seasoned leaders to trust them.

As if tired and hopeless of the protracted and routine demonstrations, leaders, including KP Oli, seem to term the movement useless and irrelevant in the present context.

There's news of a clash between the Maoist and the CPN (UML) cadres, the first accusing the latter of intending to derail and foil the peace talks.

Arranging a press conference, the leaders of the Maoist talk team elucidated their rationale for limiting the Army within the perimeters of 5kms from the barracks.

A new report has revealed that the CPN (UML) is having a crisis as its cadres are joining the Maoists in big numbers.

Political leaders seem to have reached the conclusion that as long as the Maoists are engaged in the peace talks, the success of their mass movement is nowhere in sight.

Monday, May 19, 2003

Talk team facilitator Narayan Singh Pun called me today to say that there's no need for the Army and the government to be scared by the Maoists' position to restrict the Army within the 5km perimeters of the barracks as it was their tactical ideology. He further observed that they declared the demand just to create the impression that they aren't weak, or just to save their face.

Pun also added that, as asked by the PM, he approached the Japanese Ambassador to Nepal to construct the Kathmandu-Hetauda Highway and to name it King Gyanendra Highway. He said the Ambassador was positive about it.

According to a source of information, Maoist leader Matrika Yadav, one of the Maoist facilitators and interlocutors at the peace talks, behaved rudely at the Army checkpoint at Kapurkot while heading to Salyan and then to Rolpa for the Central Committee meeting.

Orders have been issued for the Army to stay alert and keep their surveillance on the Maoists' activities in these regions.

With the Maoists sitting for dialogue without laying down their arms, the general public is skeptical about the future of the peace talks.

A senior leader of the CPN (UML) disclosed the Party's secrets, indicating a deepening factionalism therein. He said Ishwor Pokharel and Pradip Nepal have earned disrepute as corrupt leaders in the Party, and also suggested actions against them by the CIAA on corruption charges. He disclosed that leaders like KP Oli and Radha Krishna Mainali would help such investigations.

The government's introduction of the system of transfers, postings, and promotions to the Secretaries of the government on the basis of their merits, seniority and performance will create very good impressions in the public, that the government is committed to a system of meritocracy in the bureaucracy.

I've got a tipoff that Dr. Baburam Bhattarai is set to head for Delhi soon to consult with the Indian PM's advisor, Brajesh Mishra.

There's information that the Maoists have planned mass meetings in major places across the country, presumably to publicize the exact status of the ongoing peace talk. Top leaders are going to address the mass meetings in those places.

Wednesday, May 21, 2003

At an informal meeting with me, Prem Raj Karki, a Joint Secretary at the CIAA, frankly admitted to his dissatisfaction at the CIAA for dillydallying in the investigation of the involvement of political leaders in corruption. He also disclosed that the Chief Commissioner of the anti-graft body has been influenced by the CPN (UML). He sounded greatly distressed, saying, 'Forget about stringent actions against the corrupt by the CIAA when judges themselves are bogged down in corruption.'

It's known to all that taking into consideration the sensitivity of its security, the pedestrian sidewalks along the Royal Palace had been blocked for public use. Now that the construction of the Observatory Post is nearly complete, it's now time to lift the restriction. As is being proposed and received positively, it'll be better to replace the Nepal Police by the APF in the outer perimeter of the Royal Palace.

Consulting with the heads of the security forces and the members of the expert team, the PM gave direction to the government talk team to consult the expert team before the beginning of the formal dialogue with the Maoists on the important agenda.

Prof. Dr. Mohan Prasad Lohani, one of the members of the group of eminent persons to facilitate the peace talks, has expressed his inability to remain on the team.

I learned that the PM has consulted with the Nepali Congress leader, Ram Chandra Poudel, at his residence and asked him to cooperate to remove the long-drawn political impasse.

Thursday, May 22, 2003

The fourteen-member Nepali and Indian joint expedition team comprising four Nepal Army soldiers, three Indian Army soldiers, and seven Sherpas scaled Mt. Everest at 3.15 a.m. today.

The CoAS is arriving at the Everest Base Camp to express his congratulations to the team for their feat and victory.

Two Lancer helicopters provided by India to Nepal as assistance landed on the Tribhuvan Airport. The date and time for the formal handing over of the helicopters are yet to be finalized.

Friday, May 23, 2003

Khum Bahadur Khadka, in an informal meeting with me, repeated his old grievances against the CIA for allegedly being vindictive towards him. He asked me if the King would be kind enough to send a signal to the Judge at the Appellate Court, Tapa Bahadur Magar, to liberally handle the case of corruption charges labeled against him.

He also advised that it would be better on the part of the King to appoint Sher Bahadur Deuba as PM and form a broad-based government, selecting leaders with a good public image from the Nepali Congress, the RPP, and the CPN (UML).

His advice sounded indeed useful when he said that senior advocate, Ganesh Raj Sharma, can prove instrumental to convince GP Koirala to come to a consensus.

Entrepreneur Raghu Shah made important disclosure of his reciprocal financial assistance to Bam Dev Gautam, and the CPM (UML) leader's repaying the assistance by supporting the King's move.

A youth leader of the Nepali Congress, Chandra Bhandari, wondered at the widening chasm between the King and GP Koirala. He quoted

Shailaja Acharya, another leader of the Nepali Congress, as saying that it isn't right on the part of the political parties in general and the Nepali Congress Party, in particular, to exert unnecessary pressures on the King at a time when the Maoists are fighting to take over the entire state authority.

He also added that Acharya is convinced that the ongoing movement is increasingly becoming rudderless and directionless while the former PM GP Koirala is flirting with the Communist agenda.

It's heard that Sharad Chandra Shah is going to hold a press conference where it'll be no surprise if he criticizes, which he's been doing for quite some time now.

In the wake of the comments of all hues and colors made against him, the Minister of Health, Dr. Upendra Devkota, interacted with over a dozen doctors in the PM's office at Baluwatar.

There's news that a program attended by some four thousand people is going on at Gharti Gaon of Rolpa where the Maoists have blocked the entrances to the place for non-Maoist people.

The King asked me to give him a name to head the government in case a new Cabinet is to be formed.

Monday, May 26, 2003

There's news that the second joint expedition team of Nepali and Indian army personnel scaled Mt. Everest. The team included six Nepal Army personnel, six Indian Army personnel, and the rest were five Sherpas to assist the mountaineers.

I heard that the PM is making tremendous efforts to have the support of the CPN (UML) for his government. In a series of his meetings with the leaders, he consulted with KP Oli and Bamdev Gautam for the purpose.

A reliable source says that the two leaders, in view of the precarious situation of the country, have made up their mind to support the King to maintain peace for the greater good of the country and the people.

One of the leaders of the Nepal Peasants and Workers' Party has informed me that the CPN (UML) is receiving funds from INGOs for the continuance of the movement.

There's news of Khum Bahadur Khadka quitting the Nepali Congress (Democratic) to join the Koirala faction.

Sources say that GP Koirala, after a long wait, has visited the Indian Ambassador to Nepal. The content of the discussion between the two hasn't been made public.

It's learned that Shekhar Koirala, as the emissary of GP Koirala (his younger paternal uncle), went to Delhi and visited AR Raghunathan, a senior official at the foreign department. Shekher reportedly brought Raghunathan's message to GP Koirala, proving that Shailaja Acharya's line of thought is right.

The Maoist facilitator and interlocutor at the talk team, Krishna Bahadur Mahara, in an interview to the editor of the *Prakash* Weekly said it'll be no surprise if the peace talk fails, and the Maoists revert to war again.

Addressing Army officers on the occasion of the convocation held at the Shivapuri Army Staff College, the CoAS expressed his conviction that the ongoing peace effort is guided by popular wish, and at the present time, as it is, it's unfortunate to hold political demonstrations to stage street agitation by the political parties. He advised the graduating officers of all ranks to work towards quashing all such activities intended to disrupt peace and law and order.

The King commanded me to make all security arrangements for his inspection visit to the Eastern Division.

He also instructed me to take a soft line with the demonstrators of the political parties while keeping vigil on Maoist activities.

Friday, May 30, 2003

Amidst the speculations on the Cabinet's reshuffling, the five-party joint meeting preemptively decided to forward the name of the General Secretary of the CPN (UML) to head the new government.

With this political cook up going on in the five parties, former PM Surya Bahadur Thapa, himself confident of his assumption of premiership once again, seems to be preparing a list of those to be included in his Cabinet.

The Maoists, on the other hand, are reportedly circulating a confidential circular to their cadres to stay alert in case they are required to go underground once again.

Asked to clarify it, talk facilitator Pun expressed his confidence that the talk will continue; but even if it failed, it wouldn't occur to him that the Maoists would go back to the jungle.

Sunday, June 1, 2003

Compared to the recent mass meetings held by various political parties, which had drawn negligible numbers of people to listen to their speeches, there was a much bigger turnout of people at the mass meeting held by the Nepal Samata Party.

Speaking to the mass, Narayan Singh Pun, the founder leader of the Nepal Samata Party, made it clear that he was a staunch royalist, and there was no alternative to monarchy to hold the country together in view of the geostrategic location of Nepal.

He also challenged the ambiguous Maoists as well as the other forked-tongued political parties to make their position clear on what solutions they have for Nepal and its present dilemmas.

According to a confidential source, Prof. SD Muni and the former Indian Ambassador to Nepal, KV Rajan, who arrived in Kathmandu recently, are lobbying for Surya Bahadur Thapa as the next Prime Minister.

Top Indian leaders, such as Chandra Shekhar and George Fernandes, are also reportedly in parleys with the political leaders in Kathmandu.

GP Koirala seems to have made a secret plan to propose Madhav Kumar Nepal as the next PM, have the Parliament reinstated through Nepal, and wait for the most opportune time to remove him tactically, and with a certain agreement with the Maoists aim for holding the general elections.

According to a new piece of information, GP Koirala and Dr. Baburam Bhattarai have reached an agreement to have their respective parties converged under the same rhetoric of the 'Army under the Elected Government, and the royal honorific of 'Shree Paanch' (His/Her Majesty, and His/Her Royal Highness) reserved only for the King and the Queen (Majesty), and the Crown Prince and Princess (Royal Highness).

Madhav Kumar Nepal and Prakash Chandra Lohani want to meet me.

Meanwhile, Dr. Baburam Bhattarai has visited the Nepal Sadbhavana Party (NSP) leader Hreedayesh Tripathi at his residence.

Tuesday, June 3, 2003

Surprisingly, Maoist leader Dr. Baburam Bhattarai has made his position clear, that his Party would rather like the continuity of the Lokendra Bahadur Chand Government than its replacement by the CPN

(UML). He has even threatened of the resumption of the civil war if things go otherwise.

GP Koirala has indicated his support for the General Secretary of the CPN (UML), Madhav Kumar Nepal, as the next PM, a clear case of his doublespeak.

There's been an overwhelming increase in the trend of neutralist leaders visiting me, offering advice to and seeking an audience with the King. I tell them to visit the Principal Secretary for that purpose.

With the political affairs at present increasingly getting unstable, the scenario looks like this: fluidity lending space for foreign meddling here, and the leaders of the political parties routinely holding their mass meetings, joint meetings, entering into convenient covenants, threatening to gang up on the monarchy, all without concrete agenda, and seemingly guided by the lure of personal gains and to satiate their hunger for power politics.

For a couple of days now, there have been frequent and numerous political meetings, visits and consultations being held in Kathmandu.

Dr. Baburam Bhattarai, Ram Bahadur Thapa, and Krishna Bahadur Mahara are known to have visited the former Speaker of the House, Daman Nath Dhungana.

With its due information and notification given to India, the RNA has decided to purchase Armored Personnel Carriers (APC) from China for its UN Peacekeeping training purposes. The CoAS called me with regards to releasing the necessary amount from the AWF to pay for the purchase. The King asked me to convey his message to the CoAS to correspond in due time.

The CoAS reminded me to arrange for the conferring of the First Class Medals to the Army Expedition team to honor their valor in scaling Mount Everest.

In view of the mass movement called by the political parties dragging on, I consulted with the security heads about it.

The King commanded me to immediately start the internal preparations to confer the Crown Prince with the royal title of 'Prithuladhish' in 2004 and to hold the King's coronation in 2006.

Friday, June 6, 2003

The 41 members of the National Assembly and other leaders, after holding their mock meeting, decided to pass a 'resolution of regret' as the police blocked them at the entrance to the Singha Durbar, Nepal's Secretariat in Kathmandu.

The Military Attaché of the Embassy of Pakistan to Nepal conveyed the ISI Chief's wish to visit Nepal to the Chief of the NID today. Sensing that a formal invitation at this period wouldn't be suitable, the Military Attaché asked the NID Chief to host a dinner for his reception when he comes to Nepal on a private visit.

The King gave his consent when approached for permission.

According to a source, some individuals close to Lokendra Bahadur Chand, in a well-calculated manner, advised him to recommend to the King for the reinstatement of the Parliament. It was obvious that the King would forward the proposal to the Supreme Court which, in turn, would reject it on the grounds of it being unconstitutional.

Monday, June 9, 2003

Some reports alleging that the Royal Palace agreed to the Maoist proposal to restrict the Army within 5km of the barracks is utterly wrong, and therefore, deserves outright refutation.

Because the media referred to the PM, the Minister of Physical Planning and Construction, and the Minister of Information and Communication as the authoritative agencies to have taken such a decision, it's only appropriate that full denial and rectification should be issued by them in the first place.

In order to consult with and take fresh reports of the expert team to facilitate the peace talks, the PM called them to his office.

Shailaja Acharya has expressed her resentment at the recent press statement given by a leader of her own party, Ramchandra Poudel, accusing him of his inability to understand the sentiments of the Party leadership and supporters.

Another leader of the Nepali Congress, Chakra Prasad Bastola, made it clear that the time had come for the political parties to reconsider the relevance of the mass movement. It's also reported that he has demanded of GP Koirala not to take any action against those cadres and leaders who opposed the idea of the movement.

The Special Emissary of the British Government, Sir Geoffrey James, currently visiting Nepal, attended the dinner hosted jointly by the CoAS, Gen Pyar Jung Thapa, and the Joint Coordinator of the NSC, Maj. Gen Rookmangud Katawal.

The dinner took place at the residence of the Honorary ADC, Maj. Gen Bharat Keshar Simha.

Wednesday, June 11, 2003

The PM has today proposed the establishment of an "Academics' Cell" with the view to providing the King with timely advice and suggestions by intellectuals and the civil society for his moves and decisions. The cell is to have Gita Bhakta Joshi, Madhav Ranabhat, and Prof. Ramesh Raj Kunwar as members.

The Minister of Labor and Transportation, Kamal Chaulagain, spelt out his view that it's the most opportune moment for the CPN (UML) to split so that its hardcore Communists will side with the Maoists, and those with socialistic inclinations will join the democratic line, leaving the space open for those non-partisan members free to join neither camp but to remain in the middle to form a third political force.

Upon the request made by Her Royal Highness Jyotshna Basnyat to avail of the Army Officers' Club for one day as the venue for holding her son's initiation celebrations, Lt. Gen Victory SJB Rana asked the CoAS for his permission.

The King rejected the request, adding that royal relatives shouldn't be encouraged to use the Army Officers' Club for their personal purposes, and suggested to hold such family events elsewhere.

Friday, June 13, 2003

Peace talk team facilitator Narayan Singh Pun sounded distressed at being trapped between the Cabinet and the dialogue team. He shot a tirade, saying perhaps it's the reward for his honesty and sincerity.

Pun requested me to arrange an audience with the King as he was soon flying to London.

GP Koirala, accompanied by the former Chief Justice, Laxman Prasad Aryal, visited the former Chief Justice, Vishwo Nath Prasad Upadhyaya, to ask him whether or not his Party's proposal for the reinstatement of the Parliament is constitutionally valid.

Monday, June 16, 2003

Regarding the application filed by the German national, Dieter Kups, at the Kathmandu District Administration Office for the issuance of a Nepali citizenship certificate in his name in the category of 'Descent' (*bangshaj*) and spreading confusion in the Ministry of Home Affairs about its legality, the King issued his special command to instantly release and deport him to Germany.

It's to be noted that he was involved in the processing of herbs in the Gorkha District for quite a long time, and with his Nepali visa having expired, he was in jail for the last two years for overstaying in Nepal.

The members of the fresh talk team to negotiate with the Maoists, the Minister duo, Dr. Prakash Chandra Lohani and Kamal Thapa, visited the former Minister of Physical Planning and Construction and talk facilitator, Narayan Singh Pun, to learn about the status of the dialogue, the point where they halted, and where they should restart the dialogue with the Maoists. Sources say they also have a plan to consult with the experts' team of facilitators.

Corresponding to the Ministry of Home Affairs, the Indian Embassy in Nepal has again asked for its permission to station at the Embassy 40 Special Security Bureau (SSB) soldiers equipped with arms and ammunition for the Embassy's internal security.

With the Ministry of Foreign Affairs in a mood to permit it, it's better, for now, to withhold the permission until a decision to this effect comes from the NSC.

The heads of the security forces, at a meeting, briefed the PM on the overall current security situations in the country. It's worth noting here that it's the first briefing made to the PM since his assumption of the high office.

Madan Pyasi, the Station Chief of the Indian Intelligence Agency, RAW, at the Indian Embassy, referred the Maoist leader, Matrika Yadav, to one Mathur in Patna of India. It's learned that the officer asked Yadav to reveal the status of their relations with the PWG (People's War Group) and the MCC (Maoist Communist Center) of India. In response, Yadav is known to have stated that there was no nexus or any working relations with those groups except mutual ideological affinities.

With the additional duties of the Customs Patrol given to the Army due to expire in July, the RNA Headquarters has proposed to the Ministry of Finance to arrange its own monitoring team from the new fiscal year.

I advised the King to accept the role of patron of the HIV/AIDS Control Campaign Mechanism. It's a UN-sponsored program that receives substantial funds as grant and assistance to use in the treatment and rehabilitation of HIV patients.

It's been some months since the death of the Raja of Jajarkot, Prakash Bikram Shah. Since he has no sons borne of his ritually married queen to legitimately succeed him to the title, some men of the local Thakuri clan have come forward with their candidature to the kingship. They are Tej Bikram Shah, Binod Bikram Shah, and Rajiv Bikram Shah.The media has covered the news rather widely.

I advised the King that no immediate decision be taken on a delicate issue such as this.

The King only laughed at my briefing.

According to an informal source of information, the same five political parties have decided to spearhead their mass movement against the monarchy.

When I briefed the King on it, he commanded me to inform the Principal Secretary about it.

Friday, June 20, 2003

To the best of my knowledge, the Maoists have been indifferent to give momentum to the dialogue following the formation of the new Cabinet; instead, they seem to be concentrating on weaving their strategies in the eventuality of the failure of the talk.

A series of Central Committee Meetings of the CPN (UML) and the Nepali Congress has been convened in their respective party offices.The General Secretary of the CPN (UML), Madhav Kumar Nepal, and a senior leader of the party, Bharat Mohan Adhikari, talking to the foreign media made every effort to justify the relevance of their mass movement.

In the wake of the peace march organized by the Maoists, a special directive has been issued to the security forces to stay alert and watch their activities. They are also ordered to function strictly in a chain-of-command mode should the peace talks fail.

Monday, June 23, 2003

When I asked for his permission the King allowed me to attend the dinner hosted by the Chinese Ambassador to Nepal.

Leaders of the CPN (UML), KP Oli and Bam Dev Gautam, have sounded positive about the Lokendra Bahadur Chand-led coalition government. But lately, I'm hearing murmurs of discontent regarding the manner of the performance of the government within the Cabinet itself and among the coalition partners.

Sharing his career experiences, the retired Deputy Chief of the NID, Lal Bahadur Thapa, promised all his possible help to the King. It's worth noting that he can help generate the people's faith in the monarchy along the border areas between Nepal and India because of his popularity in the Tarai.

The King, therefore, commanded me to keep him in touch.

Shortly after receiving the 11-point proposal, meant especially for breaking the longstanding political impasse, from Surya Bahadur Thapa, GP Koirala held a closed-door confidential meeting with the PM and Amod Prasad Dhungana of Nepali Congress, with a copy of the proposal forwarded to the former CJ, Vishwo Nath Prasad Upadhyaya.

The US envoy to oversee its 'Interest and Property' in the world is currently visiting Nepal. Shortly after his arrival, he met the Joint Coordinator of the NSC, Maj. Gen Rookmangud Katawal.

Speaking at the dinner hosted in his honor, the envoy expressed his keen interest to learn about Nepal's further plan and preparedness in both the scenario of success or failure of the peace talks with the Maoists.

In between his speech, he made it clear that the US has no other interest than the resolution of the present problems and the prevalence of peace in Nepal. He reiterated the US policy of seeking an end to terrorism in any form and in any part of the world.

The security heads unanimously requested the US envoy to see Nepal with American eyes rather than through Indian lens, as is surmised in Nepal.

Calling me from London, Narayan Singh Pun requested me to do him a favor by delivering a carton of useful books to the King.

For some time now, it's being realized that there's been an increase in the Maoist leaders' visits to several persons and places.

According to a confidential source of information, a team from the South Block quietly visited the PM with their prescriptions, among which is to invite both the Nepali Congress and the CPN (UML) to join the government, and all to mount their moral pressures on the King.

Its also learned that the visiting group's suggestions include holding the proposed general elections to constitute the Constituent Assembly (CA) and make massive amendments to the Constitution.

The source also quoted the Indian entourage as saying that the ouster of such leaders as KP Oli and Bam Dev Gautam won't make any difference to the CPN (UML).

The King today commanded me to search for important documents concerning the Panchayat System's relations with India and China during the 1970s.

A friend of mine, who lives in China and now in his private visit to Nepal, came to meet me at my residence. He said he found the Chinese Ambassador a bit distressed and dissatisfied with the indifferent attitude of the Nepali officialdom and the ineffective performance of the political machinery.

He said he wasn't happy with his own works, either, to enhance the relationship between the two countries. He said he found excessive Indian influence and interference even in domestic political issues in Nepal.

He said he was moved to see the ruinous conditions of the Hetauda Cement Factory, constructed by China for its friend, Nepal.

Friday, June 27, 2003

It's known to all keen observers in Nepal that when he has to create a political stir, GP Koirala goes to Biratnagar, his hometown.

This time, too, he went there, ostensibly to attend a program arranged by the Press Union at Siraha, and during his stay, he initiated the 'fast-onto-death' relay protest staged by some leaders, a gentle and a Gandhian way to mount pressures on the person in power.

Speaking at the occasion, he compared the 1950-51 democratic movement with the one going on at present; that the first movement was directed against the Ranas and the present one against the King. He also announced that he won't rest until the present movement takes momentum, and he made it clear that the movement won't stop until the King corrects his October 4 move.

Sources say that the political parties have made an alliance with the Maoists to embark on street agitation, with a core minimum common agenda.

It's been reported that the Swedish Ambassador in Delhi has reiterated Sweden's full support for multiparty democracy and constitutional monarchy in Nepal.

I advised the King on honoring former PM Lokendra Bahadur Chand and former Minister Narayan Singh Pun with awards for their pioneering efforts to peacebuilding.

Information recently received here says that at the 11th Communist Conference, the CPN (UML) and the Maoists have principally been divided over the question of holding general elections to CA.

Taking into consideration the fast-unto-death relay hunger strike taking place at the Ratna Park, to be followed by a torch rally in the same evening, a special directive has been issued to the security personnel to avoid the use of force in those public places but take precaution that the situation doesn't go out of control.

The political parties have started a signature campaign for the nationalization of the entire property of the late King Birendra and his family.

With formal permission from the King, I attended the dinner hosted by the Chinese Ambassador. Speaking on the occasion and recalling the age-old Nepal-China relations, the Ambassador referred to the 45-minute conversation between the Chinese President and the Royal Nepalese Ambassador to China. This justifies the importance China accords to Nepal.

The Ambassador also highlighted China's security concerns and insisted that the government of Nepal control, even by using force, anti-China activities seen in Nepal.

In response, I assured him that Nepal is committed to its 'One China Policy' and will never allow its soil being used to hurt China.

I also tried to remove China's fears that the weapons and logistics provided to Nepal by the UK and the US are purely for combating terrorism.

In view of the steadily improving China-India relations for some years now, I asked the Ambassador what their attitudes to Nepal will be like.

Monday, June 30, 2003

Narayan Singh Pun is constantly making efforts to garner support for the King's move at home and abroad alike.

Over a dinner, he and Prof. Dr. Surya Subedi discussed the current political situation of Nepal. He said he found Prof. Subedi positive about the future of the ongoing peace talk.

Similarly, as commanded by the King, he's soon to visit Lord Camoys and discuss and update him on the current politics of Nepal.

Lord Camoys has been a constant friend of Nepal's Royal Family since the time when he was the local guardian to the late King Birendra, Crown Prince Dipendra, and Prince Nirajan as well when these Nepali royals studied in Britain, especially at Eton.

As a followup program since the visit to Nepal by the Indian CoAS, Gen Nirmal Chander Vij, a military team has recently arrived in Kathmandu with a couple of truckloads of arms and ammunitions and logistics.

Similarly, an IED team from Britain has arrived with equipment and to conduct training to the No 14 Brigade personnel. The program to handover the supplies and carry out a light training to the RNA soldiers will soon follow.

A friend and neighbor of mine, Ashok Raj Pandey, quoted the Chinese Ambassador as expressing his displeasure at the apparent lack of clarity in the government's policy over the question of Tibetan refugees in Nepal. The Ambassador allegedly demanded that the government probe seriously the inflow of Tibetan refugees and deliver them to the care of the Office of the United Nations High Commissioner for Refugees (UNHCR). Referring to the manner of dealing of the Secretary at theMinistry of Home Affairs, Tika Datta Niroula, he said such activities [being reportedly supportive of the Tibetans in Nepal] will only amount to bolstering anti-China activities in Nepal.

At an informal meeting, the Japanese Ambassador to Nepal, Zenji Kaminaga, and I discussed the politics of Nepal, because when highly placed officials of two or three countries meet, the conversation obviously turns to politics and issues related to foreign relations.

He expressed surprise why the General Secretary of the CPN (UML), Madhav Kumar Nepal wasn't trusted to head the government.

Turning the conversation to development issues, he made the Japanese policy clear, that it isn't interested in making investments in large-scale hydroelectricity projects in Nepal where the operating costs are obviously high due to several hassles, and it can't sell the generated electricity to India at competitive rates.

In a succession of visits, I met the Norwegian Ambassador who asked me why we were objecting elections to the CA?

In reply, I tried to drive home the point that the Royal Palace understands the CA elections as akin to Nepal embarking on a path to republicanism. Nepal can, if it cares, to resolve the long drawn out problems, and afford to opt for the CA elections, too, provided that there's a guarantee of unalterable provisions for multiparty democracy and constitutional monarchy.

According to a reliable source of information, the Indian Ambassador to Nepal, Shyam Sharan, hastened to meet and hold consultations successively with former PM Girija Prasad Koirala, and the General Secretary of the CPN (UML), Madhav Kumar Nepal.

Apparently, for a couple of weeks now, the two leaders, Koirala and Nepal, have been making rounds of the foreign power centers in the capital much more than minding those at home.

PM Surya Bahadur Thapa has given his strong instructions to the Ministers, and the Secretaries of the Ministries and the Departments thereof, to work in perfect coordination when it comes to giving interviews or issuing press statements.

He has also asked the government peace talk team to routinely consult with the group of eminent personalities appointed as facilitators before meeting the Maoists across the negotiation table.

I heard that the Maoist leader, Krishna Bahadur Mahara, has tried with the Chief of the NID, Devi Ram Sharma, to arrange his audience with the King.

Earlier, they were repeatedly asking me for the same appointments.

Thursday, July 3, 2003

While operating their search-and-seize recces, the security forces today discovered 180 kgs of explosives and roughly 100,000 bullets, reportedly delivered to the Maoists by a Bangladeshi gang of criminals.

As per sources, despite their marathon and multiple attempts, the political parties are still far from reaching an agreement on the question of holding the general elections to the proposed Constituent Assembly (CA), with the Nepali Congress and the CPN (UML) remaining unmoved on their stand for heavy amendments to be made to the existing Constitution, while the Nepal Peasants and Worker's Party, the Nepal Sadbhawana Party, and the Joint People's Front stand firm on holding the elections to the CA.

The General Convention of the ANNFSU (Revolutionary) has begun at the Royal Nepal Academy. It's known that the three-day convention will elect its leaders in its cameral session.

A report has it that the Maoist Party has removed Rekha Sharma from the ANNFSU's Women's Front.

The hotly debated issue of the restriction of the Army within the 5-km perimeters from the barracks has been put on hold for now.

In the name of their pro-people works, the Maoists are resorting to coercion which calls for an immediate halt.

A special directive has been circulated by the RNA Headquarters down to the Army units to exercise restraint as a normal course of deportment even when the RNA soldiers and the Maoist combatants come face to face.

Sunday, July 6, 2003

A program for the King to inspect the new sample uniform for the Army has been set at the Royal Guards Brigade Officers' Mess.

Following its tradition, the Principal Military Secretariat of the Royal Palace, on the eve of the King's birthday, hosted an honor dinner which the King and the Crown Prince graced.

It was a rare occasion where both the King and the Crown Prince were present.

The CoAS, former ADCs, former Military Secretaries, and several Major Generals also attended the dinner.

Monday, July 7, 2003

The King's 57th birthday celebrations, amidst joys and happiness, concluded today.

To mark the celebration with pomp and ceremony, a steering committee headed by the PM had been formed.

Another civic celebration committee headed by the former PM Kirti Nidhi Bista was also formed for the same purpose.

A long waiting line of people, from highly placed officials down to the common individuals, and from diverse walks of life, walked all the way from the Royal Military Pavilion to the Royal Palace to offer their best wishes of 'Happy Birthday!' to the King.

On the auspicious occasion of the King's Birthday, and adhering to the long observed tradition in its history, the Nepal Police, on its part, hosted a grand dinner and a cultural show at the BICC.

Thursday, July 10, 2003

Former Minister Kamal Chaulagain shared with me his critical analysis of the former PM Lokendra Bahadur Chand's Cabinet. He pointed out the several lapses of the Cabinet: Poor coordination in it, and the posturing of 'I know best because I'm closest to the King!' attitude of some of the Cabinet Ministers.

One consequence of this comeuppance was the confusion which the official talk team created in the Maoists' minds: the absence of unanimity and the dabs of individual styles in the state's talk team's approach flummoxed the Maoists.

According to Chaulagain, the lackluster performance of the Cabinet can also be attributed to the fact that many of the Ministers are more technocrats than politicians.

In the absence of Narayan Singh Pun as facilitator and intercessor, the Maoists, on their own, are trying other channels to arrange their secret audience with the King. Sources say that DR Lamichhane, Birendra Jhapali, Damodar SJB Rana, Birendra Bahadur Shah, and Prabhu SJB Rana have been approached to use their good offices to have the appointment made.

Coinciding with the arrival of the Foreign Secretary of India in Kathmandu, the Indian Ambassador invited the CoAS to a lunch.

The CoAS informed me that he hosted a dinner last night in honor of the outgoing Israeli Ambassador after the expiry of his tenure in Nepal.

Friday, July 11, 2003

Radha Krishna Mainali, one of the veteran leaders of the CPN (UML), who often offers me advice, met me today to discuss the current political situations. He told me candidly that the previous PM, Lokendra Bahadur Chand, was, by all means, an acceptable figure in the PM's Office. But he wasn't given a free hand to select his choice Ministers in his Cabinet, and therefore, the government performed like a marionette.

He then worried that if a situation like the present one lingers any further, it's obvious that the Maoists will draw closer to the political parties, which, for the monarchy, is going to be counterproductive.

Mainali added that neither holding elections to CA nor a referendum will resolve the ongoing crisis; it'll rather aggravate the situation and deepen the crisis even further.

Therefore, he added, the best course of action for the King is to go along with the political parties, else the monarchy may suffer the fate like that of Norodom Sihanouk of Cambodia.

Mainali requested me to arrange an audience with the King. I told him that the Principal Secretary was the right official for his purpose, so he better contact him. He said he finds me more approachable than the Principal Secretary!

It's generally accepted that Nepal's Leftist leaders are studious, so they make a subtle analysis of everything. Therefore, on the basis of the observations and analyses of the situations, I briefed the King on the current state of affairs.

His analysis: It's crucial to weaken the Maoists through dialogue or by any other means at a time when India is readying itself to weaken the Royal Palace by bringing the Maoists and the mainstream political parties to agree on a common minimum agenda against the monarchy. His diagnosis contains much truth.

According to some fresh bits of information, Dr. Harka Gurung has been making some provocative remarks against Hinduism and the monarchy for some time now.

It's learned that the small political parties don't want the Maoists to inch closer to the mainstream political parties.

Sources say the former President of Bangladesh, General Hussain Muhammad Ershad, and the former PM of Pakistan, Benazir Bhutto, are arriving in Kathmandu to attend the wedding ceremony of Charles Mendies, son of Thomas (Tom) Mendies – a known Protestant missionary and a pioneering 1950s hotelier in Kathmandu.

According to Maoist Parth Chhetri, Dr. Baburam Bhattarai is soon going to meet the BJP leader, Lal Krishna Advani, in New Delhi.

Sources say the Maoist Supremo Prachanda and GP Koirala have agreed on a joint movement against the monarchy, and press for the reinstatement of the Parliament.

Wednesday, July 16, 2003

The King had the Chinese Ambassador to dinner tonight. The Ambassador reportedly complained about the Secretary at the Home Ministry, Tika Datta Niroula, and his negative role which displeased China.

According to the information, the Secretary of Defense, Madan Kumar Aryal, is to fly to India to attend the Nepal-India Joint Consultative Meeting which will be attended by the top officials of the Ministries of Defense, the Foreign Affairs, and Home Affairs from Nepal and India.

The Master General of Ordnance, Maj. Gen Balananda Sharma, had his interview in New York for the post of the UN Peacekeeping Force Commander in East Timor. However, Malaysia bagged the post as its side of the East Asian balance weighed heavier in its favor in the lobbying.

The Working Committee Meeting of the Nepali Congress will unveil its future course of the mass movement. Shortly after his return from Biratnagar to Kathmandu and voicing his opinion about the current political situation publicly at a press conference, GP Koirala expressed his confidence in the success of the movement which will open the way for the reinstatement of the Parliament.

Legal experts have opined that failure to hold general elections within six months obviously leads to the restoration of the Parliament.

Maoist leaders such as Ram Bahadur Thapa heading to Nepalgunj after attending the mass meeting of the Tharuwan Liberation Front and Prachanda and Baburam Bhattarai already in the border areas can't be just a mere coincidence. Sources say that with the peace talks on the brink of collapse, the Maoists have circulated a directive to its leaders and cadres to move to safer places outside Kathmandu.

Visiting the CoAS at the Army Headquarters, the Indian Ambassador showed his keen interest to learn about the status of the peace talks with the Maoists.

The NSC meeting, chaired by the PM, proposed to bring the APF under the NSC, and the NID under the PM Office.

The same meeting also decided to ask friends of Nepal to use their diplomatic and moral pressures on the Maoists not to leave the peace talks.

Monday, July 21, 2003

It was last Wednesday when His Majesty the King had granted his audience to the Chinese Ambassador to Nepal, Wei Hwang Kwong, followed by a dinner, where the envoy complained against the Secretary at the Ministry of Defense, Tika Datta Niroula, for his dubious role [his empathy for Tibetans in Nepal] and had suggested that he be transferred from the Ministry.

Reminding the King of his royal conferment of the rank of Honorary General of the RNA on the CoAS of India, the Chinese envoy suggested that similar honor could well be considered for the Chinese CoAS, too.

According to recently received information, the Politburo Meeting of the Maoists, held at the Nepal-India frontiers, has concluded that the peace dialogue has reached a point of no return, largely due to the psychological trap set upon the King by the US. Cleverly enough, they tactically gave out the impression to the people at home and abroad as though the failure of the peace talks is attributed to the government.

The sources further said that they have circulated a firm directive to the leaders, cadres, and combatants to stay in defensive position.

The hypothesis that the June 1, 2001 Royal Palace Massacre was engineered by an external agency is now duly justified, as the new King is scared of the same doom befalling him: the minute of the Maoist meeting has concluded.

As though it were a routine and a casual ritual, GP Koirala, while talking to people close to him, has said they are about to stir the seventh phase of the decisive popular uprising until it attains its success. He also disclosed the plan to ignite the first phase of the uprising to be launched by the government employees' organizations, followed by a broad-based 'noncooperation' campaign.

Given the King's medical checkups at the Military Hospital taking place in the next fortnight, the CoAS has canceled his visit to South Korea, but it won't affect his participation in the conference of the CoASs taking place in Hawaii, the US.

Amidst the uncertainty looming large over the future of the peace talks, the PM discussed the strategies to be adopted to ensure the success of the talk with the heads of the security forces.

KP Oli of the CPN (UML), currently visiting North Korea, will visit China on his return trip and will consult with the leaders about the political situation in Nepal, as though he's seeking to minimize the image of Madhav Kumar Nepal, the General Secretary of the Party.

Meanwhile, the rifts and differences seen in the Party have been reconciled, for the present.

Tuesday, July 22, 2003

GP Koirala has said that the political parties seem to be scared of the prospect of the peace talks going nowhere, and the situation leading to the declaration of martial law and then civil war. The King must initiate dialogue with the Maoists and instantly find a solution to the prolonged political deadlock.

The Colonel Commandant and Lieutenant General of the British Army and the Brigade of Gurkhas, Trousedell, is soon to visit Nepal to submit his annual report on the British Gurkhas to the King, a regular formality followed every two years.

It's advisable that the Crown Prince, as Colonel-in-Chief, interact with the ranking officers of the Army for him to have the basic knowledge of the events and developments taking place there.

Friday, July 26, 2003

The agitating parties have expressed their fears of Army rule in case the peace dialogue derails.

Talking to people close to him, Govinda Raj Joshi of the Nepali Congress Party said he met the US Ambassador to Nepal, Michael E. Malinowski, and found out about the strong policies adopted by the US against the Maoists.

Meanwhile, Maoist leader Rabindra Shrestha has reportedly explained to the Party cadres that their next moves will be to concentrate entirely on urban guerrilla warfare.

The underground leader of the CPN (Mashal), Mohan Bikram Singh, has attributed the October 4 move of the King to the admixture of bad governance of the parties in power, and the violence wreaked by the so-called People's War waged by the Maoists.

He has also criticized the ways in which the alchemists in the political parties are projecting the CA elections as though it were a panacea.

India, after all, has been able to obtain permission from the Surya Bahadur Thapa Government to deploy 40 armed Indian security guards for the internal security of the Indian Embassy. No advance information of it was provided to the Royal Palace. The embassy was seeking this particular permission but was held up by the previous governments, successively terming it as 'still under consideration.'

I met the UNDP Nepal Representative, Henning Karcher, and the Special Envoy of the Secretary General of the UN, Samuel Tamrat, for a working dinner at the Hotel Shangri-La. We discussed the ongoing peace talks and the ways in which the UN can help Nepal in its peace efforts.

Bharat Mohan Adhikari of the CPN (UML), with the consent of the General Secretary of his Party, came to talk with me about changing the present government with certain negotiations of rights and privileges reserved for the King if their Party's respectable entry and powersharing is ensured.

Birendra Jhapali, one of the facilitators in the Maoist talk team, in an informal meeting with me expressed his Party's regrets over the exclusion of Narayan Singh Pun as a facilitator in the talk team with the change in government. Jhapali also disclosed Pun's own plan and ambition to become Prime Minister on the success of the peace talks as his platform. He referred to Damodar SJB Rana and Prabhakar SJB Rana being privy to Pun's strategy.

The Minister of Environment of the Arunachal Pradesh of India, on a visit to Nepal for his participation in a conference hosted by the ICIMOD, was caught with counterfeit Indian currency notes with the denominations of 500 and 1000 by the police at the Tribhuvan International Airport (TIA). The case was hushed up, and the apprehended was let off without inquiry and investigation at the special request of the Indian Embassy.

The Maoists still persist on their earlier precondition of the limitation of the Army within the 5-km perimeters from the barracks, and the release of their detained members. The government is going to reciprocate their demands by making public the whereabouts of the Maoist leaders, such as Birendra Shrestha, Bamdev Chhetri, and Muma Ram Khanal.

Sources say the facilitator of the Maoist talk team, Birendra Jhapali, and the Minister of Communication and Information, Ramesh Nath

Pandey, went to Palung by helicopter where Pandey reportedly met the Maoist leaders.

Maoist supremo Prachanda is to meet the officials of the South Block.

According to a source of information, the RPP leader of Hetauda, Dipak Bahadur Singh, has been found poaching endangered rhinos and other protected wildlife in the Royal Chitwan National Park. The King has ordered an inquiry into it.

Wednesday, July 30, 2003

With the government going a step forward in displaying its flexibility by releasing three of their central-level leaders, the Maoists are reportedly confused as to what to do next.

Sources say that India is mounting its pressures on the Nepali Congress and the CPN (UML) to join the government without further delay.

The King has commanded the CoAS to take detailed notes of the political parties' attitude in their indecisiveness and make arrangements to have a telephone monitoring system installed in the Royal Palace.

According to a reliable source of information, the coordinator of the Maoist peace talk team, Dr Baburam Bhattarai, asked the facilitator of the government talk team, Karna Dhoj Adhikari, about the King's present thinking in the changed context.

Bhattarai made it clear that the Maoists will happily accept the UN's intermediary role to facilitate the peace talks in Nepal. It's learned that the Maoist leader called on the Special Envoy of the UN Secretary General, Samuel Tamrat, asking him to support their idea.

There's news of a policewoman on duty at the TIA Customs of stealing Euro 2,000 from the bag of a Greek female tourist. The money was discovered from the person of the policewoman when the tourist made a scene and an instant investigation was made on the spot. Such activities only erode the credibility of the Customs officials and darken the image of the country as well. Pilferage of tourists' unaccompanied baggage is also frequently reported at the only international airport of Nepal.

Reminding the PM of the insufficient budgetary allotment made to the RNA this fiscal year, the CoAS asked the PM for the release of an additional Rs. 2.5 billion for the Army to purchase immediately needed logistics.

There's news of distasteful behavior of the spouse of Murari Raj Sharma, Nepal's Ambassador to the UN. Sources also report of the envoy himself being involved in demonstrations against the government. It calls for immediate actions from the Cabinet and the Ministry of Foreign Affairs against him for breaching the code of conduct and diplomatic decorum.

Sunday, August 3, 2003

Today, Bamdev Gautam of the CPN (UML) told me about his financial crisis and reminded me of his repeated requests to meet the King. He said he wanted to help the King this time, too, even at the cost of action against him by the Party.

Accordingly, in a regular briefing, I advised the King that some monetary assistance to Bamdev Gautam would be a big help.

Monday, August 4, 2003

In an interaction with journalists from home and abroad, GP Koirala said there's no alternative to the King, and for the political parties and the Maoists as well, reaching some kind of an agreement, and the best course of action, for now, would be to bring the Constitution back on track and return the basic rights to the people.

Terming the peace talks a sham, Koirala stressed on the immediate restoration of the Parliament as the best option available to end the political impasse.

According to some sources of information, the employees' union at the Hotel Yak & Yeti took to smashing furniture and vandalizing vehicles in a protest.

A confidential source of information has it that the Maoists have issued a directive to their combatants to eliminate the top officers of the security forces and the Army if they defy their demand for limiting the Army within the 5-km periphery from the barracks.

None of the four successive Chief District Officers (CDOs) has reported to the District Administration Office of Surkhet. It's worth noting that four CDOs had been transferred and posted there but not a single officer has reported for his official duties yet, all fearing attacks from the Maoists. It paints a bleak picture of the present state of the administration across the country.

The Chinese Ambassador to Nepal called me to inform in advance that he's going to send an invitation to the Crown Prince for inaugurating and Their Majesties to enjoy witnessing the Chinese acrobatics performing at the Chinese Trade Fair soon to begin at the BICC. He also told me that one RP Pradhan and some Chinese investors are to start a factory in Nepal in a joint venture.

On being advised to resign rather than bear the blames for being unable to bring the other political parties to his government, PM Surya Bahadur Thapa reportedly thundered at the men close to him, 'I'm Surya Bahadur!' He added that he was appointed PM in accordance with Clause 35(2) of the Constitution. 'So, do you expect me to leave just like that, eh?'

But he's often reported of dozing off during meetings.

The CoAS, back to Kathmandu from his inspection of the Army units in several districts, shared his experiences with me over the phone. He said he has promised to send the soldiers of all ranks on duty in Khara of Rukum District to the UN Peacekeeping Mission to recognize their display of valor in the operations in the district. He said he also listened to each soldier and has made up his mind to facilitate timely transfers and relief from their hardship and deployment.

Wednesday, August 5, 2003

In a regular briefing to the King, the Chief of the NID, Devi Ram Sharma, complained of the drastic cuts in the budget for his department which will cause adverse effects on the performance of the staff without adequate training and necessary equipment.

Referring to the fact of national intelligence agencies operating directly under the head of the government elsewhere, Sharma said its placement under the Ministry of Home Affairs has further complicated its works and performances.

In reply, the King assured him to talk to the PM to rectify it soon enough.

Monday, August 11, 2003

The CoAS today told me that the PM had asked him to recommend the names of some retired Colonels and Brigadiers for the purpose of political appointments. He asked me to suggest the names of those retirees with good track records.

The government talk team held consultations with the Secretaries of the Ministries of Defense and the Home Affairs and the heads of the security forces. They are going to consult with the members of the technical team of experts, too.

The Maoists have proposed Nepalgunj as the venue of the next dialogue, coinciding with the gathering of many Maoist leaders in nearby Lucknow, in India.

According to a confidential source of information, some Maoist leaders, like Rabindra Shrestha, Bamdev Chhetri, and Post Bahadur Bogati have headed to Lucknow for the assembly, presumably to discuss in advance the extents to which they can be flexible in the peace talk negotiations.

Even a layman can sense that the indian intelligence agency RAW is providing covert support to the Maoists and overlooking their movement across the Nepal-India borders and their open activities in Lucknow. Taking all of it into consideration, it's pertinent for the Army intelligence to increase its monitoring of such uncanny occurrences.

In my briefing to the King, I said that the Surya Bahadur Thapa Government has failed on both counts: to take the political parties in confidence and to give full shape to the Cabinet.

The political appointments made by the government have generated a host of negative criticism from many quarters.

Wednesday, August 13, 2003

Colonel Nepal Bhushan Chand, on being promoted to the rank of Brigadier General, has been transferred from the Principal Military Secretariat of the Royal Palace to the RNA Headquarters.

The King congratulated him for his contributions to the Principal Military Secretariat.

Chand advised me on the timely updates on the Situation Center set up at the Principal Military Secretariat.

In his briefing to the King, the Joint Coordinator of the NSC, Maj. Gen Rookmangud Katawal, highlighted on all the developments taking place in the NSC since his assumption of the responsibilities in that office.

He also mentioned the assurance of the continuity of financial assistance received from the British Government for the construction of the NSC office building and its operations in the days ahead.

Maj. Gen. Katawal complained that he wasn't invited to the meetings of the PSOs that could hinder the policy formulation of the NSC which lately has been preparing the National Security Policy to study and analyze the information collected and suggestions provided by experts and intellectuals. He requested the King for the approval of the policy. According to him, it's pertinent to have the National Security Policy ratified by the Parliament, too.

The King said it will be better for the NSC officials to keep a low profile so that he has less workload for him and more time to interact with the politicians, journalists, and intellectuals. He commanded the NSC Chief to circulate this directive to the concerned NSC departments and sections.

Narayan Singh Pun called me to inform that the Maoists feel uncomfortable with the new peace talk team.

The joint conference of the Division Commanders has been postponed to August 18 and 19 with the Crown Prince to grace the special occasion to start at the Tribhuvan Army Officers' Club where he can interact with the top Army officers.

With the possible failure of the peace talks in sight, now is the opportune time for the transfers of the Division Commanders.

Thursday, August 14, 2003

The logic behind the Maoists' proposal to have the venue of the peace talks in midwestern Nepal was thought to be the easy access to safe havens across the Nepal-India borders the instant the peace talks failed.

However, it's mutually agreed to hold the next round of the dialogue in Hapure of Dang District.

The government talk team has declined the Maoists' idea that they and the government talk team fly in the same airplane to the venue.

The government made it clear that the venue of the dialogue in Hapure shall be guarded by the security forces.

Pashupati Shumsher JB Rana of the RPP, currently in India, reportedly visited, one after the other, the Indian Minister of Defense, and then the Chief Minister of Madhya Pradesh, Digvijay Singh.

Monday, August 18, 2003

The coordinating conference of the Division Commanders discussed the current political situations and found themselves with similar analyses: a likely doom of the peace talks.

The participants pointed at the severe lack of logistics and vehicles in many Army Divisions. They suggested that the RNA Headquarters purchase helicopters and provide logistics from the funds of the AWF as depreciation funds received from the UN.

One report that some Divisions don't even have enough space to house detained Maoists was really a touching one, calling for immediate actions.

Wednesday, August 20, 2003

Today, the IGP of the Nepal Police, Shyam Bhakta Thapa, informed me that the Counter Terrorism Cell of Nepal Police has monitored the telephone conversation of various leaders, personalities, and people under surveillance. Inviting me to the Nepal Police Headquarters, he played the audio tapes of some of the exchanges: Narayan Kaji Shrestha (the then underground leader of the Unity Center), Lila Mani Pokharel, Amik Sherchan, Drs. Mathura Prasad Shrestha and Devendra Raj Pandey (independent intellectuals), Shyam Shrestha (independent Leftist journalist), and a certain man called Mahesh.

In the telephone talks, most Maoists were critical of GP Koirala.

There was a talk between Prachanda and Narayan Kaji Shrestha.

On most tapes, the Maoists offered their reciprocity by willing to carry all the mutually differing parties along with them to participate in the formation of the Constituent Assembly provided the belligerent parties supported the Maoists who, in return, would also help them revive the Parliament. They were even heard to reject the idea of a conditional CA, and if the elections to the CA took place, they would freely canvas for republicanism in Nepal.

This series of conversation clearly indicate the political strategies of the Maoists.

One audio tape had an interesting talk between Krishna Bahadur Mahara and Matrika Yadav while meeting Madan Pyasi, the agent of the Indian Intelligence Agency, RAW, at the Indian Embassy in Kathmandu.

The remarks made by Narayan Singh Pun to a Maoist leader were very constructive.

Keshar Bahadur Bista was heard soliciting support to become Home Minister.

Some tapes revealed the Maoists' detailed discussions on the question of choosing either war or the Constituent Assembly and propagating their manifesto far and wide.

One Maoist leader talked to Brajesh Mishra, the security advisor to the Indian PM, even while a certain round of the peace talk was going on.

On being briefed on the above classified information, the King commanded me to share it with all the security heads and have them formulate their security policy and plan counterstrategies to that effect.

My briefing also included the Chinese Ambassador's wish to have the Nepal Television Studio Building, constructed for Nepal by China, handed over to the government as soon as possible as he was soon to be relocated to Bahrain. The Embassy had to contact me because the Ministry concerned continued to be unresponsive to the idea of taking over the new facilities.

The government employees' unions, associated with their respective political parties which are presently agitating, have unveiled their phase-wise protest campaign against the government. The civil servants working at the Employees' Provident Fund, the Rashtriya Banijya Bank, and the Nepal Bank Limited are already wearing black bands in office as their token of protest. This is the first round of the planned protests. This negative development must be taken seriously.

Speaking at an all-party interaction program, the former IGP of the Nepal Police, Dhruba Bahadur Pradhan; the CPN (UML) leader, Jhalanath Khanal; a leader of the Nepali Congress, Narahari Acharya; and the firebrand Communist leader, Lila Mani Pokharel, unanimously declared that the King was the real cause of the problems in the country.

Friday, August 22, 2003

Being frustrated at the abuses of the core values of the Nepali Congress Party and currently campaigning to purify it by bringing Girija Prasad Koirala and Krishna Prasad Bhattarai together, Shailaja Acharya is reportedly staging a fast-unto-death at the Party Office from the coming Monday. She has accused Koirala of debasing the core values of the NC, causing it to deviate from the very ideology of BP Koirala, the Party's founder.

Remarking on Acharya's planned hunger strike, Koirala is known to tell his close associates: 'Shailaja's going mad!'

The government is contemplating the formation of a Central Political and Security Command comprising Kamal Thapa, Buddhi Man Tamang, heads of the security forces, and the Secretaries at the Ministry of Home Affairs and Defense.

Tuesday, August 26, 2003

PM Surya Bahadur Thapa and the government talk team are reportedly holding consultations with the five agitating political parties to end the ongoing political impasse.

Citing the deterioration of the security situation as the reason, the CDOs of Chitwan, Panchthar, Ramechhap, and Siraha have respectively imposed curfew in these districts.

Colonels of the RNA, Kiran Bahadur Basnet and Ramindra Chhetri, were rushed bleeding to the Military Hospital as a band of assassins shot them at a point-blank range within a brief interval in Kathmandu. Col. Basnet didn't survive while Chhetri is undergoing treatment in the hospital.

The King sounded serious when I briefed him about the incident. And the instant he heard it, he commanded me to have the PM brief the Ambassadors and friends of Nepal on the Maoists' resumption of terrorist activities.

The shooting of the two Colonels this morning in the capital has forcefully taken the country in another direction, and it's likely to adversely affect the future of the peace talk that's been continuing for the last eight months.

Dealing with these issues, we, the Secretaries of the Royal Palace, held our meeting. We proposed that the CoAS issue a press statement on today's incident, for the reason that it's the first direct Maoist attack on the Army after they broke the truce.

We also discussed the future course of action to be pursued by the government and decided that the Royal Palace Secretaries and the heads of the security forces hence jointly draw the future security blueprint.

Wednesday, August 27, 2003

The Maoist Supremo, 'Prachanda', issued a press statement, declaring the end of the ceasefire, further stating that its relevance ended when the peace talks bore no fruit.

The statement further read, that if the government announced forward-looking programs of reforms, with the assurance of holding elections to the CA, the Maoists would return to the negotiating table anytime and anywhere.

The PM, on his part, called an emergency meeting of the Cabinet and the heads of the security forces, reaching a consensus to issue a counter press statement, with due address to the mainstream political parties, calling them to objectively evaluate the gravity of the situation, halt their street protests and cooperate with the government to firmly counter the terrorist acts of the Maoists.

The heads of the security forces advised the government that, in cognizance of the suspension of the peace talks by the Maoists, the government put them on the terrorist watchlist and announce a State of Emergency in the country without delay.

A firm directive has been issued to the forces to strengthen their security system in the Kathmandu Valley.

Friday, August 29, 2003

Talking to those close to him, and recounting the King's stay in Doha for five days on the excuse of his medical treatment, GP Koirala termed the failure of the peace talks and the political developments of the last couple of days as part of the rehearsal for a 'grand design.' He added that there isn't any alternative to a strong protest movement.

With the sudden increase in the Maoists' activities following the collapse of the peace talks, we sent a representative to the General Secretary of the CPN (UML), Madhav Kumar Nepal, to seek his views on what he thinks will help resolve the current crisis. It was suggested to him that it'll be best to call off the present political protest for the fear that, if the situation further deteriorated, the State of Emergency will be declared and the country will under military rule.

Nepal bluntly asked, 'What do I get for stopping our Party's agitation?' His bargaining reflects the decision of the CPN (UML) as well as that of the Nepali Congress to continue their protest programs.

In a collective briefing, the security heads advised the PM to declare a curfew in the Kathmandu Valley and other security-sensitive districts, and use the security system to its full capacity to quash the Maoists.

The Maoists have made yet another murder attempt, this time on the former State Minister of Home Affairs, Devendra Raj Kandel, who suffered minor injuries and is undergoing treatment in the hospital. His car was hit at Kusunti.

In yet another collective briefing to the King, the Secretaries of the Royal Palace advised the government to take a quick decision to impose curfew in the Kathmandu Valley and other 11 security-sensitive district administrative headquarters, and have countermeasures ready in those areas of Kathmandu deemed susceptible.

Employees at the Royal Palace, too, should take precaution for their safety and security.

The government would do well to promote posthumously the Colonel Kiran Bahadur Basnet who was shot dead by the Maoists.

It's also important for the Ministry of Information and Communication to introduce a policy of selective censorship on dissemination of news related to the Maoists.

Wednesday, September 3, 2003

Some political analysts have observed that it isn't the right time for the political parties to confront the state. Even when the American, the British, and the Indian Ambassadors have asked them to discontinue their agitations, it's unwise of the political parties not to reconsider their decision.

The sister organization of the Nepali Congress, the Tarun Dal, and the students' wing of the CPN (UML), ANNFSU, took out their silent rallies, the former from its Party Office and the latter from the Tribhuvan University Auditorium Hall.

The police took senior advocate Sindhunath Pyakurel under their control as the lawyers took out their silent rally.

For a couple of days now, some district-level Maoist cadres have been arrested; but despite tremendous searches, the assassins of Col. Basnet are still at large.

A meeting of the top-level Army officers convened by the CoAS roundly discussed the current security situations, offering timely suggestions to equip the Military Hospital with advanced facilities, raise the perks and allowances of the Army to keep the morale high and give an impressive TV interview by the CoAS.

Let alone drop their demonstrations, the five political parties have unveiled a series of their future course of programs to peacefully defy the curfew and court collective arrest of the leaders.

Describing that some strangers have been suspiciously and frequently making rounds of his house on the pretext of looking for rooms to rent in the neighborhood for some time now, the Royal Palace's Principal Secretary, Pashupati Bhakta Maharjan, called me to have security forces deployed in the area. I instantly informed the IGP of the Nepal police about it.

Saturday, September 6, 2003

According to information shared by the DMI of the Army, the Maoists have dispatched a task force of sharpshooters into the Kathmandu Valley to assassinate the top officers of the security forces to exhaust and weaken the Army and the Royal Palace.

In an informal meeting with me today, the former Minister of the Panchayat era, Dil Bahadur Shrestha, admired the move of the King to appoint Surya Bahadur Thapa the Prime Minister. The King, he said, has three options: Take over the entire state authority, hand it over to the political parties, or expand the Cabinet and go for elections.

Sources have it that some leaders like Bamdev Gautam and Radha Krishna Mainali are ready to quit the CPN (UML) with a huge number of fellow partymen leaving with them to form a third political force, provided they are entrusted with important Ministerial portfolios in the Cabinet.

A serious thought needs to be given that Communists can never be reliable partners of the monarchy: Cambodia is a glaring example and lesson.

The former ADC, Maj. Gen Tara Bahadur Thapa, has talked of himself as a credible man with a proven example of service and expertise. Knowing that the King now needs people to sustain the monarchy, he offered to serve the monarchy once more as Nepal's Ambassador to Saudi Arabia or Thailand, expressing confidence that he can use his talents and influence to further strengthen the relations between the two nations and also attract the inflow of foreign investments in Nepal.

When I briefed the King on Maj. Gen Thapa's wish to serve, he appreciated his gestures and commanded me to remind him as and when needs arise.

A friend of mine, with his friend from Singapore, came to my residence. Relating the economic progress of the city republic, all made possible by its monitoring system in a relatively short period of time, he suggested that Nepal, too, apply the same system of monitoring the activities of its people to control disruption in development works, as it's one of the major hindrances Nepal is facing for quite some years now.

The Secretary at the Ministry of Home Affairs called me today to inform that his visit to India concluded successfully. He said he found India willing to help Nepal by supplying arms, ammunition and logistics for the RNA.

The Acting Chinese Ambassador rang me up today to update me with fresh information on the newly appointed Chinese Ambassador to arrive in Nepal on September 16, prior to the Chinese National Day, on October 1. He asked me to arrange the time for the Ambassador to present his Letter of Credentials to the King.

The Chinese unmistakably observe the tradition of the outgoing Ambassador briefing the new Ambassador on the country he's going to assume office in.

The Acting Ambassador reiterated China's wish to provide the RNA with logistics only, taking into consideration that Nepal's other friends aren't offended by China providing Nepal with arms and ammunition.

The IGP of the Nepal Police, Shyam Bhakta Thapa, sounded determined to apply for a key position in the Interpol meeting to be held soon. He asked me to use the government machinery to start lobbying for him from right now.

Sources say that the former PM Deuba has started talking about options being open to the CA elections and support for the General Secretary of the CPN (UML), Madhav Kumar Nepal, for prime ministership.

Retired IGP Dhruba Bahadur Pradhan, now a leader of the RPP, broke an interesting and encouraging news to me: that a conflict expert from Italy, closely studying the protracted conflict in Nepal, has offered to broker peace efforts by flying off representatives from the mainstream parties and the Maoists to a confidential place in Italy for talks and success. Pradhan added that the Italian facilitators have offered to bear all costs of the peace talk.

In response, I promised to brief the King on it first and let him know about the developments.

The American and the EU ambassadors have opined that the political parties should cooperate with the King to properly deal with the Maoists. They have promised help of all kinds in this regard.

Friday, September 12. 2003

The policemen feared to have been killed in crossfires have been found extorting money from those in search of *yarchagumba*, a local medicinal herb, in the Darchula District.

Now that the Nepali Congress, the CPN (UML), and the Nepal Sadbhavawana Party (NSP, Anandi Devi) are ready to join the government, the Joint People's Front and the Nepal Peasants and Workers' Party have decided to break away from the alliance, calling the three parties as indulging in 'mere power politics.'

Sources say that PM Surya Bahadur Thapa and former PM GP Koirala have agreed that the PM will recommend to the King for the reinstatement of the Parliament, and if the Supreme Court gives its verdict on it as unconstitutional, the Nepali Congress will join the government.

According to a confidential source of information, KP Oli and Bamdev Gautam of the CPN (UML) will block the General Secretary of their Party, Madhav Kumar Nepal, from becoming PM.

Reports recently received here says that bombs went off in the Hetauda Municipality Office, the Office of the Election Commission, and in a businessman's warehouse shortly after the Maoists issued their press statement that they have a plan for carrying out serial bomb blasts across the country to spread a sense of terror among the people in order to exert pressures on the government.

Monday, September 15, 2003

The volatility of politics continues to drag on. Amidst rumors of a change of government, the Nepali Congress and the CPN (UML), each desperate to seize the day, are concentrating all their efforts on forming a new government.

On the other hand, Surya Bahadur Thapa is still unable to expand the Cabinet. It would be wise of him to make political appointments of those with a clean image and proven expertise in their respective fields.

The Nepal Police has decided to change the date of holding the 'Beating the Retreat' pomp and ceremony from the Queen's birthday to

Nepal Police Day. Also, taking into consideration the sensitivity of the security situation, and as per the decision of the IGP, the program has been set to take place in the morning.

For a couple of days now, GP Koirala has intensified his consultations with legal experts to create an environment for the revival of the Parliament.

Wednesday, September 17, 2003

With the CPN (UML) displaying its soft bent for joining the government for quite some timenow, the Indian Ambassador to Nepal, Shyam Sharan, has quietly held consultation with GP Koirala.

Govinda Raj Joshi has said that he has got wind of the King going to make his move any time, and suggested that the Nepali Congress, therefore, set its program to preempt the King's move.

The Cabinet meeting has decided to place bounties on the heads of the Maoist leaders by putting them on the terrorist watchlist.

The Minister of Finance, Dr. Prakash Chandra Lohani, has reportedly complained of misdeeds of the RNA soldiers deployed in the Customs Patrols. The CoAS, on his part, made it clear that the instant the Cabinet decides on the withdrawal of the soldiers from its patrolling for the Customs Department, they will be back in their barracks.

The IGP of the APF, Sahabir Thapa, has requested the King to fix the time for the inauguration of the APF Headquarters Office in Halchowk from his hands. The construction of the building is reaching its completion, and will soon come into operation.

The IGP also made another request to the King, to increase the strength of the APF and make provisions for a separate hospital for the AFP's wounded personnel.

He also pleaded with the King to provide the deceased and the injured with relief package to boost the morale of the policemen battling the insurgents.The King assured the IGP of fulfilling his requests soon and also commanded him to observe strict transparency in the recruitment, transfer, and promotion of the personnel in the APF.

He also suggested that the APF can take the help of the Army as and when required.

Friday, September 19, 2003

Responding to the PM's request for a telephone tapping operation system, the US Ambassador assured him to provide equipment worth US$ one million for the APF very soon.

Asked by journalists why he's visiting China amidst speculations of a change in the Cabinet, former PM Deuba gave such a curt reply that only indicates the disinclination of the political parties for the restoration of the Parliament.

There's frontpage news of the arrest of the notorious criminal, Charles Shovraj, a serial killer on the 'Most Wanted" watchlist of the police for the last 28 years. He was arrested in Thamel.

According to a source, some Maoists have been shot dead during operations launched by the security forces in the districts of Rolpa, Panchthar, and Tehrathum.

The CoAS is scheduled to visit those areas for inspection of the Army operations.

The Secretary at the Ministry of Defense, Padam Kumar Acharya, gave a briefing to the King on the discussions with his Indian counterpart on his visit to India.

He found India mounting its pressures on Nepal to purchase its Insas rifles, and Advance Light Helicopters, with 70% of the prices as grant and 30% as loan assistance to Nepal.

Sunday, September 21, 2003

The IGP of the APF told me over the telephone that he can't make any sense as to what India wants to do by providing training to Nepal's APF personnel as well as Nepali Maoists in Chakrata near Dehradun of the Uttar Pradesh of India by just two months apart.

It's indeed a sensitive issue for Nepal to obtain India's clarifications on it through diplomatic channels.

The CoAS today briefed me regarding the Indian CoAS frequently inviting him to visit Delhi.

He also shared his plan to send his representative, Maj. Gen Rookmangud Katawal, to participate in the Armed Forces Chiefs' Conference organized by the US Asia Pacific Command.

Talking to Sher Bahadur Deuba, the British Ambassador, Keith Bloomfield, reportedly said that he found the King contemplating to restore the Parliament.

It's reported that the leader of the Gorkhaland of Darjeeling, Subash Ghising, plans to arrive in Kathmandu at the invitation of the Vishwo Hindu Parishad, and visit Lord Pashupatinath. Sources also say he's scheduled to condemn the Maoist insurgency at a press conference.

Monday, September 22, 2003

In a closed-door confidential meeting, the PM is speculated to have pleaded with Madan Pyasi, the RAW agent stationed at the Indian Embassy, to coax the Nepali Congress and the CPN (UML) into joining the government, and the latter is said to have promised his help in this regard even by leaving tomorrow for New Delhi for talks with the Indian leadership.

Meanwhile, the Indian Ambassador has agreed to receive the Nepali Prime Minister at 7 am tomorrow for a meeting.

In a series of efforts to use several channels, the Maoists have approached the UN Deputy Secretary General, Kulchandra Gautam, who has reportedly promised them help to convince the government to agree on the elections to the CA provided the Maoists shun violence and resume dialogue.

The Indian team to exchange frontier and security-related information with their Nepali counterpart arrived in Kathmandu today.

The head of the team, a Joint Secretary at the Indian Ministry of Home, remarked on the JNU (Jawaharlal Nehru University, in Delhi) being under the influence of the Leftists.

He also stated that the views held by SD Muni don't necessarily represent India's views of Nepal.

His concluding comment is that the proposed concept of a Constituent Assembly means republicanism, a fact which must be clear in Nepal.

Tuesday, September 23, 2003

At a joint interaction arranged by the student organizations of the agitative political parties at the City Hall, the speakers shot their scathing remarks and shouted derogatory slogans against the monarchy. They also disclosed their plan to disregard the curfew declared by the Kathmandu District Administration.

Calling his visit to India as purely personal and just to see his daughter in her college, former DPM Ram Chandra Poudel has headed for New Delhi where he's speculated to meet Indian leaders. When the Nepal Embassy in Delhi didn't respond to his request for accommodation at its guesthouse, he has managed his stay on his own.

The former CoAS, Gen Satchit SJB Rana offered the RNA a very useful advice: to have ex-servicemen in the security committees in the local areas.

Maj. Gen Rookmangud Katawal quoted Colonel Sutherland, the Military Attaché at the British Embassy, as saying that it was very wise of the King to make the October 1 move. But without proper action plan and timing, it couldn't yield the desired results, hence the political disturbances at present.

For the peace talks to succeed, the successive governments should've included former Minister Narayan Singh Pun.

He also said that Nepal will do well to go for 'liberal democracy.'

Responding to the Military Attaché, Maj. Gen Rookmangud Katawal replied that since the Maoists are terrorists with prohibited arms and ammunition, and the Army's operations and actions against them are legal.But it's unfortunate that certain people, in the name of human rights, exaggerate it as a violation of human rights during the military operations, making unsolicited noise.

According to the Military Attaché, the British Ambassador, Keith Bloomfield, is displeased at Sir Geoffrey James' appointment as the Special Envoy of the British Government 'Because there's already a British Ambassador in Kathmandu!,' Bloomfield has said. Therefore, he's happy that the Special Envoy wasn't granted an audience by the King.

Sutherland also said that the Chief of DFID, David Wood, seems to be influenced by the Leftists, such as the Pyakurel brothers, the Dixits, Shyam Shrestha, Malla K Sundar, and Arjun Karki. According to Katawal, Wood is a difficult man to work with and would be a hindrance when it comes taking initiatives. Reportedly, Woods didn't even see off the Special Envoy at the airport.

According to confidential information, the Maoists, in thousands, are grouping in the Chure Range between Dang and Kapilvastu districts to avenge the Doramba killing of their cadres by the security forces.

Even as mounting pressures from various sectors insist that there be a ceasefire during the big festivals of Dashain and Tihar, the Maoists don't seem to bother to reconsider it.

Regarding the Police Headquarters' policy of transferring its police officers, the PM has shown his interest in Dipak Ranjit as the SSP (Senior Superintendent of Police) for Kathmandu.

But the IGP told me on the telephone that SSP Ranjit isn't the right person for the placement and won't be able to make any positive difference.

Because the Assistant Military Attaché at the US Embassy, Major Peter Fowler, can't be present to receive the award, conferred on him on the occasion of the King's birthday by the government for his contributions to the Army, the decoration will be sent to him in the care of the Embassy. Although the US Ambassador is believed to be interested to receive the award for Maj. Fowler, who is away in the US, it'll be sent to the Embassy because of the protocol issues.

Brig. Gen Dilip Rayamajhi of the DMI shared a very interesting piece of information with me that a couple of CPN (UML) leaders have tipped off Maoist-related information. The Brig. General also told me that the lure of the portfolio of DPM to KP Oli will set the CPN (UML) aflame, threatening to split the party apart.

APF Hawaldar Dan Singh Bista, accompanied by Asst. Hawaldar Dipendra Basnet, decamped with weapons. During a search operation, Hawaldar Bista was shot dead as he refused to surrender whereas the other one is at large as yet.

Amidst the government's preparations to fill up the vacant posts of local representatives from the political parties, particularly so because local-level elections haven't been held for a very long time and the posts of people's representatives have been lying vacant, the General Secretary of the CPN (UML) didn't just publicly decline the government's offer but also declared that the Party will force its leaders and cadres to withdraw from the entire process. It seems that the parties fear that if they follow the process of filling up the posts of representatives, the government will resort to force.

Senior advocates, such as Ganesh Raj Sharma and Shambhu Thapa, have said that the government's move to fill up the posts at a time of the issuance of the ordinance openly violates the spirit of the Constitution.

The Attorney General has issued his directive to the law enforcement agency to administer the oath of office to the representatives by the district court.

In the political circles, it's being speculated that the King is contemplating to appoint the former Chief Justice, Keshav Prasad Upadhyaya, as PM, and Bishwo Bandhu Thapa as DPM after the Dashain-Tihar festivals.

The CoAS, Gen Pyar Jung Thapa, shared with me confidential information about the Indian CoAS repeatedly inviting to New Delhi even on a 'working visit.' He also told me about the invitation made by the CoAS of Pakistan, slated for November.

Saturday, October 4, 2003

The newly appointed Chinese Ambassador, Sun Heping, accompanied by his wife, joined me for a dinner at my residence. Appreciating the climate and the people of Nepal, he expressed his confidence in and commitment to further strengthen Nepal-China relations during his tenure. Recalling the moments of presenting his Letter of Credentials to the King, he said he was impressed by the King's admiration of China and the Chinese people. Speaking over the dinner, the Ambassador added that the Chinese Deputy Foreign Minister is slated to visit Nepal when the issues of development programs in the northern parts of Nepal will gain prominence. He expressed his desire to visit Mustang. The Ambassador also said he finds the residence arranged for him at Kamaladi too large and expensive. We wished his stay and tenure to be comfortable and successful.

Nepal expressed its assurance to provide the UN with 3,000 soldiers of all ranks for the Peacekeeping Missions. With respect to the UN's wish that Nepali soldiers go for the Peacekeeping Mission well equipped, Nepal finds the APCs of the UAE (United Arab Emirates) cheaper in prices, according to a report submitted by a team of technicians. Of the 40 APCs, 20 units are in good condition. The purchase includes overhaul, training, spare parts and ammunition, all not exceeding US$ five million.

To avoid misunderstanding, it's advisable to talk to the Indian CoAS about Nepal's plan to purchase the APCs from the UAE for cost-effective factors.

Friday, October 10, 2003

An advance request for an audience with the King has been sent to the Royal Palace as the British CoAS, Sir General Michael Jackson, is visiting Nepal in the second week of November.

Bamdev Gautam of the CPN (UML) has said that the Royal Palace missed an opportunity of splitting the Party to form a non-Communist force by replacing the earlier government formed by PM Surya Bahadur Thapa. He disclosed an interesting scheme that if Oli can be convinced into joining the government with the portfolio of DPM, it will cause a splinter in the Party.

He said he envisions the development of the country by modernizing its agriculture sector by augmenting the irrigation facilities.

Speaking at a dinner hosted at the Oxen Chinese Restaurant in Hattisar, actor/director Neer Shah proposed that Nepali citizenship be granted to certain pro-Tibetans with proper scrutiny of their background. He questioned: since several hundred thousand Indians can procure Nepali citizenship from the Terai region, why close the door to the Tibetans who have genuine reasons for living in Nepal ?

Monday, October 13, 2003

NC President Girija Prasad Koirala met Madan Pyasi, the RAW agent stationed at the Indian Embassy and functioning as Minister.

The meeting took place at Sujata Koirala's house at Mandikhatar where Koirala also met the PM in the afternoon.

Tuesday, October 14, 2003

The CIAA has arrested former Minister Govinda Raj Joshi of Nepali Congress, former Secretary Padam Prasad Pokharel, and former Secretary of the Ministry of Defense, Chakra Bandhu Aryal, on corruption charges.

The arrest of Joshi has created a stir in the Nepali Congress Party. An informal meeting at the Party at its Teku headquarters has taken a decision to duly protest the arrest of its leader at a time when the case is already subjudice at the court of law.

For quite some time now, several individuals with political and non-political backgrounds have been asking me to arrange their audience with the King. Those individuals include leaders of the CPN (UML), Radha Krishna Mainali and Bamdev Gautam; cineaste Neer Shah; engineer Ranjan Singh Shah; former Ministers Narayan Singh Pun, and Bhim Prasad Gauchan; the former CoAS, Gen Gadul SJB Rana; and the former IGPs, Dhruba Bahadur Pradhan, Motilal Bohara, and Hem Bahadur Singh.

A two-day meeting to address the security situation and the problems facing the security forces began at the Army Officers' Club today. The joint meeting roundly discussed the concept of the Unified Command of the Army. Most of the speakers sounded critical of the political parties, the media, and the human rights activists for their indifference to the concerns of the security forces.

Speaking at the meeting, the PM lauded the efforts of the security forces, exhorting them to move ahead with the same spirit to fight the disruptive elements.

The King granted his audience to the Colonel Commandant of the Brigade of Gurkhas, Lt. Gen Philip Trousedell, in the Royal Palace. Present on the occasion was the British Ambassador, Keith Bloomfield. Lt. Gen Trousedell submitted to the King the annual report of the Brigade of Gurkhas, a tradition and practice being observed since very long.

Expressing his concerns about the demands made by the GAESO, the King asked the British Government to direct its help towards the development of the RNA Academy rather than help in a piecemeal manner. Speaking on the occasion, the King asked the British Ambassador how the British Gurkhas viewed the Maoists and their so-called 'People's War'.

Responding to the King's concerns, Lt. Gen Trousedell expressed his assurance of legal settlement of the demands of the GAESO very soon and made it clear that the UK always accords high importance to the Army's welfare and development programs in Nepal.

Wednesday, October 15, 2003

The demonstration flight of the Lancer Helicopter provided to Nepal by India concluded successfully at the TIA this morning.

The Member of the State Council Standing Committee and

former CoAS, Gen Satchit SJB Rana called me today to inform about his departure to Dhangadhi to attend the program hosted by the State Council there. To his request that I arrange an Army helicopter for his return trip to Kathmandu, I asked him to talk to the CoAS to see if there's one available in Dhangadhi.

The King rang me on my cell phone to share his idea of taking fresh reports of the current security situations and progress of the operations at a dinner. The instant I received his call, I talked to the IGP about the King's wish, and the IGP gave words to arrange it at the Mahendra Police Club and send a request to the King for his gracious presence.

I had only just reached my office when my cell phone beeped again. It was the King's call. 'How did the PM learn about the confidentiality between you and me about the dinner?' He sounded angry. Exercising restraint, I pleaded to the King that it can, in any case, be known to the PM through the channels of the IGP, the Minister or the Secretary of the Home Affairs, even if I maintain it as a secret on my part.

The King only said, 'You screwed up!'I heard the phone slammed down on the other side.

As scheduled, the working dinner at the Mahendra Police Club took place, and the King arrived punctually at 8 PM, by which time all the heads of the security forces were already there to brief the King on the security situations.

Monday, October 20, 2003

It's a shame to admit, but it's true. The camouflaged material selected for the uniform of the Army, and freighted to Jhapa, has been lost. This incident, happening at a time such as this, is really a sensitive matter for a commoner to poke fun at the security forces. From detailed searches, it was discovered that the thief had handed the consignment to DSP Upendra Kanta Aryal who explained that it came to him from the garment manufacturer at Sita Paila. I told the IGP that such a sensitive mishap mustn't just be passed off without placing it to an intensive inquiry and investigation.

Sources say the agitating political parties are set to step up their mass movement shortly after Tihar.

Arjun Narsingh KC of the Nepali Congress Party has spread his remarks that GP Koirala has gone crazy for making statements for some time now. It seems he wants to meet me and keep in touch with the Royal Palace.

It would be a very good idea for the Royal Palace to arrange a collective audience with the King for the Mayors of the Metropolitan Cities who are currently attending a conference in Kathmandu.

Simultaneously, the Maoists are known to have threatened the participants of extortion, urging them to raise their voice in support of the CA elections.

It's also learned from a confidential source of information that the Maoists have issued a circular to their cadres to assassinate high-ranking civil servants and political leaders.

A program has been set for the King to grant his collective audience to the heads of the security forces and the PSOs at a dinner shortly after the Tihar Festivals to discuss the current security situations and the security measures to be adopted to cope with the steadily increasing insurgency.

Tuesday, October 21, 2003

The leader of the RPP and former Minister, Rabindra Nath Sharma, on his return from the US, has been taken into custody at the TIA by the CIAA.

It's hard to believe but true that it's come to light that some security officers, even during a casual conversation with friends or relatives, tend to divulge the secrets of the security operations against the Maoists. It calls for immediate identification of the loose mouths and the rectification of the problem in time.

According to some recently received information, the Maoists have kidnapped Colonel and the Chief of the Selection Team of the British Army, Griffith, from Baglung and taken him to an unidentified place. The Colonel was in the area for the selection of Nepali youths for recruitment in the British Brigade of Gurkhas.

The Military Attaché at the British Embassy has asked the government to withhold any military action until the British national is released by the abductors.

On the other hand, the Maoists, in a press statement, have clarified that the abduction is a symbolic political move, and the abductee will remain safe during the period of his detention. There's also the speculation among the people that it's a makebelieve nobbling enacted by both sides, as the British Government is on the watch out for an opportune moment to establish their channel of communication with the Maoists. It's also suspected that the British team has left with the Maoists a communication set during the dramatization of the capture.

The houses of former Minister, Sharvendra Nath Shukla, and his brother, in Marchabar of Rupendehi are gutted down to ashes as the Maoists set fire on them last night.

It's learned that the Maoists, for some time now, are seeking to expand their contacts with the friends of Nepal, diplomats and other people to gain their support for their revolution.

There's a report that decaying corpses of Maoist combatants are strewn in the rivers and jungles on the borders of Pyuthan and Baglung when they were shot dead by the security forces during the rebels' attack on their camp.

There's report that Maoist combatants in thousands have assembled in Bhalubang, Kusum and Channauta for their counterattack on the Kalikot District Administration Headquarters and the Police Station simultaneously.

Following a ceaseless search and inquiries, those involved in the shooting of Colonels Kiran Bahadur Basnet and Ramindra Chhetri have been identified. In this connection, for further inquiry, the Maoist Valley-In-Charge, Himal Sharma, has also been taken under control.

There have been gossips about and complaints against Crown Prince Paras frequenting hotels at night and behaving roughly and rudely with the people there.

I told the King that the Crown Prince's rash behavior at as critical a time as this will not reflect well the monarchy.

The King said the Crown Prince himself should realize the sensitivity of such things.

In the meantime, he asked me to find ways how the leaders of the political parties can be dragged for action by the CIAA on corruption charges.

Friday, October 24, 2003

I was surprised when Maj. Gen Rookmangud Katawal referred to the British Military Attaché as asking Katawal whether the Army killed the

Maoist cadres and activists in Doramba who had already surrendered. The way the Military Attaché expressed his concerns about the Doramba incident justifies that the Maoist abduction of the British Colonel in Baglung is just pretense.

The RNA Headquarters instantly clarified to the Representative of the Secretary General of the UN that the abduction was preplanned and a sham.

The RNA has been notified by the UN to apply and be prepared for the interviews to be held at the UN Headquarters in New York for the position of Force Commander in Cyprus.

Tuesday, October 28, 2003

It seems that the Maoists are now reactive and quietly active at inciting the civil society and administrative machinery to turn them against the security forces. It's advisable that the security forces on their part strive to suppress them (the Maoists) by means of psychological and other forms of operation. The King has commanded the security forces to 'hit them where it hurts them the most.'

It's been increasingly realized that there must be a provision for the inclusion of the Secretaries of the Ministry of Home Affairs and the Ministry of Defense when it comes to the King's audience to the diplomats and the heads of the security forces.

Satchit SJB Rana asked me, again, to arrange a helicopter for him and one Shakti Amatya for their trip to and from Doti and Dhangadhi. I expect him to know about the sensitivity of the situation. But when he pressed his request, I promised him to ask the CoAS regarding his needs.

In an impromptu meeting with the CoAS and the IGPs of the Nepal Police and the APF, the PM demanded a list of the ex-Colonels and senior police officers, the reason being the King's wish that there be a 'Village Defense Force at the local levels across the country.'

Wednesday, October 29, 2003

Sources have it that the former PM, Lokendra Bahadur Chand, traveled on bus to Dehradun and Haridwar. The purpose of the journey has been kept confidential.

The PM has given special instruction to the heads of the security forces to tone up security in the Terai, particularly along the highways, to discuss and explore the modalities of the Village Defense Forces and to arrange for the rehabilitation of those displaced by the Maoists.

Monday, November 3, 2003

According to information received from reliable sources, the Maoist Commander, Nanda Kishor Pun, was seriously injured during his attack on the Kusum Base of the security force in the Banke District.

Similarly, another report has revealed that several Maoist combatants died by falling down a cliff while fleeing from Bhawang of Rolpa as the security forces cordoned off the area and launched their operation.

It's also learned that several Maoist combatants, injured in the Arghakhanchi attack, are undergoing treatment at Hukam and Maikot of Rukum District.

A joint meeting of the Nepali and Indian intelligence officers concluded in Lucknow. It's learned that the meeting couldn't achieve its objectives as the Indian side raised their concerns over the increase of Pakistan's ISI activities and transaction of counterfeit Indian currency notes in Nepal. The Nepali side also counter-accused India of providing safe havens to Nepal's Maoists in Indian territories, demanding for their arrest and having them extradited to Nepal.

As requested by the Chinese Ambassador for his audience with the King, I confirmed the appointment to the envoy. I advised the King to discuss with the envoy Chinese assistance in the development of the northern parts of Nepal bordering on Tibet.

According to a revelation from the Intelligence Department, the Maoists are preparing their massive attacks on Army and APF bases, keeping the police engaged in minor skirmishes. It seems that the rebels want to panic the government with their attacks to force it to talk with them.

Tuesday, November 4, 2003

As planned, the King granted his audience to the British CoAS, Gen. Sir Michael Jackson, who shared his experiences of serving in Bosnia Herzegovina for nearly two years as Commander of the NATO force. The General said he found similar problems in Nepal, too. Offering his suggestion, he said Nepal should seek to settle the Maoist question more in terms of social and economic development than by mere military operations. The rehabilitation and reconstruction after conflict management are tougher, he said, recounting his knowledge and experiences. Lauding the efforts of the RNA for its bravery in fighting the Maoists despite the difficult mountain terrain, he advised the Army to respect human rights during its military operations. Acute poverty is the breeding ground and it set the stage for Maoism to grow to the scale it is now, he analyzed.

Expressing his thanks to the British Army for its continued support to the RNA, the King reiterated his conviction that what the Maoists are perpetrating are terrorist activities, and that Nepal should take measures to bring them to talk by the use of force and other means. He recounted his conviction that the poor and the marginalized have joined forces with the Maoists to vent out their fury, with no alternative means to life in sight. The King asked the British Government to help establish an intelligence training school for the security forces in Nepal.

Thursday, November 6, 2003

This is the weekly evening for the Prime Minister to have his audience with the King. Therefore, as required by the norms of our offices, we, the Secretaries of the Royal Palace, normally remain in our respective places when the PM visits the King for briefing, and until he leaves. We're on standby if and when the King needs us for his immediate consultations and feedbacks.

However, I didn't remain in my office this particular evening, as I was scheduled to attend the dinner the Chinese Ambassador had invited me to, and I had the advance permission to do so from His Majesty King Gyanendra Bir Bikram Shah Dev.

CHAPTER: THREE

THE END OF MY JOURNEY

Scenarios after I left the Royal Palace

As my service at the Royal Palace was terminated so suddenly, I strove to shake off the 'Palace Hangover' to live a normal life, resuming my usual morning walk which I had discontinued quite for some time due to the Maoists' policy of physically eliminating highly-placed officials, including officers in uniform.

Some personal works with pressing needs to be done, but left pending for quite some time, now kept me busy, and I gradually got used to normal life. I started doing things that I wanted to do; in fact, I was finally living my own life.

However, being a true soldier of the RNA, I constantly sought an audience with the King to express my gratitude for all the trust, assignments and honors bestowed on me during my 30 years of service at the Royal Palace.

But to this day, I haven't been obliged.

But I harbor not the slightest grievances; although later on, I learned that snoops were set on me by the Royal Palace to monitor me. I let it pass as my 'positional hazard'.

Barely had a week passed after I had left the Royal Palace when the Maoists launched a massive attack on the Kali Prasad Battalion stationed in the Beni Bazaar of Myagdi District, causing much damage to the RNA. But luckily, a Company of the Gorakh Bahadur Battalion defended the barracks with valor and prevented it from being overrun by the Maoists.

It must be remembered that the Gorakh Bahadur Company had spent the night there on its way to Pokhara from its special training exercises. It seems that the Maoists had no knowledge of this Company spending the night there.

I was deeply moved by the unpleasant news which I heard on an FM radio news update the next day. But there was nothing I could do.

Had I still been Military Secretary at the Royal Palace, I would instantly be privy to such information, and I would immediately be coordinating emergency actions. But now was a different time for me, and I felt some relief at not having to relive what I would abnormally be going through had I been in office.

It was the first ever attack of this scale mounted by the Maoists against the strength of a Battalion of the RNA. Until that point, only smaller Army units had been engaged by the Maoists. It was later known that the Maoists could well have overtaken the Kali Prasad Battalion, had it not been for the fateful presence of the Company of the Gorakh Bahadur Battalion.

The Gorakh Bahadur Battalion was my mother unit when I joined the Army. This fact, and the bravery of the Battalion, indeed made me proud, as I was also the Colonel Commandant of this very Battalion.

I learned that four days after the Beni debacle, the King, being invited to a dinner at Lt. Gen Victory SJB Rana's residence, said, 'I sacked my Military Secretary!'

I can make sense now that some of those present rejoiced at it while others were taken by surprise.

Had I myself been present there, I would perhaps sing this folksong for the ears of the King:

Ramrosanga jindagi chinnu
Bainsako kheti garera nahidnu.

Know (your) life fully well
Don't loaf around reaping a younker's harvest.

With some partners, I had started a plastic factory in Banepa; but due to the lack of attention from us, the enterprise had remained neglected for a long time. So I decided to take over its management. All the while, from the moment I was unwarrantedly fired by the King as his Military Secretary at the Royal Palace, Kathmandu's media had been after me for exposes and scoops in their newspapers and on TV channels. I resisted the temptations, being aware of my hallowed profession and its lofty decorum. I remained quiet for the next four years, and spoke to the media only on the condition of anonymity and on off-the-record preconditions.

Peace Efforts in Nepal, and the Carter Center

It had already been two months and I was living the life of a retiree when Dr. Duman Thapa came to my house.

He said, 'Jarsap, because of the mayhem, murder and terror unleashed by the Maoists, the Nepali people have become miserable. So, if possible, we must do something to make peace in the country.'

'This is indeed noble,' I said. 'But what can we do about it, and how?'

He explained: 'I'm with the INGO, the Mountain Resource Management Group ((MRMG), and it's aided financially by the California-based McKnoll Foundation to develop small-scale development schemes in Nepali villages. There's an American, named Carron Basset, at the Foundation, and he's in Nepal at present. He said that the Carter Center can help in Nepal's peace initiative. He says he has already talked with them about the idea, and the Center is interested in making the overture.

'But the Center wants the approval, even a verbal one, of the Royal Palace. If such an understanding is obtained, an undercover team will be sent to Nepal by the Center for making preliminary studies. The study team will hold its consultations with all Nepali agencies concerned. Based on the subsequent report submitted by the visiting group, the Carter Center will then forward its proposal for concrete plans for the peace talks.

'In this case,' Dr Thapa asked, 'Can you possibly obtain the King's consent?'

'Well, I'm no longer with the Royal Palace,' I said, 'but I have excellent relations with many of those who are there. I also closely know many in the government. I was also involved in the state-Maoist peace talks from the very beginning. So I can send a feeler on this proposition to His Majesty the King.'

Later, thinking of who could do this, a name sprang up in my mind – the King's business associate, Prabhakar SJB Rana. I knew him well enough to call and discuss the proposal with him. I rang him up and explained to him the whole idea of the new peace efforts. He promised to call me back after his talk with the King about it. After a fortnight, Rana called me with the news of the King's approval for the Carter Center to initiate its groundworks. The King wanted the Center to seek the views of all the political parties and stakeholders concerned in a quiet manner without any publicity.

I told Rana that Dr. Thapa and I would like to meet him and discuss the entire gamut of the proposal. He called us at his office in the Soaltee Hotel.

We learned that the former US President, Jimmy Carter, had already approached the King about his wish to work towards peacebuilding in Nepal, and the King had corresponded with him with a letter of thanks for his interest. However, nothing had progressed from that point.

Nepal's Honorary Consul in the US, Richard Blum, who is also a member of the Board of Directors at his companies and well known to the King, might be instrumental in our peace efforts. Thus, we decided to correspond with the Carter Center through Caron Basset at the McKnoll Foundation to take the initiative to broker the peace efforts in Nepal.

The idea took shape as Matthew Hodes, the Chief of the Conflict Management Program and his Assistant, Bryan McQueen, from the Carter Center arrived in Kathmandu to start their consultations with the PM, the leaders of the political parties, civil society activists, and the CoAS, as well as Maoist interlocutors. Shortly after meeting Padma Ratna Tuladhar, the contact person to the Maoists, and with the understanding to launch the peace process, the team returned to the US to prepare a report on it. It was also agreed that the Carter Center will have Dr. Duman Thapa, Padma Ratna Tuladhar and me in the US for further consultations on the peace efforts.

We updated the King and the PM through our contact persons. Dr. Duman Thapa, Padma Ratna Tuladhar and I were invited by the McKnoll Foundation for consultations at the Carter Center in Atlanta, Georgia, the US. The Prime Minister, Sher Bahadur Deuba, wanted to include Sharat Singh Bhandari, a former minister and a Congress MP, in the entourage as his representative.

It was very heartening to know about the successful track records of the peace efforts, peace brokering, conflict management and election monitoring achieved by the Carter Center in the different troubled parts of the world.

The descriptions of the various sensitive aspects of peace-brokering, peace talks, and conflict management by Professor Dr. William Wiry of Harvard University and an expert in peace efforts were encouraging.

Next, we roundly discussed the report prepared by the US team which had visited Nepal to study the nature of the conflict in the country.

During the discussion, I mentioned that although the problem was in Nepal, the solution could be found in New Delhi; so it would be wise of the Carter Center to read New Delhi's mind, otherwise, the entire enterprise will be useless.

I think all those attending the discussion liked my advice, and the Carter Center decided to consult with India about the Center's initiative to broker peace in Nepal.

The conducive environment for the Carter Center to be in Nepal for the aforementioned purposes was created when the PM of India, Dr. Manmohan Singh, appreciated the Carter Center's idea after its representatives consulted with him in New York during his visit at the UN. Thereafter, the Carter Center's study team arrived in Nepal and met with the stakeholders, the Ambassadors of the US, the UK, India, and China. The team also met the officials of the Ministry of External Affairs of India, the Government of Nepal, the leaders of the political parties as well as the Maoist leadership.

It was at that time that Dr. Duman Thapa and I were associated with the Carter Center. Dr. Thapa later became the Country Director of the Center in Nepal when they opened its office in Kathmandu, and I assisted it as and when I was needed.

The former PM, Sher Bahadur Deuba, came to power again, on June 9, 2004, and the CPN (UML) joined the government, terming it as 'regression as half corrected'. A high-level committee to monitor the plans and activities of the government was formed with senior leaders of the parties in the government and outside it.

The General Secretary of the CPN (UML), Madhav Kumar Nepal, NC leaders Dr. Minendra Rijal and Chakra Prasad Bastola, and Dr. Duman Thapa and I flew to the US to discuss the plans, procedures, and modalities of our peace offensives in Nepal. Jimmy Carter, in the discussions, expressed his concerns as to what role Nepal wanted the Center to play, what the complications were, and whether or not and how the Maoists would join the dialogue for peacebuilding in Nepal.

Responding to the former US President's concerns, Madhav Kumar Nepal made it clear that Nepal would welcome the Center's peace initiative, and in the same breath, he said that every possible help would be extended by the Nepal Government.

The Center, on its part, set a program to give a weeklong orientation to some leaders of the political parties at the Carter Center and Harvard University to form a body for 'Negotiation and Coalition Building.' It was agreed that the peace process, in three phases, would include highly placed government officials and some Maoists leaders in the negotiating teams.

Jhalanath Khanal and Bhim Rawal of the CPN (UML), visiting the US on their own, arrived at the meeting, requesting that they also be included as participants in the discussions. When the Carter Center refused to oblige because there was no earlier information of their participation Khanal quietly walked out while Rawal sat at the back just as an observer and to ostensibly help the General Secretary of the Party.

The participants in the meeting weren't amused at this kind of ludicrous attitude of these Nepali political leaders.

When the Carter Center set an orientation schedule on 'Negotiation of Coalition Building' at Harvard University, Chakra Prasad Bastola and Dr. Ram Sharan Mahat from the Nepali Congress, Dr. Minendra Rijal and Bimalendra Nidhi from the Nepali Congress (Deuba), Jhalanath Khanal and Ashok Kumar Rai from the CPN (UML), Dhruba Bahadur Pradhan from the RPP, Dr. Prakash Chandra Lohani from the RPP, Anil Kumar Jha from the NSP, Pari Thapa from the RJ (Mashal), Padma Ratna Tuladhar as human rights activist, Dr Duman Thapa and I left for the US to participate in the Harvard University workshop.

All the participants found the week of the orientation program very useful. The week was followed by formal and informal meetings and discussions with US Congressmen and others, and a briefing at the UN Security Council. A brief discussion followed in the Office of the Carter Center in Atlanta which Jimmy Carter also attended.

The political confusion here in Nepal deepened further by the King's second putch on February 1, 2005.

However, the Carter Center, pursuing its peace efforts, sent a request to Nepal to send the names of the top-level officials of the government. Dr. Duman Thapa visited and consulted with Lokman Singh Karki, the Chief Secretary of the government, who wanted to lead the team himself, with the Minister of Foreign Affairs, Ramesh Nath Pandey.

Belatedly, the Minister of Foreign Affairs sent the names of 12 individuals, including his son, Nischal Nath Pandey. I was surprised to see the names of such individuals who weren't the types the Carter Center had expected to serve the purpose of the peace efforts. Dr. Duman and I found the list crowded mostly by the Minister's favorites and relatives, many of them just eligible to go on their pilgrimages. We duly advised the Carter Center accordingly. We suggested that the Center ask the government to send another list with ranking officers of the Army, the Police, and top bureaucrats of the Administration – in fact, as per the Center's very previous requirements, made in advance – who would effectively help the negotiation when it would be held.

As expected, the Carter Center asked Nepal to send another list of officials to involve them in the preliminary discussions of the modalities of peace efforts. Unfortunately, Foreign Minister Pandey didn't bother to respond to the Carter Center's request, and thus the peace initiative experienced the first obstacle before it would actually begin.

The King was preparing to attend the UN General Assembly, being held in September 2005. Dr. Duman Thapa was trying to arrange a meeting between the King and Jimmy Carter. However, in a flagrant breach of official protocol to set the program of the head of state by the Secretary of the Ministry of Foreign Affairs or the Chief of Protocol, to our surprise, it was the other people who had chalked it. Even the Nepali UN Mission in New York didn't have any knowledge of the program.

Going ahead, the Carter Center offered to charter an airplane to fly the King all the way to Atlanta, arrange his meeting with the former US President, Jimmy Carter, and escort him (the King) back to New York for the General Assembly.

But the Minister of Foreign Affairs, Ramesh Nath Pandey; his son, Nischal Nath Pandey; and the operator of the Temple Tiger Jungle Resort in Chitwan, Basanta Raj Mishra; and his brother, Dr Hemanta Raj Mishra, who was working New York, had already prepared the King's program and they didn't want to follow the Center's itineraries.

However, an attempt was being made to have Jimmy Carter meet the King in Frankfurt on his return trip.

Dr Duman Thapa told me, however, that he was stunned to hear Minister Pandey's cold response to his request that Nepal not miss this opportunity for making peace even at the extent of the King going to Atlanta and meeting the former US President, Jimmy Carter, who was so keen to build peace in Nepal. He told him that peace in Nepal was more important than just the question of protocol. Dr. Duman Thapa was stunned to hear Minister Pandey's curt reply: 'The King *is* King! Who is Jimmy Carter?'

Knowing that the US President George W Bush wouldn't receive him and some nonresident Nepalis there had set a protest program, the King's trip to the US to participate as head of the state in the UN General Assembly was canceled at the last minute.

We received information that Jimmy Carter was even desirous to talk to the King on the phone anytime. Therefore, we – Dr. Thapa, Maj. Gen Tika Dhamala, and I – hastened to make it happen as soon as possible. But this last chance also slipped off our hands as the King didn't call him, nor was Jimmy Carter informed to call the King.

All of this conveyed the impression as though the Royal Palace wasn't serious about the peace talks.

Thus, the efforts of the Carter Center couldn't make any headway. It was due to a mixture of situations: the King setting out on his own road map, the overwhelming influence of the Maoists' so-called 'People's War', and the political parties embroiled in their mass movement.

The situation took a new turn as the political parties and the rebels, under the overt or covert influence of India, reached an agreement in New Delhi. Carter and his Center officials arrived in Nepal to observe the general elections held to form the Constituent Assembly (CA) after the successful conclusion of the Mass Movement II.

I heard comments of various hues made by diverse people regarding my association with the Carter Center. I had no ulterior motive other than peacebuilding in the country. Maybe the Royal Palace thought I reaped a measure of economic benefits from the Carter Center.

On April 3, 2008, a cooked-up news story appeared on the front page of the *Janadharana* weekly. Headlined 'The Maoists and China: Next Target after the Royal Palace', the story tried to assassinate my character by alleging me of being a CIA agent. This was baseless, fake and deplorable. But only yellow journalism would have the audacity to write such nonsense. I was too stunned to even read it. I couldn't believe that the Royal Palace could slander me and the Carter Center through a newspaper. The Palace Secretaries whom I knew very well feigned ignorance and took to trading charges against one another when I enquired about it. I took it as the reward for my 30 years of service to the Royal Palace! I went to court against the *Ghatana Ra Bichar* weekly for its news story that assassinated my character barely a week after I quit my service. Babita Basnet, the editor of the weekly, didn't respond to my question about the baseless news story. How would she do it on her own when she was merely a device handled by the then Press Secretary of the Royal Palace? I came to the conclusion that the whole article was a creative imagination of the Foreign Minister, Ramesh Nath Pandey.

The Royal Move of February 1, 2005

The whole of Kathmandu was abuzz with speculations just the day before the King's move, on February 1, 2005. Addressing the nation that there was no alternative to this move, the King announced the dissolution of the Deuba Government because the October 4, 2002 takeover had bolstered his confidence. A mix of factors, such as the psychology that everything can be set right by the use of force, and similar advice given to the King by the Secretaries, and the heads of the security forces, particularly that by Lt. Gen Rookmangud Katawal, led to taking the move.

But the background for the takeover was already on the making as the parties couldn't launch any effective mass movement, and the foreign embassies also adopted a strategy of 'wait and watch if things could turn for the better.' There was also speculation of the arrangement of the King's meeting with the Maoist leadership on February 4 somewhere in a confidential place of Rolpa, Baglung, or Arghakhanchi. Maybe the psychology that the takeover would create an even more congenial environment for dialogue with the Maoists also worked. The advisers of the King were in a fool's paradise to think that Article 127 was the blessed provision in the Constitution to make an opportune move to undo the tangles of the political impasse. In times past, the King had special prerogatives; but King Gyanendra had become very weak after the Royal Palace Massacre. The King would do well to think twice before making such a big decision that was akin to holding a two-edged sword. It didn't seem to be a well-calculated move on the part of the King to think of the late King Mahendra (the King's father) as his role model and disregard the political parties no matter how corrupt they were. The advisers didn't seem to accurately gauze the rise in the level of the people's awareness by now to give wrong advice to the King because the education level and the Information Technology had improved a great deal in the country from King Mahendra's time in the 1960s.

The King thus dissolved the government and took over the entire executive authority and sovereignty of the country in his hands until another provision came to replace it. It seemed that the people, loathful of the misrule of the political parties, quietly welcomed the King's move in the hope that things would be better. It also seemed that the people didn't think it otherwise even when the right to rule by the people was annulled. Instead, they saw a better future in the King's rule.

When the King formed a Cabinet, himself as the Chairman of the Council of Ministers, and when it failed to deliver the expectations of the people, the environment gradually turned otherwise. The selection of the individuals made by the King was of those favorites who were already tried and tested as failures and already outdated. This, too, added to the disenchantment of the people, and it also resulted in the loss of trust of the friends of Nepal.

The move brought the political parties together to launch a bigger protest movement. Simultaneously, the King's proclivity for advisors-courtiers, such as Ramesh Nath Pandey and Sharad Chandra Shah, also added fuel to the fire that would shortly incinerate himself.

The ADC, Maj. Gen Tika Dhamala, came to my house the very next day after the February 1st move. Over a cup of tea, and recalling the two mass movements of 1980 and 1989-90 for reforms in politics, I told him that should there be a third popular uprising, it would be, this time, directed against the monarchy itself as against the corrupt politicians' misrule. The King's risktaking, if it failed its mark, would result in disasters. It did not, by any means, seem to be a well calculated move because the conviction of the Royal Palace, that the political parties and the Maoists could no longer join hands to cause harm to the monarchy, was too outdated and no longer relevant because of the Indian role, which I came to know about while I was associated with the Carter Center, which overwhelmed every angle of thinking this time.

It is hard to believe how the Royal Palace did not give serious thought to the fact that if any two of the three forces – the Royal Palace, the political parties, and the Maoists – come closer to forge ahead for a common goal, it can easily eliminate the third one. Therefore, the best course of action for the King would have been to bring the Maoists to the talk, deliver good governance, and control corruption because people were desperate for peace and free and fair delivery of services. The need of the hour was to give the people a feel of change, and end of bloodshed.

Maj. Gen Tika Dhamala promised to find time for me for an audience with the King and talk to him in person.

There were good people in the King's team, but it looked as though the bad advisors' influence prevailed all over to impress the King. I consulted with the Vice Chairmen of the Council of Ministers, Dr. Tulsi Giri and Kirti Nidhi Bista, about the Carter Center's interest in Nepal's peace initiative. They said the King did listen to them but acted on his own. Kirti Nidhi Bista even said he did not know why the King chose them. If he did not listen to them, Bista confided in me, it was useless to be there, and he was thinking of an exit. Also, Dr Giri told me that the King did not listen to their inputs; he kept his cards too close to his chest and acted on the contrary. Perhaps Ramesh Nath Pandey prevailed upon them.

Despite all these, Dr. Duman Thapa and I were doing our best to approach influential personalities at home and abroad and the Maoist leadership as well to bring peace in Nepal because we knew very well that if the King did not bring the Maoists to the negotiation table, India was contemplating a move against Nepal any time, which was likely to be detrimental to the monarchy. We worked out the steps to be taken, consulted with the ADC, Maj. Gen Tika Dhamala, in detail, for him to brief them to the King as soon as possible. Coincidence, one may call it, that there were 12 points in the work plan for the King to move along on. We would elucidate each point if an audience was granted to the two of us. Maj.Gen Tika promised to inform us once he made the appointment. In between, we corresponded with the Carter Center and waited for Maj. Gen Tika's reply. Belatedly, though, I was informed that I could meet the King and brief him on the matter at the dinner hosted by the RNA Headquarters on his birthday.

Unfortunately, no such opportunity presented itself as there were many people, and the King only asked me, 'How are you, Vivek? Somehow, I managed to say it out, 'I'm okay, Your Majesty, but the country isn't. If we don't take timely steps, India is sure to come up with certain steps that are going to be detrimental to the monarchy.'

The King did not seem to take it seriously. It was barely five months ago when I had cautioned him that when India arranged its 12-point agreement between the political parties and the Maoists on November 21, 2005, a clear pointer was the two forces were coming closer to eliminate the third: the monarchy.

As the people could not get delivery of services to the scale of their expectations, they gradually lost trust in the King's February 1st step. The King should have started his move from buttressing the Commission for the Investigation of Abuse of Authority (CIAA) to take legal actions against the crooked leaders on corruption charges; instead, he formed the Royal Council for Corruption Control. As the Supreme Court called the Royal Council unconstitutional, the February 1st move lacked luster.

The King's visit to Africa turned out to be another counterproductive jaunt, particularly when the Minister of Foreign Affairs, Ramesh Nath Pandey, remarked that it marked a departure from and added a new dimension to Nepal's foreign policies.

The King's sojourn to Jomsom and the good time spent with the artistes there did not contribute to calm the situation, either.

Additionally, Minister Pandey's overstay of 11 days in China to meet the President to garner China's support also did not prove fruitful.

Moreover, Nepal's proposal to have China as an observer in the SAARC Summit in Dhaka came with inadequate homework, a design by the Minister of Foreign Affairs, Ramesh Nath Pandey, to trap the King, and offend India to read it as though the King played the China Card to thwart India's interests in Nepal. As a result, India brokered the 12-point Comprehensive Peace Agreement (CPA) between the political parties and the Maoists in New Delhi.

The failure of the February 1st move also marked the King's neglect of his diplomatic strategy, so much so that even the closest friends of Nepal expressed their surprise at the mismatch of the overstatement and the underperformance of the head of the government. Unable to have his audience with the King and visit the CoAS, the Special Envoy of Britain, Sir Geoffrey James, vented out his ire of his humiliation with Col. Nirendra Aryal of the Human Rights Cell of the RNA.

Tipped off that the King of Nepal was set to make his move on February 1, 2005, the Chief and Deputy Chief of the Indian Intelligence Agency, RAW, hastily flew to Kathmandu in order to discuss it quietly with the King in the residence of Prabhakar SJB Rana at Tahachal. To the best of my knowledge, the RAW chiefs advised the King to drop his idea. But the King moved ahead anyway to have a shot at the move. This only triggered the 19-day mass movement, and culminating in the said

12-point CPA, signed mutually by the political parties and the Maoists, and authored by India in New Delhi.

Even with the political parties' mass movement going on, Karan Singh, a leader of the Indian Congress (I), as a special emissary of the Indian Government, and accompanied by the Secretary of the Ministry of Foreign Affairs, Shyam Sharan, arrived in Kathmandu to hold consultations with the warring parties.

I knew that Shyam Sharan even sternly cautioned the CoAS, Gen Pyar Jung Thapa, that an army operation against the warring parties would amount to the RNA having to answer for the violations of human rights in the International Court of Justice.

Things seemed to be slipping off the King's hands. Compelled by the pressures of the mass movement, the King, in an announcement, addressed the political parties on April 21 to cease their movement to form an all-party government. But the parties declined the King's offer.

In yet another address, on April 24, he announced the reinstatement of the Parliament which would open the way to holding the elections to a Constituent Assembly, and that would actually mean the abolition of the monarchy.

The first act of the revived Parliament was to pass the Citizenship Bill which was withheld by the late King Birendra, later to be declared unconstitutional by the Supreme Court. It is apparent how India took its move to the abolition of the monarchy to serve its interests in Nepal. Perhaps it was part of what Girija Prasad Koirala rightly read and repeated so often in his later years as 'Grand Design.'

The Comprehensive Peace Agreement (CPA) was followed by the promulgation of the Interim Constitution. The Maoists joined the government which held the elections to the CA in which although no party obtained a clear majority of votes, the Maoists emerged as the largest party against all odds. Communities and interest groups at home and abroad were stunned at the Maoists' emergence as the largest elected party and tried covertly to prevent it from ascending to power. The political forces in Nepal clearly got divided into two camps: the status quoists and democrats versus the revolutionaries and totalitarians. I think the popular expectation sided with those who represented forward-looking changes.

The very first meeting of the Constituent Assembly on May 28, 2008, at midnight, declared Nepal a Federal Democratic Republic by terminating its monarchy with the history of the reign and rule of the Shah Dynasty for 240 years. The interplay of the Maoists' 'People's War', the former PM Koirala's wish to become the first President of Nepal, and India's aim to remove the monarchy from the country contributed to the announcement of Nepal as a Federal Democratic Republican State.

The monarchy in Nepal has become history now, although, given the country's geopolitical and strategic location and linguistic diversity, it was the most potent glue to bind all the disparate caste and ethnic groups together in Nepal.

The Delhi Settlement of 1951 among the then three political forces – the Shah Monarchy, the Ranas, and the Nepali Congress – was replayed in the 21st century by the Maoists, the political parties, and India in the form of the 12-Point Agreement signed by Nepal's seven political parties and the Maoists in New Delhi in 2006. That history repeats itself came true in 2006. As had two of the three forces in 1951 –the monarchy and the Nepali Congress – teamed up against the Ranas, in 2006, the political parties and the Maoists joined hands to eliminate the monarchy through the Mass Movement II of that year. Coincidentally, India brokered both the settlements of 1951 and 2006, and both accords were signed in New Delhi.

Perhaps it is no exaggeration to say that if King Gyanendra had chosen to enter into a secret pact with India, the monarchy would obviously have continued to exist. It is particularly so because at that very fragile time, India had forwarded some proposals serving its interests which were of far-reaching implications to Nepal. India had actually forwarded a draft proposal of the same intents in 1989 to King Birendra which he had declined and joined hands with political forces in the country. In the bargain, he was demoted to a constitutional monarch.

King Gyanendra, it must be admitted, most admirably preferred to walk out of the Royal Palace than to concede to the Indian proposals to the perpetual costs of Nepal.

It seems that Nepal was declared a Federal Democratic Republic in a hurry. The historically eventful decision, whether Nepal should remain as a Kingdom or opt to become a Federal Democratic Republic State, could have been put to popular vote, in a referendum.

It obviously leaves room for questions for the future, as the Mass Movement II of 1989-'90 concluded with an agreement between the monarchy and the political forces.

But this time, the end of the monarchy in Nepal was due neither to the Maoists' People's War nor the political parties' Mass Movement in the first decade of the new millennium, per se: The crucial ax fell from somewhere else.

The Republic of Nepal, the Ground Realities, and the Nepal Army

Unfortunately, on the very day of May 28, 2008, while Nepal was being declared a Republic inside the BICC (Birendra International Convention Center) at New Baneshwor, now used as the Parliament, I was attending to the last rites of my friend, Brig. Gen Daman Narsingh Rana, at the Aryaghat of Pashupati.

The CoAS, Gen Rookmangud Katawal, arrived when the Constituent Assembly, in its meeting inside the BICC, was declaring the demise of the long-entrenched monarchy in Nepal, though there were already speculations on a 'Baby King' or a 'Ceremonial King' as an alternative.

Well, how would the RNA – the Royal Nepal Army – take it? We, the serving officers and ex-service Generals, were surprised to find the CoAS did nothing about the issue. As the CoAS, he could have taken a stand on having a referendum on the future of the monarcy. He let it go as though it was a decision of little importance. It was his 'foresight' that had prevented any 'undesirable incident' from happening.

As already said, the first meeting of the Constituent Assembly hurriedly declared Nepal a Federal Democratic Republic. But there is a long journey in front to institutionalize it. For already in the few years since then, many fissures have surfaced, threatening to complicate the way ahead. Although it is already three years since the Constituent Assembly has been in place, the writing of the Constitution and the logical conclusion of the Peace Process is far from achieving its due success, with the political parties still swaying back and forth in their power struggle, unable even to elect a Prime Minister. With the political parties mired in their political ambitions and power politics, there has been an alarming increase in foreign meddling and interference in the country's internal affairs. With the political parties embroiled in petty partisan interests and self-seeking, the national interests have been pushed to the corner.

There are fears of ethnic and regional shibboleths pushing the country into a quagmire of disintegration; the demand for the entire 'Madhesh as one Pradesh' (province) is a case in point. Now people don't know what it is like to have law and order, a sound economy, a clean administration, an efficient service delivery, and employment opportunities for the youths in the country. Insecurity and a poor economy have added to myriad complications.

To see the problems confusing the future course of the country, a folksong rings in my mind, and I feel like singing it for the lawmakers:

Kahile hola timro man palaune ?
Mardaina ash gharbar chalaune.

When will your mind sprout?
Hope won't die to have a home and family ......

The RNA, since its inception, remained, for quite a long time, under the patronage and guidance of the Royal Palace, so much so that even after the final Mass Movement I, of 1990-1, it was not trusted as the army of the people.

But this construct has undergone a change since the institutionalization of the Constituent Assembly. Even the political leaders seem to be trying to cultivate their rapport with the Army, which is a good sign, particularly when the Maoists still have their own combatants, called the 'people's army'.

The Maoists still seem to have a vindictive attitude toward the Nepal Army because the Maoists-led government not only refused to extend the tenure of the service of eight Brigadiers General but also sought the immediate ouster of the CoAS, Gen. Rookmangud Katawal. Things turned even worse when the President sent a letter reversing the decision of the executive Prime Minister, Pushpa Kamal Dahal, alias 'Prachanda', clearing the way for the CoAS to remain in office.

The CoAS waited for the letter to arrive from the President, at 11 pm, and had remained in the Army Headquarters for two days, unclear of what fear haunted him, an event the first of its kind in the entire 250-year history of the RNA. It will be wise for the leadership of the RNA to think in advance how the Army will insulate itself from undue interests of political players to drag it into controversy.

It concerns the event of June 2006 when Major General Pradip Pratap Bom Malla went to the Supreme Court to seek justice for the extension of his tenure of service when the same was not done. The CoAS, Gen Rookmangud Katawal, and the ADC, Maj. Gen Tika Dhamala, both asked me to have Maj. Gen Malla withhold his case from registering it at the Supreme Court. He won the case, but the CoAS did not reinstate him in office. But when it came to the case of the eight Brigadiers General, the CoAS reinstated them without even bothering to wait for the Supreme Court's verdict. The two cases of identical nature being treated differently reflect on the CoAS's dual character; whether he did it of his own accord or a strong force was behind it will come to light in the days ahead.

He could not remain immune to his own personal controversy, either, as he put his whole energy to have his tenure of service extended for another term, even to the extent of making rounds to the leaders of the political parties. But all his efforts went in vain. Finally, he was compelled to go on the customary one month's leave before his automatic retirement date. It is interesting to know that he took the incoming COAS, Lt Gen Chhatra Man Singh Gurung, to introduce him to the Nepali Congress Party President, Girija Prasad Koirala, a custom reserved only during a king's reign. This action of his was by no means in conformity with the gravitas of the position of the CoAS. Such overt and covert activities on the part of the high-profile CoAS only contribute to the infiltration of politics into a hitherto professional army. It is by no means a good omen for Nepal's armed forces.

Nepal's political instability, coupled with poor economic conditions, increasingly render Nepal vulnerable to foreign meddling in the country's internal affairs. By now, it is no surprise that the Maoists, who waged their so-called 'Jana Yuddha', took blessed shelter in India while their killing of civilians and others went on in Nepal. And the unprecedented events, be it the 12-point agreement, or the Maoists' ascendance to power, or the termination of the monarchy in Nepal, the Maoists were able to keep India in good humor.

Maybe it was because of the Maoists' character itself or a shift in the strategic interests of India, the relations between the two seem to have strained for some time now, particularly since the Maoists formed their own government; and the Madhesh uprising was a clear pointer to the divorce. All speculations and analyses aside, when the Maoists emerged as the largest party in the CA elections and when they formed their own government, however, the number game of politics and diplomacy turned its new face: the Maoists inching closer to China. It will be immature to speculate that China has now a newfound love for the Maoists of Nepal; it has its own interest in eliminating anti-China Tibetan activities from the soil of Nepal. The statement of the Chief of the Foreign Affairs of the Chinese Communist Party that China does not interfere in Nepal's internal affairs and wants the same from others, too, is a clear indication of China's Nepal policy.

It is obvious that, since Nepal shares its borders with both India and China, both giants do have their interests of influence, too. What is difficult to understand, however, is why India does not demonstrate its good neighborliness with Nepal, playing ignorant of Nepal sharing 1,400km borders with the next superpower. History is a testimony to the fact that the Nepal Army has valiantly fought battles and daunted the British and Tibetan forces. It is not outrageous to expect the retention of the NA's glorious history with foresight from its leadership.

Some Unforgettable Episodes

Of the 35 years and four months of my career in the RNA, I served for 30 years at the Royal Palace as ADC and Military Secretary to His Majesty the King, spent about four or so years in training, and the rest in the barracks.

While serving as ADC and Military Secretary, I got the opportunities to visit all the 75 districts of Nepal and several countries.

It was not as though all the events and experiences were sweet, but a blend of bitter and sweet memories which, when I recall them, both cheer and haunt me now.

Let me describe only those that seem most memorable to me.

Arms Diplomacy vis-à-vis Nepal

Arms and ammunition is the second biggest industry in the world, along with oil, motor, and tourism. The trade and transactions in armaments entail complicated game plans, black marketing, and huge kickbacks, and commissions, too. It is said that there is a Mafialike network of arms dealers the world over. The Bofors Scam of India and the Lockheed Scandal in Japan are just a couple of many such sensations. It is obvious that all countries, rich or poor, spend a significant percentage of the annual budget for their defense needs by equipping their armed forces with the latest and modern arms and ammunition.

The RNA replaced its obsolete Three-Nought-Three (303) rifles with SLR (Self-loading Rifles) purchased from India and the UK, giving the 303s to the Nepal Police. In time, and on the grounds of the SLR also being outdated, and the manufacture of its models and spare parts discontinued by the owning company, the need for replacing them with new weapons was pressing.

In addition, the NA (renamed as Nepal Army) troops participating in the UN Peacekeeping Missions also needed weapons of international standards.

In this connection, the NA had obtained some sample weapons from different companies in 1995/6. A task force was formed for the testing of those samples with a special team of experts and technicians, representatives from the DMO and MGO, and in coordination with the R&D Directorate of the NA Headquarters.

Obviously, many foreign companies were interested to sell and supply their weapons to Nepal. The sample weapons tested by the team included the American M-16 A2, the Israeli Galil and Tavor, the HK G-36 of Germany, the Stayer of Austria, the Indian Insas, and the South Korean K2 rifles.

The Russian AK47 was dropped from the trial testing because it was not a 3-G weapon, and it was also reportedly used by the Maoists in various places who presumably procured it from weapon smugglers in India.

Only the American, German, Israeli, Austrian, and South Korean sample rifles met the quality and standard tests. The M-16 A2 rifles purchased for Nepal's UN Peace Keeping Forces by the then CoAS, Gen Dharmapal Bar Singh Thapa, in 1996-97, were also found satisfactory. The Bhairabh Nath Battalion was using the Israeli Galil, and there was no complaint heard against its efficiency, range of fire and make. The sample weapons test team had recommended either of the M-16 A2, the Galil or the K2 as fit for purchase from the standpoints of guarantee and cost efficiency. The Stayer and Tavor were also found good but were rejected due to the "bullpup" system (magazine behind the pistol grip).

Compared to other rifles, the 3-G German HK G36 was found to be the best, but its price was slightly higher. But the offer by the German Company, Heckler & Koch, owned by the British Aerospace looked well because the manufacturer would supply 5,000 units directly from its factory and the rest would be assembled in Nepal. Perhaps the company's plan to assemble its arms in Nepal and sell them in the Asian countries, if successful, would not only create jobs but also introduce foreign technology in Nepal which, in the long run, would contribute to strengthening the economy and self-reliance on basic weapons for the soldiers.

I found the late King Birendra positive about the company's idea. It was his earnest wish to make Nepal self-reliant in every way, which was why he began the practice of observing the firing of several kinds of weapons on the annual occasion of the Hindu Dashain Festivals. I had noticed his fondness for the HK G36 which he keenly observed during its firing.

The then Crown Prince Dipendra himself had tested the firepower of the sample rifles on the Nagarkot and Thankot firing ranges and found the German model the best. As if that was not enough, the model was put to test in the coldest (Kaisang) and hottest (Amlekhgunj) places, along with the other sample models, and the HKG qualified itself as the best in every way.

As has already been and often referred to earlier in this book, India always mounted its pressures on Nepal to purchase its Insas rifles by, among other methods, sweettalking the then Royal Nepalese Ambassador to India, Dr Bhekh Bahadur Thapa, and even using the good offices of the previous Prime Minister, Inder Kumar Gujral, who advertized the suitability of the sidearm even to the King and the Prime Minister of Nepal. India also offered easy terms on the purchase: 70% of the costs as grant, and only 30% as down payment. The hard sell rose to such a level that India even supplied 1,200 units of the rifles, gratis.

But the team set to test the quality of the weapons did not recommend it. India's posture was that since the Indian Army itself was using the Insas, why should the Nepal Army turn to other sources? But the Indian soldiers' own frustration with the rifle's inferiority has had been amply covered in the Indian press itself.

On the basis of the rigorous process of tests and recommendations of the team of experts, the Royal Palace gave its approval to the RNA Headquarters to start the process of purchasing the HK G-36 rifles. Thus the process to purchase the weapons got the legal nod as the CoAS, Gen Prajwalla SJB Rana had the PM pass the decision from the Cabinet and subsequently the agreement was signed.

I was Military Secretary then.

Following the agreement, the process to open the LC (Letter of Credit) for the purchase of 5,000 units of the German rifles progressed. But a complication arose for the company to obtain the export license to supply the weapons to Nepal both from Germany and Britain because the company, Heckler & Koch GmbH, had been sold to the British Aerospace. Thus the supply seemed to freeze.

At this very juncture, the Royal Palace Massacre took place. Two months had passed before the manufacturing company corresponded with the RNA Headquarters, stating its inability to supply the arms, as per the contract.

It did not take long for one to sense that an influential person was behind the scene to effect the cancellation of the agreement. Apparently, India was displeased at Nepal's decision to import qualitative weapons, plan for its assemblage and also possible future sale in the Asian countries.

Obviously, it was unfortunate that Nepal could not procure the best possible weapons at a time when the Army was mobilized to fight the Maoist rebels. Even the US dillydallied to supply its M-16 A2 rifles in small quantity when it was approached after the failure of the first arms deal. Thus Nepal was left with no alternative than to purchase Insas rifles from India.

The possibility that there was a mix of game plan and foul play of some sort in the purchase of weapons in Nepal cannot be completely ruled out and written off.

I find one incident related to the purchase of weapons worth of reference here. The arms trade is a big business for middlemen and very lucrative for stakeholders in high offices.

Here I recollect one incident of earlier times when the Nepal Army bought 4,500 M16 A2 Rifles from the USA. It concerns with one night in 1997 when I was on duty in the ADC Office in the Royal Palace when the Principal Military Secretary, Lt. Gen Shanta Kumar Malla, called me at around 10 o'clock at night. He sounded as though he was giving me a very important piece of information. He said he was surprised to see the assembly of the types of people at the dinner hosted at the PM's residence at Baluwatar. The invitees included the then CoAS, Gen Dharmapal Bar Singh Thapa, the Chief of the ADCs, Maj. Gen Tara Bahadur Thapa, the Royal Nepalese Ambassador to the USA, Dr. Bhekh Bahadur Thapa, PM Sher Bahadur Deuba's relatives, and one Bipin Karmacharya, the middleman to supply the M16 A2. Gen Malla wanted this piece of news to be reported to the King immediately.

It reminded me of another incident of the Director of the Army Welfare Fund at the RNA Headquarters, Brig. Gen Tilak Bahadur Kunwar, sharing his experience of having to write a pad of cheques worth Rs 42 crores to pay for the purchase of the American M-16 rifles for the troops going for a Peacekeeping Mission.

The two references: Principal Military Secretary Lt. Gen Shanta Kumar Malla's description of notable VIPs gathered at the PM's quarters at Baluwatar, and Brig. Gen Tilak Bahadur Kunwar having to write a book of cheques resembled some connection between them.

The row over the purchase of arms was already there even before I took over the ADC Office as its Chief since my promotion to Maj. Gen on April 14, 1998.

Concurrently, the issue of the purchase of an RJ-100 airplane with funding from the Army Welfare Fund to pay for it for the flights of the VVIPs of the Army as and when needed and at other times lease it out to the Royal Nepal Airlines was also going on. I came to know that an agreement of some sort was reached between the then CoAS, Gen Dharmapal Bar Singh Thapa and the representative of the British Company regarding the deal on the HKG-36 rifles and the RJ airplane which was by no means appropriate in view of aircraft's price and technical aspects. I wondered at the way an impromptu Cabinet meeting hastily decided to purchase the airplane with half of the sum to be arranged from the Army Welfare Fund and the rest from other financial sources.

On my individual initiative, I briefed the King on the haste made for the purchase of the airplane from the fund raised by the blood, sweat, and toil of soldiers, without giving a thought to the exorbitant price and technical aspects of the aircraft. I pleaded with the King to have this matter reviewed because the blame for the wastage of the fund may eventually and ultimately come on him as the Patron of the Army Welfare Fund. I also told the King that the aircraft in question has to fly 365 days a year and eight hours every day to break even, which was impossible.

King Birendra looked serious while listening to my pleading and said that not even a single Rupee would be spent from the Army Welfare Fund; he would have no objection if the government managed the funds from somewhere else.

When CoAS, Gen Dharmapal Bar Singh Thapa's tenure of service expired, Gen Prajwalla SJB Rana succeeded him, and all of a sudden he appeared positive towards the purchase of the airplane which he had objected to earlier.

But the Army Welfare Fund had refused to invest in it, so the process came to a standstill. It saved the Army Welfare Fund, and when I look back, it gives me a great pleasure.

The Social Services National Coordination Council, to Me

Towards the end of the Panchayat Era, the role of the Social Services National Coordination Council was a leader in the sector of primary healthcare and children's education. The members of the Royal Family, each one in his or her sector, contributed to the spread and development of the Council; Queen Aishwarya Rajya Laxmi Devi Shah was its patron and chairperson.

The Council shone with the slogan of 'Let us dedicate our service with (our) heart, mind, and soul as witness', coupled with its objective of coordinating with the numerous I/NGOs with the assurance of transparency to contribute to the development of the country. It enjoyed its heydays for over ten years, from 1979 to 1990.

Nepal being the only Hindu Kingdom in the world, the Royal Palace striving to glow with its image as a unique and sovereign state was obvious.

But just around the same time, the mushrooming growth of I/NGOs was gradually ripping apart the social and cultural fabrics of the country. Christian missionaries actively working in Nepal were using their whole energy to turn Nepal into a secular state although Hindu philosophers, scholars, yogis and therapists from all over the Indian subcontinent were able to make disciples in the millions all over the world. Hindu swamis, scholars and preachers have had their force of scholarly preaching and intellectual and spiritual discourse, but the Christian missionaries had no such attributes other than their economic clout to lure the people to religious conversion. Because converting a Hindu into a Christian was illegal, anyone found doing so would be arrested and put to legal action during the Panchayat era. Conflict of interest was obvious: the Royal Palace was for checking the huge flow of foreign funds into Nepal, designed to promote conversion by the Christian missionaries and wearing the garb of I/NGOs. It is obvious that the monarchy to the Christian missionaries felt like an eyesore. In the name of emancipation and equality, the missionaries spared nothing to incite the issue of ethnicity to buckle the sociocultural fabrics of Nepal. If incidents and evidence are anything to go by, the missionaries and the Maoists had a covert nexus to weaken the twin binding factors of Nepal: its religion and monarchy. The Maoists, in the name of their Cultural Revolution, damaged and demolished temples and imposed secularism on Nepal. The Maoists' effort to undo the age-old system of the appointment of the Principal Priest (Mul Bhatta) at the Lord Pashupatinath Temple is a testimony to prove their nexus with the missionaries because they needed funds for their armed struggle.

There had been occasional complaints against the Queen for appointing her kinfolks in the Council, and those appointees often displayed their power and influence to show off their association with the Royal Palace. People often gossiped about large amounts of money being laundered in the name of the Council. However, to the best of my knowledge and long experiences of working in the Royal Palace, I can vouchsafe that the Queen was not the type to have any greed for money and appointments; in her social sector activities, she was rather generous and would give money to institutions and programs from her personal purse and account. For its functions, dedication to and fairness in work, the SSNCC earned a pristine status; for it was oriented mainly to service.

When it comes to the Social Service National Coordination Council, one incident springs to my mind. It happened in 1986 when the Council was holding its Social Service Week with a two-day workshop inside the City Hall. Kumar Khadga Bikram Shah and Prof. Ganesh Raj Singh presented their papers on the occasion of the Council Day.

While commenting on the papers, Prof. Asharam Shakya mockingly thanked Chandrabir Gurung, the Member Secretary of the Council, for recommending his name as a nominee to receive the award of the Gorkha Dakshin Bahu for his social services since 1951. He continued lampooning by adding that one does not get to heaven by getting an award. The entire hall reverberated with laughter and applause to hear the biting satire.

To everyone's surprise, the Queen looked angry and walked out of the hall, as if to ease the_situation as the meeting had already exceeded the stipulated time.

The next day, addressing the august gathering, she said that the awarding of national honors passed through a rigorous process of selection and was accorded only to the most deserving individuals on the occasion of the King's Birthday. It was not as if someone recommended and the recipients received it. In fact, social service was more of the sense of benevolence for one's peace of mind, not with prospects of rewards or material gains.

There was applause from the front seats of the hall.

Handing the ADC's Office over to Me

Normally, when an employee or the head of any department retires or goes on a long leave, it is a common practice in the RNA and in the Royal Palace for him to hand the confidential documents and keys over to the immediate officer next to him.

But I was stunned at the way I was handed the responsibility of the ADC Office. It must be the 13th of March 1998 when it happened. Maj. Gen Tara Bahadur Thapa of the ADC Office in the Royal Palace was taking his retirement, effective from April 13, 1998. There is a long-held tradition of the outgoing officer taking a leave of one month prior to retirement.

However, without any formal handing over of the confidential files and documents of the ADC Office, he asked me to take over the responsibility with only a short telephone call. He said that the keys to the file cabinets were on the table. I was stunned at the way the official handover system in the Royal Palace was carried out. In order to set a system, therefore, prior to my transfer to the Military Secretariat, I prepared a detailed manual for a proper process for handover and takeover in the future.

The Sudden Retirement of Colonel Shashi Pratap KC

It feels like just yesterday when I reached the Office of the ADC at 11 AM on March 13, 1992. Relieved of his daytime duty, Colonel Shashi Pratap KC was preparing to go home while Brig. Gen Shanta Kumar

Malla was still in his office. Jokingly, I asked Col. KC to prepare a heavy treat as he was going to get a promotion soon. 'Sure,' he said, with a glow of joy on his face.

Barely 15 minutes had passed after Col. KC had left when a call from the King (Birendra) came. Brig. Gen Malla felt a little inconvenient. I noticed a dark wave of panic pass over his face as he talked to the King. I sensed something was wrong. He asked me to see if Col. KC was around. He had already left, I said. Brig. Gen Malla then asked me to leave a message for him to call the Palace as soon as he reached home. He called back half an hour later. Brig. Gen Malla told him over the telephone not to come to the office anymore, and to take a leave instead. After a while, Col KC called me at another number from his house. It seemed strange to me. He asked me if I knew the reason behind this sudden order. I replied I had no knowledge of it.

Col. KC's summary termination of service happened just one month short of his official retirement, on completing a Colonel's five-year service tenure. I came to know that the Royal Palace had him take a month's leave prior to retirement, effective immediately on the expiry of the leave. Actually, Col. KC was dreaming of the post of the Principal Military Secretary to the King. But this sudden decision undid his aspirations. I can only imagine his shock at this bolt from the blue.

I think Brig. Gen Tara Bahadur Thapa had got wind of it in advance but he never disclosed it to any of us.

It is pertinent to mention here that Shashi Pratap and I were inducted on the same day to serve as ADCs to the King at the Royal Palace. More about him later.

My Emergency Telephone Call to the Press Secretary

This happened in 1977.

The ADCs in the Royal Palace had only two Saturdays off from work, that too if no additional duties were called for.

It was one such Saturday when I was in the ADC Office when King Birendra called me to convey his message to Chiran Shumsher Thapa, the Royal Palace's Press Secretary, to have him call His Majesty at a certain number.

Instantly, I called the Press Secretary at his house. But an unfamiliar voice, as though playing a prank, said he was not in at the moment. However, I sensed that the Press Secretary was there the entire time; he avoided talking to me, as I was only a junior officer (Captain). Shouting at the one who had picked up the phone, I ordered him to summon his boss ('*tero Bau*!') right away. The Press Secretary was obliged to come to the phone and explained, as an excuse, that the person lacked manners. I told him to have careful and attentive people in his place and added that, as I was on duty, I had called him from the ADC Office, and his casual approach to a call from the Royal Palace was unbecoming of him, to say the least. I told him the King wanted to talk to him that very instant. He sounded frantic at the mention of the King and his call and tried to explain himself to me with polite apologies.

He was always careful and correct with me in his conduct in the coming years.

The Plan to Rig the Ballot Box

In 1986, the elections to the Kathmandu Nagar Panchayat (Municipality) had only just concluded, and the counting of votes had started in the City Hall from the next day. Of all the candidates in the election fray, the immediate and fierce competition was between Ravi Charan Shrestha, a nominee of the PPEC, and Haribol Bhattarai, representing the popular democratic movement. Bhattarai was leading Ravi Charan Shrestha.

I was on duty at the ADC Office. The Principal Secretary to the King, Ranjan Raj Khanal, called me to convey Prime Minister Marich Man Singh Shrestha's message on fudging the ballot box as the opponent, Haribol Bhattarai, appeared to be winning. Khanal wanted me to brief the King on it.

I told the Principal Secretary to talk to the King in person or on the telephone himself about it. But he sounded reluctant to do so, and asking me to do it for him, he slammed the telephone down.

The trick to cause Haribol Bhattarai to lose the elections by an unfair means pricked my conscience. I dreaded the situation in case the rigging came to be known that very day or the next. It would surely defame the monarchy.

I briefed the King on it. He only listened to me, nodding his head without saying anything.

It is worth revealing here that there were factions and subfactions within the Panchayat itself, to play unfair games to have one win and defeat the other in the elections. I also knew that the officeholders of the PPEC were active in tampering with the election results. But it could not happen, and Haribol Bhattarai won the elections with a sweeping majority of votes.

Queen Aishwarya's Rage

Their Majesties, King Birendra and Queen Aishwarya, were in Bangalore for the King to participate in the SAARC Summit in 1987. I was assigned to arrange their pilgrimage to the Tirupati Balaji soon after the conclusion of the SAARC Summit. I had reached Bangalore two days earlier with an Avro (HS748) plane of the RNAC (Royal Nepal Airlines Corporation) at the ready for Their Majesties' flight to Tirupati.

Since large aircraft could not land at Tirupati, I had gone there by car, accompanied by my security detail and the first Secretary at the Royal Nepalese Embassy to India in New Delhi, Janak Bahadur Singh. We had reached Tirupati one day before Their Majesties' arrival. Earlier, the Ambassador to India, Jagadish SJB Rana, had reached there to send information to the Royal Palace that the puja for Their Majesties was arranged to be solemnized in a place a little way further from the temple as the temple authorities said there was no provision for a devotee to pray right inside the temple.

The norms of the temple also required that the King worship, clad in a *dhoti*. Because there was no telephone network available, we could not consult with the King about what he would otherwise prefer to do.

Their Majesties were accorded a warm reception at the Tirupati Airport. We escorted them straight for the *darshan* (worship at the main temple of the Tirupati Balaji) with a big fanfare amidst the chanting of *shloka*s, and after the *darshan* of all the deities, the royal couple was taken to the hotel for briefing and lunch.

At the hotel, the King's Principal Private Secretary, Narayan Prasad Shrestha, summoned me after his 15-minute of meeting with Their Majesties. He looked flustered.

Just as I entered, the Queen shot an angry look at me.

She fumed, 'Why wasn't the puja held in the mandap [of the Mandir] itself?'

I explained, 'Sarkar, the temple authorities require His Majesty to change into a *dhoti* for the prayers but we don't have telephone network there to consult with Your Majesty about it.'

She fired more angry words at me while the King remained silent. I knew that Jagadish SJB Rana had also received his share of similar royal salvos.

After lunch and rest, the King and the Queen expressed their keen interest in the Balaji Trust. The Trust Authorities explained that it operated its medical, engineering and agriculture colleges with its funds, now with Rs. 54 crores in safe deposits. This made the Tirupati the richest Hindu temple in India.

The Queen then asked why they were not allowed to worship (conduct puja) in the prayer square (mandap) itself.

The priest explained that one's presence before God with devotion was much better than holding *puja* with rites and rituals and oblations, which were merely one's external gestures. He added that the donations and gifts made by the devotees to God Tirupati went directly to the Temple Trust's account.

Only after the Queen offered gold coins and cash to the Temple did her fury calm down.

The Queen's abuse made me lose my appetite until we reached Bangalore. It still feels just like yesterday when the Queen turned her angry face to me. It would have slipped into my unconsciousness had it not taken on a foreign soil.

Soon after his return from the Tirupati Balaji, however, the King founded the Pashupati Area Development Trust (PADT), with the Queen as its first Chairperson. Unfortunately, the PADT could not serve its idealistic purposes, as it is now being increasingly abused by political interference following the end of the Panchayat.

My Consultations with the King's Principal Private Secretary

Around the beginning of 1990, the political parties and their sister organizations jointly steered their mass movements that gradually gained popular support. On the other hand, the Panchas, the stakeholders in the Panchayat Polity from the senior down to the local-level cadres squandered time and money alike, labeling the architects of the mass movement 'a handful of anti-national elements operated by foreign hands and enemies of the people.'

Just at that time, my father fell ill, and I had to take him to Bangkok for treatment where I remained for three weeks. By the time I returned to Kathmandu, on February 19, I still had some more days of leave to join my office. Yet I went back to my work at the ADC Office of the Royal Palace.

Their Majesties were on their regional sojourn in Pokhara, with still one more month of stay there. I went to meet the Military Secretary, Shanta Kumar Malla, and the Press Secretary, Chiran Shamsher Thapa, to catch up on the latest issues and developments. They expressed ignorance of how the King was being briefed on the civic disturbances continuing in Kathmandu.

There were the likes of such Royal Palace executives as the Principal Secretary, Rewati Raman Khanal; the Principal Military Secretary, Lt. Gen Rishi Kumar Pande; the Chief ADC, Brig. Gen Tara Bahadur Thapa, and the officials of the PPEC in the entourage of Their Majesties.

Next, I updated the King's Private Secretary, Narayan Prasad Shrestha, on the news and views on Nepal published in the newspapers of Bangkok, and with the fresh turn of events of the mass movement.

Dismissing the force of the movement, he called it a bunch of people seeking instability more instigated by some foreign elements than anything else. He told me that the people had accepted the Panchayat System through the Referendum, and the verdict could not be challenged. Referring to the verdict of the Referendum of 1980, I said that although the Panchayat won over 'multiparty democracy' with only a slender margin of 10% – 55% versus 45% – there had come a sea change in the people's expectations, looking for timely political reforms. The state should never overlook this truth.

One must bear in mind that the state edifice then was constructed for Panchayat.

He replied that he entirely agreed with me and I was not far from the truth, either. He added, however, that there was nothing we could do since they (the King and his circle of advisors in Pokhara) did not realize it. But the ground reality showed that the mass movement of the political parties was spreading like wildfire.

Back in Kathmandu and on a walk one morning, the King asked me, 'What's the situation outside like, Vivek?'

I described the mass movement expanding every single day, backed by democratic movements throughout the world, and of course, in Nepal's case, more so by India's interests. I ventured to caution the King to be wary of those who wanted to serve only their own interests by keeping him in the dark from the stark reality. I also advised him to reconsider the matter in addressing the people's expectations.

I sensed that the King did not like my offer of advice.

The mass movement heated up, gaining intensified support at home and from abroad, making the King back out to replace Prime Minister Marich Man Singh Shrestha with Lokendra Bahadur Chand. When that strategy also did not bring the expected results, there was no choice but to put the Panchayat in the backburner and to introduce the multiparty political system.

Thus, the downfall of the Panchayat Politics is more attributed to the refusal of the King and his minions to realize timely reforms to be made to the system than anything else. And, of course, the vested interests of the circle of his advisors also served to quicken the disappearance of the Panchayat Polity.

The Creation of Two More Vacancies of Lieutenant Generals

One afternoon in the second week of March in 1994, I was hastening to His Majesty's Office when the Royal Orderly called me to escort the King out on a drive. I found the King already in the driving seat. He gestured at me to sit next to him as he drove the car. Handing me an envelope marked 'Confidential', he asked me to open it and read the contents. It read that two vacancies of Lieutenant Generals had been added in the Army, by the decision of a Cabinet meeting.

Turning to me, the King said, 'Is the matter known to you?'

'No, Sarkar. Because it's a high-level decision, such matters don't come to my notice and knowledge,' I pleaded.

'Who do you think must've known about it?' he asked, his face darkening in fury. I sensed he was very angry.

'Apart from Cabinet members, it's mainly the PM [Girija Prasad Koirala], the CoAS [General Gadul SJB Rana] and perhaps the Principal Secretary [Pashupati Bhakta Maharjan], and the Principal Military Secretary [Lt Gen Shanta Kumar Malla] who must know about it, Your Majesty,' I pleaded frankly.

'Why didn't they tell me, then?' he said, even more crossly.

I remained silent as I did not know what else to say. I feared that

I would further enrage him if I said anything more, as the matter was serious, and I being only a junior officer to comment further.

After driving for a while in and around the Palace premises, the King walked into his office, scorn on his face. He commanded me to summon the Principal Military Secretary. He promptly appeared before the King and soon walked out, panicked. While leaving, he reminded me to receive the COAS, Gen Gadul SJB Rana, who would arrive any moment.

Next, the King called me to summon the Principal Secretary, Rewati Raman Khanal. He arrived at the King's office instantly. Khanal asked me to inform him once the CoAS arrived for his audience so that he would call PM Koirala afterward and inform the Palace Gate. Soon the CoAS arrived. He and the Principal Military Secretary went in. They wore a gloomy look when they came out after about 20 minutes. PM Koirala also arrived to meet the King, followed by Principal Secretary Khanal. It was already seven in the evening. Because ADC Tika Dhamala had already come to relieve me, I walked out of the Royal Palace. I dreaded the King's fury today because it was the first of this kind in my lifetime. I felt a sense of relief to leave my office.

I came to know that the King expressed his grave concerns over the creation of the vacancies of two Lieutenant Generals without his knowledge and prior approval of the proposal. The PM gave a straightforward answer, typical of his usual slippery kind: that the Army was traditionally taken as the King's domain; so he assumed that the King already knew about it, and he, therefore, merely seconded the proposal on that foregone premise.

It seemed that the CoAS and the Principal Military Secretary had kept the King in the dark about it. Later I knew that a high-voltage game plan was behind all that.

I thought the King would take actions against the CoAS and the Principal Military Secretary but nothing happened.

The next day, as the grand rehearsals for the celebrations of the Nepali New Year 2051 BS were going on at the RNA Headquarters, I heard many officers say 'Congratulations!' to Maj. Gen Yogendra Pratap Jung Rana who was speculated to be promoted to CoAS after the creation of the twin ranks of Lieutenant Generals.

It could not happen, however, and no promotion to the rank took place in his tenure. Instead, some dominos fell: He rather faced the

'Court Martial on Corruption Charges' brought about by the CPN (UML) Government in 1995.

In connection with the same disgrace, even the CoAS, Gadul SJB Rana, had to resign his post. Only two weeks had remained for his official retirement.

After his forced exit, Dharmapal Bar Singh Thapa and Krishna Narayan Singh Thapa were promoted to Lt. Gens; and Dharmapal Bar Singh Thapa became the CoAS.

Undesirable to the RNA Headquarters but Recognized by the Royal Palace

King Birendra was always punctilious in his observation of the norms in the Army when it came to the appointment of the new CoAS. He saw to it that the new CoAS enjoyed a free hand to make and implement his plans for the first one year so that his efficiency and foresight could be observed. The King would give him his feedback and suggestions if any flaws were noticed in his enactment of his plans.

The CoAS, Gen Dharmapal Bar Singh Thapa decided to retire Brig. Gen Victory SJB Rana. He appealed to the King for justice. The King, after consultations with the PM, the CoAS, and me, issued his order for Rana's stay in office. He was later promoted to Maj. Gen, and selected to be the UN Force Commander in Cyprus where he served for two years, and retired as Lt. General.

Similarly, the then CoAS, Gen Prajwalla SJB Rana, had contemplated retiring Brig. Gen Rookmangud Katawal, citing the reason being the correction in his date of birth, made by the special command of the King. From the minute study made of the documents of the correction of his age some 20 years ago by the royal command, the King gave his orders of stay to Brig. Gen Rookmangud Katawal who later became the CoAS.

The precedent went into effect in another case, this one relating to Brig. Gen Balananda Sharma, also relating to the facts and figures in his personal and documents. He later became Maj. General and served as the UN Force Commander in Syria, and retired as Lt General.

The Principal Military Secretariat of the Royal Palace withheld the CoAS's recommendation for the promotion to a junior officer, superseding his seniors, whereby, after studying the concerned files, the then Col.

Pawan Jung Thapa was promoted to Brigadier General, who later became Maj. Gen. and served as the UN Force Commander in Sudan, and retired as Lt General.

Thus, if and when there were flaws seen in the promotion or retirement of high-ranking Army officers on the part of the RNA Headquarters, the King, with his common sense and righteousness, would prevent injustice meted out to the innocuous and deserving ranks, which seems to be painfully absent after the abolishment of the monarchy.

Two Men of the Royal Palace on My Memories

My long service at the Royal Palace gave me enough time and opportunity to work and be associated and interact with and learn a lot from people of the highest ranks of the Generals down to its personal attendants and office assistants.

Of all the people I heard of and worked with, two men I can never forget are Hawaldar Chandra Jung Shah, a Royal Orderly of the past when the late King Birendra was still Crown Prince. It must have been some time in 1976 when the then Chief ADC, Lt. Col Shanta Kumar Malla, asked me to sort out the old files in the Shri Sadan ADC Office to burn the unnecessary ones.

While leafing through the documents in the files, my hands rested on a page where I found a one-paragraph resignation letter submitted by the Hawaldar, Chandra Jung Shah, whose wording, addressed to the Crown Prince, touched me. The resignation read that he was forced to resign and take retirement due to his family's situation, and he asked for some time for him to pay off the loan he had taken once he would receive his provident fund at his retirement.

He could have begged, as traditions have it in the Royal Palace, for many things but instead thanked the Crown Prince for his trust and love. The honesty of the soldier can be a glowing example and leave precedents for many soldiers of all ranks in the Army.

The other person I equally revere is Lt. Col Komal Bahadur Bista, ADC to the Queen Mother Ratna. At the time of his retirement, due to his age, he sought the Queen Mother's audience. She wanted to know if he wished for anything, in cash or kind, but he only asked for her permission to present her some *prasad* as a gift on his return from the pilgrimage.

That was another precedent of honesty of a soldier of the RNA.

Normally, those who serve the Royal Palace want to prolong their stay there, even after their official retirement or wish for financial or material favors when leaving. But Chandra Jung Shah and Komal Bahadur Bista are those two great soldiers to display the bigness of their heart without any covetousness for personal gains.

By the way, I have never met or talked to Chandra Jung Shah until today.

My Journey, from Rukum to the Royal Palace

There's a little and beautiful village, called Banfikot, in the northern outback between Musikot and the Sano Bheri River. Musikot is the District Administrative Headquarters of Rukum.

That was where I was born. But I spent most of my school years in my maternal uncle's house in Musikot, cared and reared by my loving grandmother. I rarely visited my family. I went home only during holidays and long vacations.

I vividly remember my first day at school. It was the Basanta Panchami, the day Saraswoti, the Goddess of Learning, is worshipped. I started going to school in 1956. I was perhaps seven years old.

There was just one school, called Bhasha Pathshala, in the district. I was instructed to offer a coin and a flower to the teacher and bow to his feet to begin my first learning, a classical system to honor a teacher. In seven years of study, I reached Grade 10, as there was a system of awarding double promotion to deserving students who performed very well in the exams in those parts of Nepal.

Having passed Grade 7 from Musikot School, I was sent with my cousin to Dang, a trekking taking four days. I sat for the entrance tests to the high school at Bharatpur in Ghorahi to qualify for entry in Grade 9. Because the school remained closed on a strike, we moved on to a high school at Khalanga of Salyan District where Marich Man Singh Shrestha was the headmaster. He later became the Prime Minister of Nepal. We came home back in Musikot to obtain our parents' permission to join that school; but by that time, my father and his brother-in-law had campaigned to open up a high school there to retain the students in the village. The school was upgraded to high school on a temporary permission. A qualified teacher from Dang was appointed to head the School. He was Bamdev Dhital. Thus the newly named Tribhuvan Janata Secondary School commenced its teaching of high school classes.

Although the school obtained a temporary certification to operate high school classes, the students of Class 10 were required to go to Dang to appear in the SLC Board Examinations. With still three months left for the SLC exams to begin, we, the students of the Tribhuvan Secondary School went to Dang to prepare for the Board Exams.

An unidentified fear and a ghostly image haunted me: *What if I failed?* But when the results came out, I had passed the SLC.

My interest to study Science at the Intermediate level kindled in my mind. The regulation required that for one to have Science as specialization at the Intermediate level, s/he had to have Elective Science and Mathematics in the SLC. But because our school was just a newly proposed and poor secondary school, it could not afford to teach those subjects. That would entail additional financial burden by having to hire another Mathematics teacher.

Immediately after the SLC exams, I headed to Kathmandu for my further education. It was also my first experience of flying in an airplane.

I happened to learn about a terrific mathematics teacher with whom I took private tuition. I completed my study of science subjects and mathematics. When I scanned through the column of the third division and did not find my symbol number, my heart skipped a beat, scared of being 'failed'. But the joy of being passed in the Second Division overweighed the scare of 'failed'. In a way, I created history in the whole of Rukum District that year as the only student able to pass the SLC. My Rukum classmate and maternal cousin, Gopalji Jung Shah, however, could not graduate.

I applied to study at the Amrit Science College, popularly called ASCOL, but did not find my name on the list of those selected. Next, I applied at Tri-Chandra College where I started my I. Sc. I applied for rooming at ASCOL where the hostel building was still under construction. Luckily, I got a chance to live in the ASCOL hostel; getting a place in a hostel in those days was like being in luck or fighting a battle and winning one.

My aunts lived in Lagantol. Occasionally, I stayed with them. Just at that time, the Rashtriya Sewa Dal (National Cadet Corps) was started with a view to giving basic military education to interested college students. I did not miss the opportunity.

I was giving my best to make my mathematics good. I found another veteran mathematics teacher who became the Chief Secretary of the government and later the General Secretary of the Parliamentary Secretariat. He was Ishwor Prasad Upadhyaya. I have not been able to meet my reverent tutor, Keshav Shrestha, of the JP High School, yet.

My Second Division score in the I Sc (Intermediate Science) exams was whetting my ambition to obtain an MSc Degree in Physics.

By the time I was finished with my BSc first-year exams, I learned that some of my classmates were preparing to join the Army. I was not interested in it. My family wanted me to obtain a higher degree, return to the village and do social works, a calling my father revered the most.

On the other hand, my friends, Dilip Karki, Suraj Shumsher Rana, Kushu Ukyab and others kept enticing me on to join the Army. I had already become the Junior Under Officer in the National Cadet Corps (NCC). I applied for the Army and somehow passed through the hurdle tests.

Around that time, a plan to construct a Military Academy for the Nepali Army Officer Cadets was planned, as there were difficulties in sending them to Britain or India for training.

When the result of Physical and Intelligence Test of the officer cadets was published on July 3, 1968, I was surprised to find that I had come second on the merit list.

I couldn't decide whether to join the Army or continue my education. I couldn't consult with my parents too, as one could communicate only by post in those days and it would have taken more than a month for a letter to reach Rukum from Kathmandu and another month for the reply to arrive. When I consulted my relatives in Kathmandu, they advised me not to miss the opportunity to join the Army.

On July 9, 1968, I went to the Chhauni for officer cadets' training with 47 other cadets where I completed the first term of the 14-week training.

That was how my service in the Army started. I sent a letter to the father in Rukum by hand, detailing everything. Because of another round of training, I could not go home to have a *tika* and blessing from my parents for that year's Dashain.

Friends congratulated me for having come first which amounted to my selection for training in the Royal Military Academy, Sandhurst, Britain.

By that time, a letter from father arrived. He said that none of the family had ever joined the Army and there was no one to turn to in the services. So he suggested that I leave the Army and pack up for home immediately.

It put me to thinking. It took me a long time to decide since I had already put in a lot of efforts to join the Army. So why leave? I wrote to the father that I was selected for training in Britain.

The second round of training started at Nagarkot. Between the training, Colonel and the Commandant of the School, Col Aditya SJB Rana called me, along with Daman Narsingh Rana and Jitendra Bahadur Bista, in his office to break the news that three seats for training in Britain were in the pipeline.

With the entire documents made ready from the RNA Headquarters and visas obtained from the British Embassy, we left for Britain. Due to the flight schedules of the British airplane from Singapore, it took us a fortnight to land in Britain.

In the journey via Barrackpore in Calcutta and the British Camp in Singapore to Britain, I happened to meet the British Gurkha Captain, CB Rai, whose sharing of his experiences as the first Nepali to undergo training at Sandhurst turned out to be very useful to me.

We stayed for two weeks at the Singapore British Gurkha Camp's Officers' Quarters and finally flew on a Royal Air Force VC-10 aircraft to the Brighton Air Force Base in England. Jitendra and I boarded a train to Aldershot in Surry. It was a new experience for us to find people either reading newspapers or sitting quiet, unlike here in Nepal.

I admired the fairness of the system's selection of cadets like Jit and me without any connections, influence or powerbase for such a kind of training in Britain. I was a rural boy, and Jitendra's father, Ratna Bahadur Bista, was a Judge. But we both were selected on the merits of our excellence in training.

On reaching the station, we found it was a holiday, being Sunday. No one was there to receive or guide us. We were confused for 15 minutes. Anyway, we took a taxi to the Mons Officer Cadets School. The duty sergeant, after looking at our papers, guided us to our room and directed us to the nearest 7-Eleven for food. We did not enjoy the food.

It was warm in the room but it was chilling cold outside. It is obvious that one feels homesick in a foreign land. Maybe we were travel-fatigued or jetlagged, so we went to bed quite early. Britain did not fascinate us so much as we had expected it would.

Next morning, we reached the office of the duty officer who advised us on things to do in the next few days before the training started. We were given an instruction manual. And then we took our equipment and uniform. Our training started at Mons.

When we met Bhagwan Singh Thapa, from the British Gurkhas, we felt a little relief because he explained to us the system there. There were altogether 33 fellow trainees in our group from different countries. The training was tough, with a lot of tasks to be done within a very short time. The trainees were required to observe strict discipline; should one commit an error, it would entail horrible punishment. Jitendra was by nature careless and easygoing, so he got the punishment of having to do additional tasks, on holidays, too.

At the conclusion, the instructor admired my performance in training; I passed both the physical and academic tests without any fault or flaw.

It snowed much in Britain at that time of the year. We were required to clear out the snow from the ground and there was no alternative to doing it. With the training in Mons over and still a little time left for the Sandhurst training, we three – Jitendra, Bhagwan Singh, and I – decided to go sightseeing in and around London.

We had limited money, but still we managed, cutting extra expenses. We spent every penny wisely or we would go broke. We received 90 pence on working days and two pounds on holidays as our training allowance. We had to manage on that nominal allowance, which is unthinkable now.

Our training in Sandhurst started one month after the one we completed in Mons. There were three colleges: New, Old, and Victory. We trainees were divided and sent to the three colleges. Luckily, Bhagwan Singh and I were sent to the New College but Jitendra was billeted in the Victory College.

At Sandhurst, there was the system of calling each trainee by the first letter of his surname. Thus Bhagwan Singh Thapa and I were called ST, a new way and practice to call the trainees. We were affectionately called ST and we became popular.

I got an opportunity to see the city of Inverness and the Loch Ness in Scotland and climb the Ben Nevis with the students of the New School. The trip was a mix of fun and adventure. In the last summer holiday, I went to the Isle of Wight. My stay in Britain allowed me to visit most of its important places, except Ireland.

During the training at Sandhurst, I contested in boxing with appreciative comments from both my trainee mates and my instructor for my performance. Once, a boxing team from Sandhurst went to compete at the Cambridge University. When I set my first foot in the ring, I found my opponent weaker. But in the second round, he landed a heavy punch on me. I lost my footing and fell down, staggering. I found myself in a Cambridge Hospital at night.

Another time, a boxing team from the Oxford University came to Sandhurst. I knocked my opponent in the first round and won without much effort. I excelled in boxing and swimming during my training.

Our training at the RMA (Royal Military Academy) in Sandhurst concluded on July, 29, 1971, with the King of Norway as Chief Guest at the passing-out parade. Unlike its long-observed tradition of three more months of training for the Young Officers' Course, we were, however, the only batch to return to our respective countries straight after the conclusion of our training at the RMA of Sandhurst.

David Allison, my training mate, offered to drive me to the airport. By the time we reached the airport, however, the plane had already taken off, an embarrassing experience for me! We panicked, but luckily there was another flight at midnight, and David arranged the flight for us. With a lot of time to spare, we moved to the Brighton Air Force Base Officers' Mess where we drank beer till 11 p.m. I boarded the aircraft, but it was a cargo plane, and I was the only passenger on board. Jitendra had left on an earlier flight to Calcutta. The next day, the plane had its transit halt at the Bahrain Airport where I again sadly found that all my connecting flights had already left, and I had to wait one more day there for my flight to Calcutta. After two days in Calcutta, I flew for Kathmandu, and thus the return trips from my training remained unforgettable and remarkable. I was coming home after three years abroad.

Back in Nepal, I went to report to the Office of the Director Military Training (DMT). The then DMT, Colonel Arjun Narsingh Rana – who later became the CoAS – welcomed us with a pleasant surprise: that another two officer cadets were selected to go to the RMA at Sandhurst, and as our posting order was yet to come, we were to train them for three weeks with morning orientation each day. Thus, Jitendra and I started training Gajendra Bahadur Limbu and Kul Bahadur Khadka. Limbu was active and a sportsman while Khadka was a little slow.

We were still at the DMT without our posting confirmed to a battalion. Just at that time, the Bhairavnath Battalion opened its Para Basic Training. So, Devendra Bahadur Shah, Jitendra Bista and I headed for it and waited for our parachute jumps. I was expecting my posting to the Kali Prasad Gana, a battalion of engineers where my study of Science subjects could be useful. But I was put in with the Gorakh Bahadur Battalion. I was placed on the frontline for the parachute jump. I jumped out without the slightest fear. At the conclusion of the training, Major General Dev Bikram Rana awarded us with our Para Wings.

I reported to the Gorakh Bahadur Battalion, then under the Royal Guards Brigade at the Royal Palace. At a Mess Night, as Colonel Rishi Kumar Pande, who later became Lieutenant General and the King's Principal Military Secretary, introduced me to the Chief Guest, Maj. Gen Dev Bikram Rana. He was surprised to see me in the Gorakh Bahadur Battalion. He also recalled the moment when he pinned my Para Wing on me. To his query why I was in that particular battalion, my answer was straightforward: 'Simply because I was sent here, and as a true soldier of the RNA, I have no hesitation to be in either Battalion. In fact, I wanted to join the Kali Prasad Battalion.'

Everyone there taunted me as though I had said something objectionable. I wondered what was so unbecoming in my conviction to cause everyone's mirth. In my lifetime, I have pursued one principle – tell the truth, no matter how bitter or unpleasant to hear it is at the moment. Truth delivers good results in the end.

As a training officer in the Gorakh Bahadur Battalion, I was assigned to conduct the Sandhurst model of training and passing-out parade to the trainees of the Gorakh Bahadur and the Kali Prasad Battalions. Despite limited resources, I did my best to provide the best possible training there.

The RNA still follows the same modus operandi in the training and passing-out parades that I set during my stint there.

I come to 1972. It was the year when King Mahendra passed away while he was residing at the Diyalo Bungalow in Chitwan. His mortal remains were instantly brought to Kathmandu. After the ceremony of his accession to the throne, the then Crown Prince Birendra was declared King of Nepal. After the crowning ceremonies were solemnized inside the Hanuman Dhoka Palace Complex, his horse carriage moved in procession to the Narayanhiti Raj Durbar (Royal Palace). I was assigned to the Guard of Honor contingent then.

The last rites of the late King Mahendra ended with State Honor, followed by a 31-gun salute. I was assigned to this Guard of Honor, too. There was a tradition of detailing a Brigade to the Last Honor where the Brigade Commander is the officer in charge of the last rites. Brigadier General Aditya SJB Rana was the Commander of the No. 1 Brigade in Kathmandu. It was when he was instructing the sequences of the last rites, the Commander of the Guard of Honor, a Major, interrupted him about the correct procedures to follow: to set alight the funeral pyre simultaneously while mounting the Guard of Honor for the Last Honor.

This caused some confusion.

While the other officers sided with the Brigade Commander's views, he turned to me and asked, '*Nani* (Son), what are the norms at Sandhurst?'

I replied that things there go according to what the Major had said. Hearing me, the General commanded that the procedures would be followed in that order.

Actually, there is no such formality for last rites at Sandhurst, and the manners in which the last rites are observed in Nepal and those in the UK do not match. Because what the Major explained sounded right, I supported him. Had he asked me if that was actually the routine at Sandhurst, I would be left answerless!

I had made arrangements for everything without any room for adverse comments. Nothing went lacking. So flawless was my arrangement that even the Battalion Commander turned to me to carry through the process of the last rites like those of the royal funerary exercises in Sandhurst.

There was a rule of maintaining two standby battalions at the Royal Palace. But for some time, a new system was introduced to relieve the battalions and send them for training. As scheduled, the Gorakh Bahadur Battalion was mobilized to Trishuli for training for one and a half years. We had a day's transit at Baneshwor.

I would like to relate something that happened in April 1972. The Battalion Commander, Lt. Col. Janardan Bikram Rana, returned to Kathmandu due to his bad health.

The training team comprised Lt. Col Shambhu SJB Rana, Major Chandra Bahadur Khatri (CB Khatri), and Captains Chandra Bahadur Rawal, Keshav Raj Sharma, and Shishir Chandra Shah.

They were all at the mess one evening when the bugle sounded. It marks the Retreat at sundown and is a military custom. I stood at attention, as is compulsory in the military. Major Shiva Bahadur also did the same.

But the other officers said there was no need to stand to attention as they were inside and under a roof and were seated.

After the last note of the bugle faded, I saluted the senior officers. 'Good Evening, Sir.'

But I was angry. I argued that the bugle at dusk signified our tributes to the departed souls and so it required everybody to stand at attention where and whosoever he was as a mark of respect.

Before I could complete my argument, however, Major CB Khatri ruled that 'those in the house needn't stand at attention during the Retreat.' When I politely reminded him of the norms, I noticed anger on his face. He said, 'I know the norms more than anyone else. A little fry of a Lieutenant like you teaching me, eh?'

I found him vainglorious and bullying. He mentioned his home base, the Devi Datta Battalion, where, he said, such observances did not apply. I said, 'But this is the Gorakh Bahadur Battalion, and I'm its operations officer, so it's my duty to have things done as per its standards.'

The row passed off that evening. It was at a party when the Major picked up a quarrel, bringing back the old issue of that evening. He heated up and flexed his muscles for a brawl. I was also ready to pay him back. I dressed him down. Other officers intervened and took him out. Capt Shishir Shah was so upset by the Major's behavior that he showed his dismay by slashing his hands with a blade, bleeding profusely. He was rushed to Kathmandu for treatment.

I later heard that Major CB Khatri got his nose broken in a scuffle with Jitendra Bista in Nepalgunj.

I got married to Geeta Shah, a daughter of a prominent Chautariya family of Gyaneshwor in Kathmandu. It was in 1972 when I was still at the Trishuli Barracks. I was not one of Kathmandu, but perhaps due to my excellent career in the Army, I got wedded to Geeta by an arranged marriage. Geeta's family was well known, with her father and uncle in the Army. Her father was Lt. Colonel Mohan Pratap Shah, and her uncle was Brig. Gen Singha Pratap Shah, who eventually became the CoAS, and after retirement, was appointed Ambassador to [West] Germany.

We have two sons, Ajaya and Kunda, born to us in 1974 and 1978.

With the training still going on in Trishuli, the Gorakh Bahadur Battalion, after one and half a years there, was to return to the Royal Palace, and I got my transfer to the Kali Prasad Battalion (Engineers) where I was promoted to Captain.

I was assigned as liaison officer of the British Royal Air force which was coming to Nepal for a mountaineering expedition, to climb the Dhaulagiri IV. Mountaineering was under the Ministry of Foreign Affairs, and it saw all expedition-related matters. I made all the necessary arrangements of all documentation from that section well in advance.

Just when I was ready to join the expedition, I heard Col. Khemraj SJB Rana, then Military Secretary at the Army Headquarters, relay a message to the Mountaineering Section not to send me on the expedition. I wondered what on earth goes on in the Army sometimes! I could not resist myself from asking Colonel Mohan Pratap Shah, my father-in-law, what the matter was. He was Military Assistant to the CoAS. He said he had no knowledge of it. But later, he called me to say that I was selected as one of the ADCs to the King, and therefore, I was called off the expedition. He also told me to contact the Royal Palace's Military Secretary's Office the following day.

When I went to the Military Secretary's Office, I found out that the Royal Palace had previously asked the RNA Headquarters to send another list of six officers. It was obvious that I was not on the first list, but I heard that I was selected as ADC to the King for two years. The second list that was forwarded to the Royal Palace had the names of Gyan Jung Thapa, Daman Nar Singh Rana, Jyoti SJB Rana, Shishir Chandra Shah. My name was the last one on the list. The first list included the names of Sadip Bahadur Shah, Chitra Bahadur Gurung, Ratna Bahadur Gurung, Nar Bahadur Gurung, Rookmangud Katawal, and Shashi Pratap KC.

My joy knew no bounds to know about Shashi and my selection as ADCs by the Royal Palace. I was extremely happy because to become an ADC to the King was not an ordinary matter in those days. It was a 'craze' to aspire to the position of prestige and potential promotions. But I had never even faintly imagined being where I would be going to.

Shashi Pratap and I relieved Capts Bhuvan Jung Shah and Niranjan Pal as ADCs. Similarly, Inspectors Ramkaji Bantawa and Sudarshan

Shamsher Thapa of the Nepal Police were relieved by Inspectors Motilal Bohara and Balendra Bahadur Singh.

We underwent our VVIP training to qualify ourselves (I stood first) as ADCs and took our appointment in April 1974, for two years. It was the time when Shanta Kumar Malla and Tara Bahadur Thapa were the Chief ADCs, and Sher Bahadur Malla was the Principal Military Secretary, and Dan Gambhir Rayamajhi as the Military Secretary.

The Principal Military Secretary, Sher Bahadur Malla, arranged our audience with the King on the very first day of our job. We saw the King with our offer of coins, a token of veneration for the monarch. He commanded us to work diligently, with the instruction to ask for clarifications when confused at anything during our service.

Shortly after taking over, I attended the phone calls and conveyed the messages to the King from the very first day. But the opportunity to escort the King had not presented itself so far.

I was still in office after my night duty on that particular day. ADC Tara Bahadur Thapa was supposed to escort the King but he had a bone dislocation, requiring him to visit the hospital. He called me with the news, and I went to the spot where and when the King was about to return from his gymnasium. The King was driving the car himself. He called out, 'Vivek, sit down.' I was a bit nervous to sit next to the King, but my nervousness did not last long as he called me by my first name. I found him a down-to-earth gentleman because he expressed his concerns whether or not Tara Bahadur Thapa visited the hospital. He drove to the Shri Sadan, his residence where he stayed till 11 o'clock, reading newspapers and necessary documents, and then came to office.

I shared my experience of accompanying the King for the first time with my wife Geeta.

King Birendra had his coronation in 1975. That same year, with only two months left for my two-year term as ADC to expire, the Chief of the ADCs, Col. Shanta Kumar Malla, asked for my wish if I would love to extend my service as ADC in the Royal Palace. When he assured me of having my preconditions met for regular military training, regular promotions, and travels abroad with the King, I agreed to continue in the Royal Palace Service as ADC.

Afterwards, I was confirmed as ADC on a permanent basis.

With my posting still in place at the ADC Office, I got an opportunity of one year of training on the Command and General Staff College Course at Fort Leavenworth, Kansas, the US, considered as one of the best military training institutions in the world. I found myself very lucky to have that opportunity, as only high-profile and caliber army officers are received there for training to acquire the knowledge of the US Army. This particular visit is considered 'the best year of my life!' by every trainee. Altogether there were some one thousand participants in the training from some 60 countries. The very training made a significant difference in my personal life and redefined my career in the Army. Meeting and training at Fort Leavenworth with the best officer material in the world was an extraordinary fact of my military life.

My Interest in Music

I have been fond of music since my childhood. I learned to play the madal (two-sided drum) and bansuri (flute) without tutoring. While at school, I often took part in stage performances. My friends and classmates often admired my singing but I never bothered to learn dancing. Although I had a passion for classical music, I could not pursue it to the level of my heart's content. I wanted to go to Lucknow to learn South Asian classical music when my uncle was there but he returned home.

My entry into the Army in 1968 and the initial three years of training put a brake on my interest and involvement in music.

After my return from Britain, I was with the Para Training School for the paratroopers' training at Maharajgunj. The Commander of the Para Battalion (now the Bhairavnath), Lt. Col Yadwan SJB Rana, himself being a music aficionado, and perhaps to keep the soldiers in good humor, set an informal routine for refreshing and fun with music in the evenings. He would love it when I played the madal and sang folksongs.

I remember one event when my unit was in transit at the Baneshwor Barracks (where the BICC stands today, at 'New' Baneshwor). Once, when the CoAS, Gen. Singh Bahadur Basnyat, hosted a dinner in honor of the visiting Indian CoAS, General Sam 'Bahadur' Manekshaw at Shashi Bhawan (the official residence of CoAS), a similar musical event was organized by Lt. Col Yadwan. I knew that he had arranged several ensembles from several units to sing songs and perform dances. He came to call me. Giving me local Nepali costumes to wear and with the approval of my CO, Janardan Bikram Rana, he asked me to accompany him to Shashi Bhawan for a cultural program to play the madal and sing. He convinced me to sing and play the madal along with some soldiers as artistes. He also put me in charge of the team. There was no question of refusing to obey the orders of the Colonel. Thus I was asked to join the musical team to perform without prior rehearsal or thought.

Without elaborate stage decoration and rehearsal, we played the music and sang songs in honor of the guest. The musical program was followed by the introduction of the performers. I was introduced as a '2nd Lt., a fresh returnee from the RMA of Sandhurst' (and the only ranking officer participating in the program).

The soirée turned out to be very popular as everyone admired the songs and dances. It was almost midnight when the program was over. Lt. Col Yadwan took me to a restaurant at Swayambhu where he forced me to sing and play, and I was still in my vaudeville costume.

I also remember the time in 1971 when a football match was organized between two battalions: the Gorakh Bahadur and the Kali Bahadur. Their Royal Highnesses, Princes Gyanendra, and Dhirendra, each played from the teams while the Crown Prince, Birendra, umpired the match.

The game was followed by a dinner in the evening. Lt. Madhu Sudan Sapkota and I presented a stage performance to entertain Their Royal Highnesses. The show, blended with wines, lasted up to 11 PM without noticing how quickly the night wore on.

Crown Prince Birendra seemed to like my folksongs very much because he called me for an introduction. I introduced myself to the future King that I joined the Army as a local lad of Rukum and recently completed my training from the RMA at Sandhurst. That was how I became familiar with King Birendra. Perhaps because he was impressed by my talents, the King chose me to serve as his ADC from among several other senior officers.

Later, when I became his ADC, King Birendra recalled the moments when I had entertained him with my songs and dances. I was happy to hear it from the King. I knew that the then Crown Prince, later as King, took notice of even the smallest things we did or said.

I remember the year 1979 when King Birendra, representing Nepal, had been to Havana for the Non-Aligned Nations' Summit. As ADC, it required me to accompany him. On our return trip, we halted in London where the Royal Nepalese Ambassador, Jharendra Narayan Singh, hosted a reception for the Nepalis living in London and some British friends of Nepal, to be followed by a dinner in honor of the King and the royal entourage. With the dinner still going on, Senior Captain Bobby Bikram Shah of the RNAC had the idea of an impromptu music program, asking two of the RNAC employees to entertain the King with songs. I also joined in, livening the evening. Captain Shah appreciated my participation and offered me complimentary air tickets to be used at my own convenience. Later, my wife Geeta felt the need of a medical checkup and treatment in Bangkok. I had only mentioned it to Captain Shah and the KTM-BKK-KTM tickets for the two of us promptly arrived. By the way, that was the first and the last time I used my free jaunts on the RNAC!

King Birendra also was fond of music, painting, swimming, and hunting. Most often on his hunting trip, I accompanied him to various places. At times, I made him dance while I played the madal.

In 1983, I accompanied the King to the Royal Chitwan National Park where the hunting would last the whole day, and the evenings, without fail, would become remarkable with the King inviting all the members in the entourage to drinks and dinner. Needless to say, drinks and dinner would lack taste without music and dance. The members would dance and play music turn by turn. At our request, the King would also dance and play the madal.

One particular evening is still fresh in my memory. The Minister of State for Forest and Soil Conservation, Badri Prasad Mandal, also joined one dinner. Drunk and eating, he rose to dance. But a bone got stuck in his throat. He struggled frantically. Rushing him to the hospital, the whole fun of the dinner was spoilt. As the doctor gave a whack on the back of his neck, the bone sprang out like a released spring. Thank God he survived the mishap.

When it comes to music, I recall the precious moments back in 1965 when there was a mela (fun fair) held at the Bhrikuti Mandap to mark the birthday of the then Queen Ratna. There was a Radio Nepal stall where I impulsively got up to sing a song.

When I finished, an employee in the technical section of Radio Nepal said, 'You qualified in the voice test. If you want, you can be a recording artist at Radio Nepal.'

It upped my confidence, giving me a measure of satisfaction. But I would never go in that direction because my destiny was leading me somewhere else.

But my love for music is still alive.

Ups and Downs

The year 2001 turned out to be very unfortunate both for my family and the Royal Family I served for so long as though all hell broke loose over us. My mother passed away in April of the same year. It came as a shock to me. I was able to shake off the shocks. But Their Majesties King Birendra, Queen Aishwarya and their entire family, with their closest relatives, met their fateful death that same year. The Royal Palace Massacre shook the entire country from the very foundation, sending shockwaves across the world. As if this was not enough, my own father also expired the same year.

I was mourning the death of my father at the Pashupati Aryaghat. Maybe because I was Military Secretary to the King, people from eminent walks of life, ranging from the Prime Minister to the Secretaries, the heads of the security forces and highly placed officials of the Civil Service came to express their condolences to me on my father's demise. When I was there to observe my mother's death rituals, a few months earlier, King Birendra had come to express his condolences on my mother's demise. But King Gyanendra did not keep up with his late older brother's legacy even though I was still serving in the post of Military Secretary to him after the demise of King Birendra.

On the ninth day of the mourning for my deceased father, the Acting Head of the Principal Military Secretariat, Shri Ram KC, came to me and broke the news of likely administrative structural changes in the Royal Palace. The proposal included the merging of the Principal Military Secretariat with the Military Secretariat, that too under the Principal Secretariat, a civilian entity. At that moment, I only said that it was a sensitive matter that deserved hard thinking with cool minds, requiring a lot of discussions and rethinking before such a move was taken.

As was instructed, I reached my office to sign my attendance on April 4, 2002. I came to know that a meeting to discuss the proposed organizational structural changes was convened in the office of Sharada

Prasad Pradhan, the Chief of the Royal Household, the same day. Pradhan had issued his notice to everyone that the new structural change would go into effect from April 14.

When I asked Pradhan about the relevance of the sudden changes, he simply passed it as 'the King's command'.

I described to him the history of the Principal Military Secretariat and the Principal Secretariat, the oldest of the offices in the Royal Palace ever since 1954, with the provision of a Lieutenant General as the Principal Military Secretary to the King. Given the growth of the strength of the Army from 15,000 then to over 50,000 now, the proposed new organizational structure was not justified by any means. It might raise practical complications for the CoAS to deal with the Principal Secretary, a civil appointment, in matters of immediate security importance.

Maybe Pradhan found my logic reasonable, so he promised to arrange my audience with the King so that I myself could brief him on the gravity of the matter.

As the King had wished, the Chief of the Royal Household and I reached Nirmal Niwas at 9 o'clock where I briefed His Majesty on the matters of national security. I presented to the King the likely hierarchical difficulties if the planned administrative structural changes were to go into effect.

As a rule, for instance, the Bills ratified by the Parliament and the Government and forwarded to the Royal Palace for the Royal Seal of Approval required to be undersigned jointly by the Principal Secretary and the Principal Military Secretary or the Military Secretary. But the proposed structural change was sure to pose legal or hierarchical complexities. Besides, the Principal Military Secretary was to answer for the overall security of the Royal Palace. According to the proposed structure, the Royal Guards Brigade, the Royal Guards Military Police and the Royal ADC Office apparently came under the Principal Military Secretariat. Therefore, the proposed new structure would obviously pose ambiguities in the proper channel with dilemmas, such as, say, who should correspond with the RNA Headquarters, the Ministry of Defense, and the heads of the security forces.

That was how I drew the likely bureaucratic ramifications that would be caused by the new structure in the future.

The King keenly listened to me and finally gave his command to operate the Principal Military Secretariat as per the existing system, with the provision of all ranks and files from the Army; and the Principal

Private Secretariat, the Principal Press Secretariat, and the Principal Personal Secretariat to be under the Principal Secretariat as per the new administrative structure.

The Crown Prince advised that there should be appointments of young officers when it came to filling the vacancies in the Principal Military Secretariat. He sounded angry when I offered my advice to fix the positions above the rank of Major in the vacancies and commanded me to settle it with the arrangement for four officers in place, with two Lieutenants and two Captains.

The established tradition of the appointment of the Principal Military Secretary made from among those with experiences as ADCs to the King was strictly observed, even more so since the mobilization of the Army against the Maoists began in 2002.

The ambit of the Military Secretary covers all military-related issues to be briefed to the King. This means that matters pertaining to the Royal Palace, the Army, the Police, and the Ministry of Home Affairs fall under the scope of the Military Secretary. It is the Military Secretary again who cosigns the Bills before the King grants them his Royal Seal of Approval.

Although it is only the King who decides who is to be appointed, transferred, promoted, or whose term of service is to be extended, I also gained the experience of the patterns and the practices of those norms at the Royal Palace. I continued my service to the King as his ADC for another one year and four months even when I was promoted to Major General. Later, he entrusted me with the additional responsibility of the Military Secretary. Even though I served in the ADC Office for a long time, I don't as yet know what delayed my promotion to Lieutenant Colonel though the post was lying vacant for years. Then again, with only a little time left for my rank as Colonel to expire, I had similarly sensed a strong force working to block my promotion to Brigadier General. Despite all that, I got the promotion in any case. There were people to wish me ill, as has already been seen in the case of Col. Shashi Pratap KC, too. Nonetheless, I got promoted to Major General and then to Military Secretary after the retirement of Maj. Gen Tara Bahadur Thapa.

All along the decades of my service at the Royal Palace, however, I always saw to it that I never let my personal and professional flaws creep in fulfilling my duties.

It was quite natural for people to criticize me for the remissness in the security system of the Royal Palace when the Royal Palace Massacre took place. Therefore, on my part, I had offered my resignation to the new King from the post of Military Secretary on moral grounds, as the Royal Palace Massacre took place on my watch because I was in charge of the overall security of the Royal Palace. The resignation was not accepted, with the King explaining his decision.

I heard that the CoAS, Gen Prajwalla SJB Rana asked the Adjutant General, Maj. Gen Sadip Bahadur Shah, to restructure the overall security system of the Royal Palace, given that it failed to the extent of the Royal Palace Massacre to happen. I had also got wind of Adjt. Gen Shah having a work plan prepared by people in his confidence to that effect.

In a series of incidents, Adjt. Gen. Shah called me one evening in 2001 to say he desperately wished to meet me. I asked him to drop at my house in the evening at around 8 p. m. He disclosed that the CoAS had asked him to make fresh security plans and stay prepared to take over the responsibility of the Military Secretary in the near future. That would mean my replacement. He said he could not help sharing it with me. Then I remembered that conspiracy still plays its games in many places in the country when it comes to power and position.

What I now heard reconfirmed the background noise I had been hearing for quite some time. So I told him that the appointments and replacements of the Military Secretary went as per the wish and command of the King, not according to the likes or dislikes or recommendations of the CoAS. That was still the system followed.

I happened to meet the former Military Secretary, Lt. Gen Shanta Kumar Malla, at a party thrown by ADC Shashi Bikram Singh at his residence on the occasion of his promotion to the rank of Colonel. Malla said he heard about my replacement by Lt. Gen Victory SJB Rana. He said he heard many people congratulate Rana for his likely placement to the new position.

I could not make sense of what was going on. One day, however, when I eventually met Gen Rana and asked him about the matter, he said, 'I'll come to your house for dinner and shall tell you everything about it.'

I invited him one day, and he came with his wife for dinner at my residence. I was more than surprised when I referred to the speculations he said there was nothing about anyone congratulating him. In fact, it was

Lt. Gen Malla himself who took him to a corner to break the news that he was soon going to take over the new responsibility of Military Secretary. Victory vowed he would never conspire against me, as he was indebted to me for my help in his retention at his post at a time when the CoAS, Gen Dharmapal Bar Singh Thapa, was set to retire him from the post of Brigadier General.

Nonetheless, I could still sense the presence of a strong force in and around the Royal Palace active to terminate my appointment as Military Secretary. My style of working in the strictest pursuance of the norms and values of my office must have hurt them in a number of ways in their pursuits of their petty interests.

Amidst these ups and downs, I put an end to the speculations one day by resigning the post of Military Secretary, on November 7, 2003. It was commanded by the King.

The End of the Road

The Royal Family had a long tradition of holding their get-together and have fun on Fridays. The dinner would invariably be followed by movies. There was no television in those days. We, the ADCs, were also allowed to join the group to watch the movies, seated a few benches at the back.

The fun and the stillness of one night in 1978 were broken by the telephone ringing at 11 p.m. I received the call, as was the rule. It was Her Royal Highness Princep Shah who asked me to pass the call to His Majesty. After a brief talk, the King turned to me and told me to get his car and security detail ready to drive to Tahachal and the residence of HRH Prince Himalaya Shah [King Mahendra's second younger brother and uncle to both Kings, Birendra and Gyanendra.] Their Majesties, along with the Queen Mother, also hastened to the carcade. HRH Prince Himalaya Bir Bikram Shah had become unwell and required to be taken to hospital. On the way to Tahachal, I informed the security detail to secure the Bir Hospital where the Prince was to be taken.

We found a team of doctors and an ambulance already there. Although everything possible was done, he could not survive the heart attack. Soon the doctor declared him dead, and an uneasy silence fell.

I saw Princess Princep placing her head on her husband's feet and say, 'How lucky of you, Hazur, to depart early enough to avoid the sight of Nepal's catastrophes!'

Was she was referring to the likely end of the Panchayat Political System, or some bad omens to befall the monarchy? I could not help being shocked by her epiphanic statement. Barely 12 years later, the Panchayat tumbled, and the monarchy, too, some years later. Her prophetic words still resonate in my ears.

It feels like just yesterday when my neighbor, Dr. Trailokya Nath Uprety [former Vice Chancellor of Tribhuvan University, and Ambassador], dropped one morning at my house to say he foresaw ill fortunes besetting the Royal Palace. It was the day after I had resigned as Military Secretary. But he, in fact, took my summary dismissal by the King as a godsend!

Barely five years had passed since I had resigned as Military Secretary when the monarchy came to an end in Nepal, something the people in the country had not even thought or imagined would happen.

HRH Princep and Dr. Uprety are no more but their individual angst proved right, and very shortly at that.

During my service in the Army and in the Royal Palace, I constantly saw to it that there was no slightest error on my part which I would regret and be ashamed of. I did my duty to the best of my knowledge and ability, and it occurs to me now that my simplicity, honesty and straightforward manner guided me ahead to success.

The training I took at the Royal Military Academy, Sandhurst, and the Command and General Staff College in the US and the divine quote of 'Sacrifice is the greatest virtue' made me bold enough to shoulder whatever responsibilities were entrusted to me and to remain unafraid to state anything wisely and objectively. I always held my moral strength high without ever getting tempted by personal gains or favors. It was my clean image that gave me the confidence to ask King Gyanendra to accept my resignation when the opportune moment came for me to do so before the environment of trust could get obscured if I lingered any longer at the post.

My service at the Royal Palace was not as easy and glamorous as people outside thought. It was akin to walking on a tightrope, as the employee was required to speak or move according to the moods of the members of the Royal Family; they would vent out their ire on the ADCs or the employees any time.

It was also not as though one had certain working hours and was relieved at the end of the shift. The duty hours would extend, and there was no particular time even for food and sleep.

Obviously, therefore, the ADC's family life would be messy. His family members would complain of him/her as being possessed more by the job than anything else. Frankly speaking, I have not taken as much care of my own sons as I have given my care and attention to Crown Prince Dipendra, Prince Nirajan, and Princess Shruti. I have even been a toilet attendant to Crown Prince Dipendra in 1982 when he was on a sojourn to Rara Lake and had suffered acute/altitude mountain sickness (AMS).

Although the principal job of the ADC was to provide security to Their Majesties, he was required to do any chore that appeared; getting taunted was part of it, too.

One's fright would obviously rise high when Queen Aishwarya drove her car, the nightmare of an accident always haunting the ADCs as they were required to sit next to her. I would silently pray to God to deliver us safely.

Frankly speaking, in retrospect, I don't find my military training and skills used as wisely and fully during the years of my service at the Royal Palace. The nature of the job sometimes deprived me of opportunities of further military training, but timely promotions and service to the monarchy gave me a sense of satisfaction.

Around 1985, I had thought of leaving the Army after three more years in it when it would reach my 20th year in the services, and I would live in retirement as Colonel to lead a free life. I even consulted with my wife, Geeta, regarding my thought of retirement. She wondered whether I was mad to quit such an important job people wish to have, suggesting me to continue with my appointment in the Royal Palace. But to live the life of an independent man to pursue my own interests always lay dormant inside me.

Hindu religious figures, such as the most revered Shankaracharyas and Swamis from home and abroad (India), had arrived in Kathmandu as invitees to the Crown Prince Dipendra's *bratabandha*. Present on the occasion were the Shankaracharya, Sworoopananda Saraswoti, and the Shankaracharya, Jayanta Saraswoti, of the Dwarka Dham and Rameshwaram, in India. Nepal's own most revered sage, the Khaptad Baba, was also present. Special arrangements had been made for their meals, rest and stay in different places, with ADCs deputed to see to their comfort and security.

I was assigned to monitor the overall arrangements to see to it that there was no lack of anything in service and reverence. My duty also included acting as a messenger between the Shankaracharyas and the King.

I had an urge to call on the Khaptad Baba who was staying at Dhrubasthali near the airport. I was growing impatient to ask him about the present and future of the country, and my working life as well. Although I also wanted to ask him his age, I dreaded not to do so. However, I did ask it in a diplomatic way.

In reply, the Khaptad Baba asked me to visit him in Khaptad, if possible, where he would deal with my concerns. In Kathmandu, he had time constraints and he was distracted by people visiting him. I found his suggestion appropriate and so promised to visit him in his abode in Khaptad soon.

In April 1988, I left for Dipayal and halted for two days at the No 5 Brigade Headquarters whose Deputy Commander was Lt. Col Prakash Bahadur Basnyat, my coursemate and very good friend.

I found the panorama of the Khaptad National Park captivating. We two walked to the dwelling of the Baba who had been living there for the last 40 years. He had come to Nepal for the first time when Dev Shumsher was the Rana Prime Minister, and ever since he lived in many places of religious importance in different parts of Nepal. Adding up the number of years of his hermitage and his stay in a particular place, or that, and the other, I guessed his age tentatively to be somewhere between 130 and 135 years.

To visit the sage in his own habitat was a unique experience and a mental solace. Without seeming to be in hurry, he started talking about the shifts of attitude in people over the long period of time. He said he found people too result-oriented without the patience to wait and make a sacrifice. The sage described the whole of humanity as oscillating between self-centrism and self-seeking, leading to unprecedented problems and diseases without names and cures. Due to the advancement of science and technology and material prosperity, time and space have shrunk. Life expectancy has drastically increased but the lack of peace of mind has troubled the individuals.

After a long philosophical musing, he turned to my two questions and said that I still had a career in the Army and the Palace, and the King and the country still needed my services. So, he suggested that I should continue in my present works.

Then he chanted a verse in Sanskrit: *'Na haste yestimaadaya dewa rakshanti sadhakam, yam tu rakshitumikshanti subudhya yojantitam.'* It means: God does not manifest Himself with a stick to protect a person; instead He does so with the endowment of knowledge and reasoning.

He thus advised me to reconsider the thought of quitting my job. I dropped the idea of resigning and returned to Kathmandu.

King Birendra knew very well that although Shanta Kumar Malla and Tara Bahadur Thapa were the ADCs assigned to him and appointed on the same date, they were not in good terms and tried to outsmart each other.

Similarly, Shashi Pratap KC and I joined the Royal Palace service as ADCs on the same date. Maybe the King feared the same ghost would also haunt the two of us, so he retired Shashi Pratap KC with the rank of Colonel and retained me at the post. I knew about it later from Brig. Gen Pradip Bahadur Singh who heard it from Crown Prince Dipendra when he said it in an informal way on his Pokhara sojourn.

King Gyanendra sought my exit, and to ease his search for another person to replace me, I left the space open by instantly offering my resignation. Being a true soldier of the RNA with a long service and experience in the Royal Palace, there was no question of my loss of respect for the institution of monarchy. I was not confused one bit that Nepal needed monarchy for the collective good of the country and its people.

Even after my ouster from the service at the Royal Palace, I was keen to see that everything went well there; that the ADCs' briefings to the King on the state of affairs were being done objectively and without flattery. If I ever happened to meet those in service there, I always told them to keep the King informed of the ground realities without any inhibition because, otherwise, briefing the King would, in the long run, be more harmful than good to the monarchy if they tried only to please him in their reporting.

I was surprised to hear Devi Ram Sharma, the Chief of the NID, telling me that those serving at the Royal Palace have all turned out to be 'yesmen' without guts to give the King correct advice even though it did not sound pleasing to report at that moment. Sincere advice normally does not sound good, just as bitter pills have blessed effects.

The former CoAS, Gen Pyar Jung Thapa, at a wedding party of the daughter of IGP Sahabir Thapa, shared more or less similar views with me. He said that no one any longer briefed the King on the day's issues quite like the way I did. Laughing, I said it was too late to talk about those things now, as he had been granted the King's audience on November 8, 2003 – the day after my resignation. He said nothing.

I also came to know that all the security chiefs missed me at the Royal Palace because I always told the King the truth without caring whether he liked it or not. I thought of a folksong:

Ratama tara jhulkinchhan sastai,
Euta pani hundaina juun jastai.

Stars shine in the night sky, dimes a dozen
None of them is like the solitary Moon.

Later I heard the then Crown Prince, Paras, tell his friends, 'I hated his guts.' Here 'he' refers to me. He was also right to say that, as the interview with him in the *Jana Astha* Weekly justifies it when he said that the Royal Palace did not lack Secretaries who spoke their minds with sound and balanced criticism.

Although I offered my resignation the instant it was sought, I still don't know the exact reason/s why was it sought.

I was a young man when I joined the Army and started training in 1968. Now it occurs to me that it was wise of me to join the services whose training and experiences have strengthened my confidence to cope with situations no matter how adverse and tough. I think of my service at the Royal Palace successful; it may be seen otherwise through others' eyes. I have rated those years of service and experience as the best and sweetest for me.

I still carry my conviction that, even though the end of the monarchy is attributed to a combination of inabilities to move with the times, my assessment is as follows: The King's failure in his accurate reading of the popular expectations, his refusal to recognize the political parties' aspirations despite their flaws, his overmuch trust in those advisors who were already failed individuals, the dearth of timely identifications and analyses of good and bad advisors, and the lack of the royal will to foil external meddlings in Nepal's internal matters that brought about the monarchy's end in Nepal.

In addition, King Gyanendra was too confident of himself; and his close advisors, such as Sharad Chandra Shah and Ramesh Nath Pandey, were actually self-centered, working only for their personal interests.

The myth that the Gorakhnath Baba had told King Prithvi Narayan Shah that the Shah monarchy would come to an end in the 13th generation of the Shah Dynasty came to be true. The monarchy of 240 years of standing went to be relegated to history in Nepal. It is said that history repeats itself and there are instances of the return of the monarchy in the world. Time alone will tell if we will see the monarchy returning or not.

I am happy that I did not have to be a part of the monarchy's end. But the thought that unfortunate things happened to the monarchy that I had an opportunity to serve 30 years pains me all the while.

Index

A

AB139 or HUEY-2 298
AB Bardhan 227
Abhishekh Shah 8
abir 22
Achham 60, 200, 201
Achyut Krishna Kharel 48, 56, 73, 74, 75, 103, 106, 113, 115, 122, 289
ADC XXI, 1, 3, 4, 5, 7, 8, 9, 10, 12, 13, 14, 15, 17, 18, 23, 24, 30, 31, 32, 33, 34, 35, 36, 37, 40
ADC Office 1, 4, 8, 9, 10, 14, 15, 17, 30, 31, 32, 107, 139, 153, 300, 403, 404, 407, 408, 409, 412, 416, 429, 433, 434
Aditya SJB Rana 421, 425
Adjutant General 154
Agni Prasad Sapkota 170
AIG 9, 21, 46, 56, 57, 123, 126
air defense 42
Air Force 146, 178, 421, 423
Ajaya Bikram Shah XXV
Alam Devi 182
Aldershot 36, 421
Amar Lama 70
Amar Singh Shah 9, 122
Ambassador 3, 8, 9, 35, 36
ambush 78, 111, 206, 257
America 110, 142, 156, 200, 201, 214, 272, 304, 309, 328
American Embassy 255, 260, 322, 328
Amik Sherchan 267, 358
Amlekhgunj 112, 141, 149, 402
Amnesty International 116
Amod Prasad Dhungana 341
Amod Prasad Upadhyaya 161, 210, 261
AMS 438
Ananda Sworoop 165
Anil Kumar Jha 388
ANNFSU 309, 316, 346
Anuradha Koirala 90
APC 226, 264, 336, 371
APF 16, 23, 55, 56, 57, 60, 61, 88, 98, 101, 118, 119, 123, 129, 135, 139, 142, 148, 149, 150, 151, 152, 160, 164, 166, 175, 177, 179, 187, 189, 195, 215, 216, 226, 276, 279, 286, 292, 309, 311, 312, 314, 322, 331, 349, 366, 367, 370, 377, 378
Arjun Jung Bahadur Singh 195
Arjun Narsingh KC 186, 374
Arjun Narsingh Rana 423
Armored Personnel Carrier 264. *See also* APC
Army Day 137, 202, 294
Army Officers' Wives' Association 181, 268
Army Training School 125, 135
Army Welfare Fund 106, 113, 116, 125, 129, 131, 162, 194, 196, 205, 233, 240, 258, 278, 279, 320, 404, 405
AR Raghunathan 334
Article 115 244
Article 127 XXII, 244, 245, 246, 391
Aryaghat 20, 21, 22, 26, 27, 143, 183, 211, 397, 432
Arzu Rana Deuba 163, 165, 227
Asha Kaji Basukala 110
Asharam Shakya 406
Ashok Kumar Rai 388
Ashok Shrestha 5, 10, 147
Ashok Singhal 101
Atal Bihari Vajpayee 216
Atlanta 386, 388, 389
Australian 99
Austria 184, 401
Austrian 401
AV Mathur 58
Avro 108, 305, 410
AY Tipnish 131

B

Babar Mahal 260
Babita Basnet 9, 281, 390
Baburam Bhattarai 28, 58, 63, 88, 133, 165, 168, 209, 210, 217, 259, 276, 278, 290, 294, 295, 296, 302, 308, 309, 310, 316, 320, 321, 323, 324, 327, 331, 335, 336, 349, 353
Baby King 397
Badri Prasad Mandal 90, 221, 290, 306, 431
Bahrain 359, 423
Baidhya 99, 157, 168, 316
Baitadi 157
Bajhang 237
Bajra Gurung 77, 109, 110, 188
BA Kumar 162
Balananda Sharma 349, 415
Balaram Gharti Magar 77
Baldev Raj Mahat 295
Baleshwor Agrawal 145
Bal Krishna Neupane 290
Baluwatar 17, 85, 285, 286, 333, 404
Baman Prasad Neupane 302
Bam Dev Chhetri 239
Bam Dev Gautam 10, 46, 53, 73, 74, 77, 254, 257, 274, 315, 316, 321, 332, 341, 342
Bam Kumari Budha 165
Banfikot 149, 418
Bangaru Laxman 140
Bangladesh 59, 60, 61, 105, 348
Banglamukhi 177, 179
Banke 107, 378
Basantapur 260

Basanta Raj Mishra 389
Bashu Rishal 79
Beijing Olympics 65
Belgium 237, 263, 277
Belgrade 34, 35, 66
Bel Prasad Shrestha 164
Benazir Bhutto 348
Beni 383, 384
Ben Nevis 422
Bhagwan Koirala 16
Bhagwan Singh Thapa 422
Bhairavnath Battalion 118, 154, 311, 424
Bhakta Bahadur Basnet 112
Bhaktapur 89, 180, 305
Bhalubang 376
Bhandarkhal Parba 65
Bharat Bahadur Rayamajhi 151
Bharat Gurung 68
Bharatiya Janata Party 101, 252, 328
Bharat Keshar Simha 36, 101, 206, 222, 329, 338
Bharat Mohan Adhikari 74, 232, 322, 340, 352
Bharatpur 128, 134, 418
Bhasha Pathshala 418
Bhekh Bahadur Thapa 114, 208, 239, 244, 402, 404
Bhendetar 118
Bheri River 120, 418
Bhim Prasad Gauchan 373
Bhim Rawal 388
Bhupendra Bahadur Thapa 170
Bhushan Dahal XXV
Bhutan 65, 174, 264, 267, 299
Bhutanese 264, 271
BICC 191, 218, 308, 347, 355, 397, 429
Bidhya Devi Bhandari 70
Bijaya Kumar Gachhadar 125, 126
Bikas Bikram Shah 137
Bikram Era 140
Bikram Sudh 235
Billiard Room 16
Bimalendra Nidhi 208, 388
Binayak Bahadur Singh 77
Binod Bikram Shah 340
Binod Raj Gyawali 28
Binoj Basnyat 298
Bipin Karmacharya 404
Bir Bahadur Balayar 134
Birendra Bahadur Shah 18, 81, 347
Birendra Jhapali 347, 352
Birendra Military Hospital 9, 15, 124, 153, 157, 162, 253
Birendra Shrestha 352
Bishnu Bahadur Gurung 151
Bishwo Bandhu Thapa 371
Bishwo Bikram Shah 229
BJP Chairman 140
Boao Conference 135
Bobby Bikram Shah 431
Bofors 401
Bonn 35
Bosnia Herzegovina 378
Brahmanand Udasin 168
Brahmin 174
Brajesh Mishra 107, 331, 359
Brazil XV
Brigade 7, 13, 18, 53, 77, 107, 109, 151, 152, 157, 169, 180, 185, 188, 189, 197, 201, 206, 209, 215, 223, 233, 242, 261, 294, 298, 305, 327, 329, 344, 346, 351, 373, 375, 424, 425, 433, 439
Brigade Commanders 109
Brigade of Gurkhas 180, 233, 242, 351, 373, 375
Brig. Gen XXV, 4, 7, 8, 9, 15, 18, 37, 62, 77, 106, 126, 139, 147, 151, 154, 159, 164, 178, 182, 188, 202, 206, 218, 230, 258, 265, 266, 297, 300, 301, 310, 314, 316, 320, 370, 397, 404, 407, 408, 412, 415, 426, 440
Britain 10, 7, 33, 36, 42, 59, 128, 131, 149, 183, 238, 243, 249, 266, 270, 283, 291, 305, 344, 394, 403, 420, 421, 422, 429
British Ambassador 230, 250, 291, 292, 367, 369, 373
British Defense Minister 131
British Embassy 124, 127, 287, 293, 369, 375, 421
British Gurkhas 99, 124, 242, 351, 373, 422
British Royal Air force 427
Bryan McQueen 386
Buddhi Goshain 329
Buddhi Man Tamang 360
Buddhist 22, 103, 266
Budhanilkantha School 31, 35

C

Cabinet 1, 5, 24, 51, 62, 76, 80, 82, 85, 89, 91, 93, 94, 104, 119, 121, 150, 158, 177, 189, 197, 200, 224, 240, 245, 248, 249, 252, 253, 255, 256, 257, 258, 261, 263, 266, 267, 286, 287, 295, 301, 313, 319, 325, 329, 333, 334, 338, 340, 341, 347, 348, 354, 356, 361, 363, 365, 366, 367, 392, 403, 404, 413
cadet 7
captains 71
Carron Basset 385
Carter Center 63, 385, 386, 387, 388, 389, 390, 392, 393
CB Khatri 425, 426
CDO 127, 185
Center for Humanitarian Dialogue 83
CGS 54, 141, 148, 155, 175, 180
Chai Haosain 137
Chairman 17, 18, 19, 94, 121, 140, 166, 182, 184, 192, 204, 227, 238, 245, 248, 254, 258, 287, 290, 293, 295, 305, 306, 307, 309, 392
Chakra Bahadur Thapa 150
Chakra Bandhu Aryal 372

Chakra Prasad Bastola 78, 79, 114, 158, 160, 198, 209, 210, 217, 227, 261, 284, 299, 337, 387, 388
Chakrata 11, 60, 61, 62, 105, 314, 367
Chandra Bahadur Rawal 425
Chandra Bhandari 332
Chandrabir Gurung 406
Chandra Man Shrestha 269
Chandra Narayan Yadav 112
Chandra Shekhar 196, 198, 335
Channauta 376
Charles Mendies 348
Charles Shovraj 367
Chaturman Singh Rajbanshi 279
Chautarias XVI
Chengdu 172
Chhabilal Khadka 209, 258
Chhatra Man Singh Gurung 159, 399
Chhatra Subba 128
Chhauni 162, 420
Chief Admiral 146
Chief District Officer 127
Chief Justice 17, 19, 26, 72, 136, 230, 255, 327, 338, 371
Chief of the Army Staff XXIII, 3, 309, 320
Chief of the Royal Household 6, 7, 17, 159, 170, 175, 289, 433
Chief Secretary 107, 143, 195, 263, 289, 315, 388, 419
China XV, 42, 43, 58, 60, 61, 63, 64, 65, 66, 67, 102, 106, 107, 115, 127, 135, 136, 137, 138, 139, 145, 148, 154, 155, 159, 163, 171, 172, 173, 177, 179, 182, 186, 188, 189, 192, 204, 222, 226, 231, 238, 240, 264, 286, 297, 306, 336, 342, 343, 344, 349, 351, 359, 364, 367, 371, 387, 390, 394, 400
Chinese Ambassador 3, 8, 138, 171, 188, 204, 225, 229, 231, 232, 240, 243, 266, 275, 293, 304, 306, 307, 322, 341, 342, 343, 344, 349, 350, 355, 359, 364, 371, 378
Chinese Defense Minister 137, 138, 139, 145
Chinese Embassy 3, 65, 264, 304
Chinese President 182, 343
Chiranjivi Wagle 208
Chiran Shamsher Thapa 15, 168, 229, 412
Chitra Bahadur Gurung 124, 127, 240, 427
Chitra Bahadur KC 168
Chitwan 18, 128, 134, 206, 353, 360, 389, 424, 431
Christian 103, 128, 213, 405, 406
Christina Rocca 195
CIAA(Commission Investigation of Abuse of Authority) 149, 220, 249, 250, 253, 256, 262, 270, 286, 294, 315, 318, 327, 331, 372, 375, 376, 394
Citizenship Bill 41, 42, 107, 110, 120, 134, 135, 136, 395
Civil Military Campaign 243, 267
CMC 267
CoAS XXIII, XXIV, 3, 7, 15, 16, 19, 25, 27, 28, 37, 43, 47, 48, 54, 76, 77, 79, 81, 84, 90, 91, 100, 101, 102, 103, 104, 106, 107, 108, 109, 110, 113, 115, 117, 118, 119, 120, 121, 122, 123, 124, 125, 127, 129, 130, 131, 133, 135, 140, 141, 142, 143, 144, 147, 148, 149, 153, 155, 156, 157, 159, 160, 161, 162, 164, 165, 166, 167, 169, 171, 172, 174, 175, 176, 179, 180, 181, 183, 184, 185, 186, 187, 189, 190, 191, 192, 193, 194, 195, 200, 201, 202, 203, 205, 206, 209, 210, 212, 215, 216, 217, 220, 221, 222, 232, 234, 237, 238, 242, 243, 250, 254, 258, 263, 265, 266, 267, 268, 271, 272, 273, 283, 287, 288, 289, 294, 297, 298, 299, 300, 301, 302, 303, 304, 308, 309, 312, 316, 320, 321, 322, 324, 329, 332, 334, 336, 338, 344, 346, 347, 349, 350, 353, 355, 360, 362, 366, 367, 369, 371, 372, 373, 374, 377, 378, 386, 394, 395, 397, 398, 399, 401, 403, 404, 405, 413, 414, 415, 423, 426, 427, 429, 433, 435, 436, 441
Col 3, 9, 13, 15, 42, 74, 102, 103, 109, 110, 113, 116, 117, 118, 136, 137, 151, 156, 162, 209, 230, 295, 298, 300, 321, 360, 362, 394, 408, 415, 416, 421, 425, 427, 428, 429, 430, 434, 439
Colin Powell 194, 195
Colonel Commandant 8, 124, 180, 242, 351, 373, 384
communique 197
Company 13, 101, 108, 115, 140, 143, 151, 154, 185, 191, 238, 383, 384, 402, 404
COMPOSA 166
Conference 34, 35, 135, 218, 308, 343, 367
Constituent Assembly XIX, 63, 65, 89, 90, 93, 161, 184, 198, 210, 272, 279, 283, 290, 325, 342, 346, 358, 359, 368, 390, 395, 397, 398
Corps Concept 117, 120, 125, 136, 189
Council of Ministers 253, 254, 258, 392, 393
Counter Terrorism 358
CPA 8, 394, 395
CPN (Mashal) 119, 168, 351
Crown Prince 9, 12, XVII, XVIII, XIX, XX, 8, 14, 15, 16, 17, 19, 20, 25, 29, 30, 31, 32, 33, 34, 35, 36, 37, 38, 39, 40, 64, 66, 99, 100, 104, 117, 136, 142, 157, 176, 179, 181, 196, 200, 209, 212, 227, 233, 237, 251, 253, 261, 285, 314, 317, 321, 335, 336, 344, 346, 351, 355, 357, 376, 402, 416, 424, 430, 434, 438, 440, 441
Crown Prince Charles 321
Crown Prince Dipendra 9, XVIII, XIX, XX, 15, 16, 17, 19, 20, 30, 31, 33, 34, 35, 36, 37, 38, 39, 64, 66, 99, 142, 176, 196, 314, 344, 402, 438, 440

D

Dadhiram Lamichhane 316
Dakshinkali Bhagwati 90
Dalai Lama 172, 232, 315
Dalit 100, 111, 174
Daman Narsingh Rana 397, 421
Daman Nath Dhungana 324, 336
Dambar Singh Gurung 103, 297
Damodar SJB Rana 329, 347, 352
Dandapani Timilsena 112
Dang XXII, 50, 53, 60, 77, 146, 178, 185, 186, 188, 214,

222, 357, 369, 418, 419
Dan Gambhir Rayamajhi 428
Danish Embassy 83, 273
Darjeeling 62, 109, 368
Dashain 62, 119, 150, 179, 180, 183, 252, 370, 371, 402, 420
David Allison 423
David Wood 291, 369
DB Lama 68, 69
DCM 299
Deepak Gajurel 11
Deepak Gyanwali 85
Deepak Singh Thangden 122
Deep Kumar Upadhyaya 162
Defense Ministry 152, 162, 182, 188, 233, 249, 270
Defense Secretary 297
Dehradun 11, 60, 61, 367, 377
Democratic National Youth Association 260
Denmark 273
Deputy Chief of Mission 109, 150, 267, 299
Devendra Raj Kandel 318, 362
Devendra Raj Pandey 358
Dev Gurung 290
Devi Prasad Ojha 85, 286, 295, 303, 305
Devi Ram Sharma XXV, 21, 46, 57, 132, 156, 197, 203, 207, 226, 230, 243, 345, 355, 440
Dev Mukherjee 173
Dev Shumsher 439
Devyani XVIII, 33, 37
DFID 206, 255, 291, 309, 369
DGMO 25, 54, 102, 155
Dhana SJB Rana 41
Dhankuta 167, 327
Dharma Bahadur Thapa 83, 252
Dharmapal Bar Singh Thapa 37, 43, 47, 133, 135, 140, 155, 160, 176, 221, 401, 404, 405, 415, 436
Dhiraj Mukhiya 62
Dhirendra Shah XIX, 16, 27, 42, 110, 139
Dhruba Adhikary 12
Dhruba Bahadur Pradhan 73, 176, 289, 359, 364, 373, 388
Dibya Mangal Joshi 38
Dieter Kups 339
DIG 1, 5, 10, 121, 123, 256, 267
Digvijay Singh 357
Dil Bahadur GC 111
Dil Bahadur Shrestha 363
Dilip Karki 420
Dilip Maharjan 170
Dilip Rayamajhi 42, 102, 106, 151, 154, 159, 164, 182, 202, 206, 230, 316, 370
Dilli Raj Khanal 232, 307, 324
Dilli Ram Oli 233
Dinanath 119, 125, 295, 299, 323
Dinanath Gautam 125
Dinanath Sharma 295, 299, 323
Dinesh Sharma 125
Dipak Bahadur Singh 353
Dipak Bikram Shah 22
Dipak Bohara 261, 276
Dipak Ranjit 370
Dipendra Basnet 370
Dipta Prakash Shah 154, 166, 238
Directorate 57, 102, 126, 186, 190, 401
Disaster Management 99, 131
Divide and Rule 284
Divisional Headquarters 53, 186, 282, 310
Dixit 253, 278, 310
Diyalo Bungalow 134, 424
DMI 127, 205, 215, 230, 243, 316, 363, 370
DMO 25, 102, 186, 205, 215, 401
DMT 141, 423, 424
Dolakha 123, 142, 219, 253, 262
Dolpa 48, 75, 98, 120, 121, 124, 134, 165, 248
Doramba 369, 377
DPKO 124, 240
DPM 20, 28, 46, 53, 73, 77, 83, 90, 101, 126, 129, 130, 132, 143, 145, 148, 158, 255, 282, 290, 306, 369, 370, 371, 372
Drabya Shah XVI
drone 179
DSP 35, 374
Duman Thapa 14, 83, 272, 273, 307, 385, 386, 387, 388, 389, 393
Dunai 48, 75, 98, 120, 121, 122, 123, 128
Durbar 8, 10, XX, XXIII, XXV, 13, 24, 45, 97, 190, 202, 206, 337, 424
Durga Subedi 161, 162
Dwarka Dham 438

E

Eastern Development Region 277, 278, 281, 284
East Timor 106, 349
election 79, 80, 100, 160, 232, 236, 242, 243, 245, 386, 409, 410
Embassy 3, 36, 62, 63, 65, 83, 109, 111, 124, 127, 133, 167, 168, 176, 186, 187, 193, 208, 216, 228, 229, 231, 234, 239, 255, 259, 260, 263, 264, 267, 273, 276, 277, 284, 287, 293, 297, 299, 304, 321, 322, 327, 328, 329, 337, 339, 352, 358, 359, 368, 369, 370, 372, 375, 410, 421
Embassy of Israel 167
Embassy of Pakistan 193, 337
Eton College 33
European Union 178

F

far western 107

Fattya Bahadur Limbu 230
FBI 260
Field Marshal 25, 43, 135, 136, 140, 153, 216
Finland 36
FM radio 118, 171, 383
Fort Leavenworth 429

G

Gadul SJB Rana 221, 373, 413, 414, 415
GAESO 242, 373
Gajendra Bahadur Limbu 7, 62, 106, 230, 423
Gajendra Bohara 34, 176, 196
Gajendra Narayan Singh 107
Gandaki 267
Ganesh Man Peace Campaign 197
Ganesh Man Singh 67, 70
Ganesh Raj Sharma 332, 370
Ganesh Raj Singh 406
Ganeshsthan 282
Garima Rana 33, 36, 37
Gautam Ratna Shakya 139
Geeta Shah 426
Genealogy of the Ranas 260
General 9, 11, 16, XX, XXI, XXII, 1, 2, 3, 5, 8, 9, 10, 25, 27, 43, 46, 47, 54, 56, 68, 69, 83, 90, 91, 93, 95, 100, 102, 109, 112, 124, 126, 133, 134, 135, 139, 140, 146, 147, 148, 154, 162, 167, 171, 173, 180, 184, 193, 196, 200, 210, 216, 217, 222, 223, 227, 233, 242, 244, 245, 249, 252, 254, 255, 257, 265, 266, 275, 287, 290, 293, 295, 297, 301, 305, 307, 309, 310, 312, 314, 320, 321, 324, 325, 334, 336, 340, 344, 345, 346, 348, 349, 350, 351, 352, 353, 356, 361, 364, 365, 368, 370, 372, 377, 379, 387, 388, 389, 390, 398, 399, 413, 415, 416, 420, 424, 425, 429, 433, 434, 435, 436, 437
General Convention 69, 126, 133, 134, 290, 346
General Secretary 9, 68, 69, 83, 91, 100, 146, 171, 173, 193, 196, 200, 210, 223, 227, 244, 245, 249, 252, 254, 257, 275, 287, 290, 293, 295, 297, 305, 307, 324, 325, 334, 336, 340, 344, 345, 351, 352, 361, 364, 365, 370, 387, 388, 420
Geoffrey Hoon 131
Geoffrey James 338, 369, 394
George Fernandes 140, 225, 278, 335
Georgia 386
German Embassy 284
Germany 35, 108, 339, 401, 403, 426
GESON 106
Ghatana Ra Bichar 9, 281, 329, 390
Girija Prasad Koirala 9, 10, 17, 22, 44, 49, 51, 53, 58, 59, 67, 68, 70, 73, 82, 83, 93, 116, 176, 186, 190, 222, 230, 252, 293, 306, 323, 345, 372, 395, 399, 413
Gita Bhakta Joshi 277, 338
GNLF 111
Gopal Budhathoki 204
Gopalji Jung Shah 419
Gopi Hamal 315
Gorakh Bahadur Battalion 7, 8, 221, 383, 384, 424, 425, 426, 427
Gorakhnath Baba 442
Gorakhpur 52, 152, 263
Gore Bahadur Khapangi 85, 165
Gorkha XVI, XX, 25, 84, 109, 143, 144, 152, 184, 229, 252, 285, 294, 339, 406
Gorkha National Liberation Front 109
Govinda Acharya 269
Govinda Raj Joshi 48, 75, 77, 122, 128, 141, 201, 242, 250, 254, 270, 312, 330, 351, 366, 372
Govinda Sharma 16
grand design 59, 361
Griffith 375
Gunaraj Luintel 9, 308
Gyan Chandra Acharya 227
Gyandas Shrestha 179
Gyanendra Raj Rai 122
Gyan Jung Thapa 427

H

Hanuman Dhoka Durbar 24
Hapure 93, 357
Hari Babu Choudhary 158, 270
Hari Bahadur Thapa 10
Hari Bibhor Karki 36
Haribol Bhattarai 409, 410
Haribol Gajurel 84, 263
Harka Gurung 348
Harvard University 386, 388
Hattiban Resort 140
Hawaldar Chandra Jung Shah 416
HDC 273, 274, 280, 285
Heckler & Koch GmbH 41, 403
Helen Shah 111
Helicopter XX, 17, 18, 19, 118, 148, 149, 154, 158, 159, 167, 178, 183, 219, 225, 226, 233, 258, 288, 304, 309, 322, 353, 374, 377
Help Nepal Group 227
Hem Bahadur Singh 373
Hemendra Dhoj Khadka 136
Henning Karcher 352
High Altitude Chamber 99
Himalaya 18, 436
Himal Sharma 376
Hindu XV, 27, 62, 101, 103, 105, 206, 321, 368, 402, 405, 406, 411, 438
His Royal Highness 16, 17, 18, 19, 20, 21, 23, 25, 38, 99, 113, 114, 115, 139
HK G36 41, 131, 154, 402
Holeri 49, 76, 77, 78, 157, 158, 159, 160
Honorary General 217, 222, 309, 350

House of Representative 232
Hreedayesh Tripathi 107
Hritik Roshan 133
Hukam 378
Hussain Muhammad Ershad 348

I

ICIMOD 352
ICRC 178
IED 193, 344
IGP 5, 8, 10, 48, 56, 57, 61, 68, 69, 73, 74, 75, 83, 88, 103, 106, 113, 114, 115, 121, 122, 123, 126, 132, 134, 135, 139, 143, 144, 149, 150, 151, 155, 160, 161, 172, 176, 179, 189, 195, 196, 213, 215, 226, 240, 241, 256, 257, 269, 273, 276, 286, 289, 292, 304, 311, 317, 322, 358, 359, 363, 364, 366, 367, 370, 374, 441
IK Gujral 129, 278
Improvised Explosive Devices (IED) 193
India 9, 11, XV, XIX, 12, 33, 41, 42, 43, 52, 57, 58, 59, 60, 61, 62, 63, 64, 65, 66, 69, 72, 73, 74, 79, 92, 99, 100, 101, 102, 104, 105, 107, 108, 109, 110, 111, 112, 114, 115, 117, 119, 126, 127, 129, 130, 131, 132, 133, 134, 137, 140, 145, 146, 148, 153, 155, 160, 162, 163, 168, 169, 170, 171, 174, 175, 177, 178, 182, 183, 186, 187, 188, 189, 190, 192, 193, 195, 198, 199, 203, 204, 205, 207, 208, 209, 210, 213, 214, 217, 218, 219, 220, 221, 222, 223, 225, 226, 227, 228, 229, 230, 231, 233, 234, 235, 236, 238, 239, 240, 241, 244, 249, 254, 257, 259, 260, 263, 264, 267, 269, 271, 272, 274, 276, 277, 278, 279, 284, 285, 286, 291, 296, 297, 299, 300, 301, 303, 304, 305, 306, 307, 308, 309, 310, 312, 314, 318, 320, 322, 323, 325, 328, 332, 336, 339, 340, 341, 342, 344, 345, 347, 348, 349, 350, 352, 353, 356, 357, 364, 367, 368, 369, 373, 378, 387, 390, 393, 394, 395, 396, 399, 400, 401, 402, 403, 410, 411, 413, 420, 438
Indian Ambassador 116, 140, 168, 173, 190, 192, 245, 272, 277, 286, 293, 297, 302, 323, 328, 333, 335, 345, 347, 349, 366, 368
Indian Army 62, 105, 140, 221, 233, 320, 332, 333, 402
Indian Communist Party (ICP) 59
Indian Defense Minister 235
Indian Embassy 62, 109, 111, 133, 168, 176, 187, 228, 229, 234, 239, 259, 263, 267, 276, 299, 327, 339, 352, 358, 368, 372
INGOs 184, 290, 333
Insas 41, 205, 223, 238, 303, 367, 401, 402, 403
Inspector 1, 5, 9, 46, 60, 61, 93, 111, 139, 159, 183, 196, 221, 266, 278
Inspector General 1, 5, 9, 46, 93, 196, 266
insurgency 11, XXI, XXII, 55, 56, 57, 81, 106, 110, 112, 138, 178, 186, 194, 198, 204, 235, 291, 292, 321, 368, 375
Inverness 422
ISDP 52, 76, 104, 127, 129, 143, 144, 146, 147, 149, 152, 154, 155, 156, 161, 171, 180, 181, 183, 184, 267, 294
Ishwori Dahal 209, 210
Ishwor Pokharel 257, 331
Ishwor Prasad Upadhyaya 420
ISI 105, 107, 108, 109, 240, 304, 320, 337, 378
Isle of Wight 422
Itahari 74, 186
Italy 85, 364

J

Jagadish SJB Rana 410, 411
Jagat Gauchan 296
Jaish-e-Mohammad 229
Jajarkot 101, 143, 233, 304, 340
Jana Astha Weekly 147, 149
Janadharana weekly 283, 390
Janak Bahadur Singh 410
Janak Basnet 252
Janakpur 125, 133, 290, 291
Jana Morcha 45
Janardan Bikram Rana 425, 430
Janardan Subedi 155, 184
Japanese Ambassador 310, 330, 344
Jaswant Singh 235
Jawaharlal Nehru University 58, 296, 368
Jaya Jetley 140
Jayanta Sarswoti 438
Jayanti Shah XIX
Jay Prakash Prasad Gupta 120, 172
Jeevan Kunj 8, 163, 174, 183, 295
Jeev Raj Ashrit 69
Jhalanath Khanal 143, 359, 388
Jharendra Narayan Singh 431
Jimmy Carter 386, 387, 388, 389, 390
Jiri 123, 253
Jitendra Bista 424, 426
Jitendra Dev 274
J. Krishnamurthy 216
JK Sinha 86, 235, 273
Jog Mehar Shrestha 77
Joint Coordinator 192, 198, 202, 204, 219, 227, 230, 243, 254, 287, 293, 338, 341, 356
Jung Bahadur 10, XVI, XVII, XIX, XX, 195, 258
Jungle Warfare School 112
Jyoti SJB Rana 427
Jyotshna Basnyat 338

K

Kabul 178
Kailash Shirohiya 28
Kali Bahadur Battalion 74
Kalikot 100, 101, 143, 157, 304, 376
Kali Prasad 383, 384, 424, 427
Kamalanayanaacharya 178
Kamal Chaulagain 83, 85, 259, 268, 274, 295, 303, 338, 347

Kamal Pandey 109
Kamal Thapa 93, 275, 280, 339, 360
Kanak Mani Dixit 253, 310
Kanti Ishwori Vidhyalaya 31
Kantipur daily XXIV
Karan Singh 395
Karl Marx 69
Karna Dhoj Adhikari 324, 353
Kashmir 59, 206
Kaski Sadan 25
Kathmandu 4, 8, 11, XVI, XX, 5, 10, 11, 13, 17, 18, 23, 51, 60, 62, 68, 74, 77, 78, 82, 83, 86, 97, 101, 104, 105, 108, 109, 110, 114, 126, 132, 133, 136, 138, 139, 140, 154, 159, 162, 164, 166, 168, 171, 173, 174, 177, 178, 180, 181, 183, 184, 186, 187, 189, 190, 191, 193, 195, 198, 208, 217, 218, 219, 222, 227, 232, 235, 238, 239, 240, 244, 245, 248, 250, 258, 260, 261, 262, 264, 265, 267, 268, 269, 275, 276, 278, 285, 294, 295, 296, 300, 305, 307, 309, 320, 323, 327, 328, 330, 335, 336, 337, 339, 344, 347, 348, 349, 355, 358, 360, 361, 362, 363, 368, 369, 370, 374, 375, 384, 386, 387, 391, 394, 395, 409, 412, 419, 420, 423, 424, 425, 426, 438, 439, 440
Kathmandu Metropolitan City 83, 187, 260
Kathmandu Valley XVI, 5, 10, 17, 23, 82, 105, 110, 177, 183, 190, 258, 267, 278, 320, 327, 361, 362, 363
KB Gurung 79
Keith Bloomfield 250, 291, 292, 367, 369, 373
Kendra SJB Rana 33
Keshang 128
Keshar Bahadur Bista 358
Keshari Bandhu 10
Keshar Jung Rayamajhi 182
Keshar Mahal 37
Keshav Prasad Khatiwada 322
Keshav Prasad Upadhyaya 26, 327, 371
Keshavraj Mahat 118, 151
Keshavraj Rajbhandari 105, 117
Keshav Raj Sharma 425
Keshav Shrestha 420
Keshav Sthapit 83, 260
Ketki Chester 19
Khadga Bahadur Bhujel 116
Khadga Bahadur Gurung 116, 117
Khagendra Bahadur Shrestha 218
Khampa 61, 64
Khaptad Baba 438, 439
Khaptad National Park 439
Kharipati 120
Kharsang 109
Khemraj SJB Rana 427
Khula Mancha 220
Khum Bahadur Khadka 57, 115, 120, 126, 150, 158, 161, 170, 173, 174, 175, 179, 193, 208, 210, 214, 222, 223, 242, 244, 250, 256, 257, 284, 286, 312, 314, 315, 318, 327, 332, 333
Khumbuwan 181
Kilo Sierra-2 47, 55, 167
King 7, 8, 9, 10, 11, 13, XVI, XVII, XVIII, XIX, XX, XXI, XXII, XXIII, XXV, 1, 2, 3, 4, 5, 6, 7, 8, 9, 10, 11, 12, 13, 14, 15, 16, 17, 18, 19, 20, 21, 22, 23, 24, 25, 26, 27, 28, 29, 30, 31, 32, 33, 34, 35, 36, 37, 38, 39, 40, 41, 42, 43, 45, 47, 48, 49, 50, 51, 52, 53, 56, 57, 60, 61, 62, 64, 66, 67, 68, 71, 72, 76, 80, 81, 82, 83, 84, 85, 86, 87, 88, 89, 90, 91, 92, 93, 94, 98, 99, 100, 101, 102, 103, 104, 105, 106, 107, 108, 109, 110, 111, 112, 113, 114, 115, 116, 117, 118, 119, 120, 121, 122, 123, 124, 125, 126, 127, 128, 129, 130, 131, 132, 134, 135, 136, 137, 138, 139, 140, 141, 142, 143, 144, 145, 146, 147, 148, 149, 150, 151, 152, 153, 154, 155, 156, 157, 158, 159, 160, 161, 162, 163, 164, 165, 166, 167, 168, 169, 170, 171, 172, 173, 174, 175, 176, 177, 178, 179, 180, 181, 182, 183, 184, 185, 186, 187, 188, 189, 191, 192, 193, 194, 195, 196, 197, 199, 200, 201, 202, 203, 204, 205, 206, 207, 209, 210, 211, 212, 213, 214, 215, 216, 217, 218, 219, 220, 222, 224, 225, 226, 227, 228, 229, 230, 231, 232, 233, 235, 236, 237, 238, 239, 240, 241, 242, 243, 244, 245, 246, 247, 248, 249, 250, 251, 252, 253, 254, 255, 256, 257, 258, 259, 260, 261, 262, 263, 264, 265, 266, 267, 268, 269, 270, 271, 272, 273, 274, 275, 276, 277, 278, 279, 280, 281, 282, 283, 284, 285, 286, 287, 288, 289, 290, 291, 292, 293, 294, 296, 297, 298, 299, 300, 301, 302, 303, 304, 305, 306, 307, 308, 309, 310, 311, 312, 313, 314, 315, 316, 317, 318, 319, 320, 322, 323, 324, 325, 326, 327, 328, 329, 330, 332, 333, 334, 335, 336, 337, 338, 339, 340, 341, 342, 343, 344, 345, 346, 347, 348, 349, 350, 351, 352, 353, 354, 355, 356, 357, 359, 360, 361, 362, 363, 364, 365, 366, 367, 369, 370, 371, 372, 373, 374, 375, 376, 377, 378, 379, 383, 384, 385, 386, 388, 389, 390, 391, 392, 393, 394, 395, 396, 397, 400, 402, 404, 405, 407, 408, 409, 410, 411, 412, 413, 414, 415, 416, 423, 424, 425, 427, 428, 430, 431, 432, 433, 434, 435, 436, 437, 439, 440, 441, 442
King Birendra 10, 11, 13, XVII, XVIII, XIX, XXI, XXII, 11, 12, 13, 14, 15, 16, 19, 20, 22, 23, 29, 30, 32, 33, 34, 35, 37, 38, 41, 42, 43, 45, 53, 57, 62, 64, 66, 98, 99, 148, 152, 160, 164, 173, 175, 182, 204, 211, 231, 305, 343, 344, 395, 396, 402, 405, 408, 410, 415, 416, 428, 430, 431, 432, 440
King Birendra and Queen Aishwarya Memorial Trust 305
King Gyanendra 8, XIX, XXI, XXII, 1, 2, 6, 8, 12, 24, 27, 40, 49, 52, 53, 60, 64, 76, 80, 81, 94, 151, 155, 169, 175, 182, 211, 231, 254, 330, 379, 391, 396, 432, 437, 440, 442
King Mahendra XVII, XX, 18, 30, 33, 39, 64, 66, 147, 391, 424, 425, 436
King of Norway 423
King Prithvi XVI, 442
King Rajendra XVII
King Tribhuvan XVII
Kiran Bahadur Basnet 94, 360, 362, 376
Kiran Bhandari XXV

Kirti Nidhi Bista 250, 251, 319, 347, 393
Kishor Shrestha 177
Kohalpur-Banbasa 64
Komal Bahadur Bista 416, 417
Koshi 267, 299
Kot Massacre XVI, XIX, XX
KP Oli 83, 171, 255, 261, 262, 273, 316, 330, 331, 333, 341, 342, 351, 365, 370
Krishna Abiral 10
Krishna Bahadur Mahara 88, 89, 91, 170, 290, 295, 299, 307, 319, 321, 323, 334, 336, 345, 358
Krishna Dhoj Khadka 215, 313
Krishna Jwala Devkota XXV
Krishna Mohan Shrestha 21, 56, 57, 88, 135, 139, 150, 179, 226, 276, 286, 311
Krishna Murari Sharma 46
Krishna Narayan Singh Thapa 145, 165, 220, 250, 254, 415
Krishna Pahadi 181
Krishna Prasad Bhattarai 44, 51, 67, 68, 69, 74, 115, 116, 223, 359
Krishna Prasad Gautam 236
Krishna Prasad Sitaula 213, 216, 299
Kul Bahadur Khadka 423
Kulchandra Gautam 368
Kumar Budhathoki 103, 151
Kumar Gorakh SJB Rana XX
Kumar Khadga Bikram Shah 406
Kumar Mahesh Singh XX
Kunda Bikram Shah XXV
Kushu Ukab 420
Kusum 10, 376, 378
Kusum Bhattarai 10
KV Rajan 140, 245, 268, 335

L

Lal Bahadur Thapa 61, 341
Lalbandi 200, 201
Lal Chand Bohara 112
Lalit Man Singh 104
Lal Krishna Adwani 229
Lal Mohar 129. *See also* Royal seal
Lamsal Commission 297
Lashkar-e-Toiba 229
Laxman Prasad Aryal 338
Laxman Prasad Ghimire 79
Lee 231, 240, 275
Lenin's Anniversary 99
Liberation Front of Kharsang in Darjeeling. 109
Liberation Front of the Freedom Fighters of Bangladesh 61
Lieutenant Colonel 103, 434
Lila Mani Pokharel 167, 168, 172, 176, 306, 358, 359
Local Development Officer 100
Loch Ness 422
Lockheed 401
Lokendra Bahadur Chand 7, 46, 53, 71, 72, 73, 82, 87, 91, 126, 215, 243, 244, 250, 251, 252, 277, 285, 304, 308, 313, 335, 337, 341, 343, 347, 348, 377, 413
Lokman Singh Karki 253, 388
London 34, 38, 105, 108, 131, 227, 254, 261, 338, 341, 422, 430, 431
Lord Buddha XV, 137
Lord Camoys 33, 344
Lt. Col 3, 74, 103, 416, 425, 429, 430, 439
Lt Gen 180, 191, 230, 399, 413
LTTE (Liberation Tigers of Tamil Elam) 59
Lumbini 137

M

M16 A2 Commando Rifle 16
Madan Bhandari 69, 70, 72
Madan Jaiswal 328
Madan Kumar Aryal 297, 349
Madan Mani Adhikari 8
Madan Pyasi 339, 358, 368, 372
Madhav Kumar Nepal 9, 13, 17, 19, 20, 24, 52, 83, 91, 92, 93, 100, 146, 168, 173, 174, 193, 196, 200, 201, 210, 213, 214, 220, 223, 224, 225, 227, 233, 236, 244, 245, 248, 249, 252, 254, 256, 257, 267, 270, 275, 276, 277, 287, 290, 291, 293, 295, 297, 300, 305, 307, 316, 322, 324, 327, 335, 336, 340, 344, 345, 351, 361, 364, 365, 387
Madhav Kumar Rimal 14
Madhav Marahattha 218
Madhav Ranabhat 338
Madhesh 62, 63, 316, 397, 399
Madhesi Jana Adhikar Forum (MJF) 182
Madhesi Liberation 181
Madhu Raman Acharya 188
Madhu Sudan Sapkota 430
Mahakali 72, 73, 267
Mahakali Treaty 72, 73
Maharaja of Jaipur 33
Mahendra Manjil 14
Mahendra Police Club 374
Mahesh Acharya 123, 137, 155, 161
Mahesh Singh XX
Maikot 378
Maj Gen 127
Makalu Base 226
Malla K Sundar 283, 369
Mangal Raj Joshi 38
Mangal Sadan 17
Mangalsen 60, 200, 201, 209
Manjul Ansari 227
Man Mohan Adhikari 71, 257
Manmohan Singh 387
Manoj Dahal XXV

Manoj Kumar Acharya 322
Mao 64
Maoist Communist Center (MCC) 59
Maoist Problem 45, 50, 51, 52, 54, 55, 57, 58, 64
Maoists 8, 10, XVIII, XIX, XXI, XXII, 5, 12, 13, 23, 24, 25, 28, 42, 45, 46, 47, 48, 49, 50, 51, 52, 53, 54, 55, 58, 59, 60, 61, 62, 63, 64, 65, 66, 75, 76, 77, 78, 79, 82, 84, 85, 86, 87, 88, 89, 90, 93, 94, 98, 104, 105, 106, 107, 108, 109, 110, 111, 112, 113, 115, 116, 119, 120, 121, 122, 123, 125, 126, 129, 130, 131, 132, 133, 134, 136, 140, 141, 142, 143, 144, 145, 146, 149, 154, 155, 156, 157, 158, 159, 160, 161, 163, 164, 165, 166, 167, 169, 170, 171, 172, 173, 174, 175, 176, 177, 178, 180, 181, 182, 184, 185, 186, 187, 189, 190, 191, 192, 193, 195, 197, 199, 200, 201, 202, 204, 205, 206, 209, 210, 211, 213, 214, 215, 216, 217, 218, 219, 220, 221, 222, 223, 225, 226, 227, 229, 230, 231, 232, 233, 234, 236, 238, 239, 240, 241, 243, 244, 246, 247, 248, 250, 252, 253, 254, 255, 258, 260, 261, 262, 263, 264, 265, 270, 271, 272, 273, 274, 275, 276, 277, 278, 279, 280, 282, 283, 284, 285, 286, 287, 288, 289, 290, 292, 294, 295, 296, 297, 298, 299, 300, 302, 304, 305, 307, 308, 310, 311, 313, 316, 317, 318, 319, 320, 321, 323, 324, 325, 328, 330, 331, 332, 333, 334, 335, 338, 339, 340, 341, 343, 345, 346, 347, 348, 349, 350, 351, 352, 353, 354, 356, 357, 358, 359, 360, 361, 362, 363, 364, 365, 367, 368, 369, 370, 373, 375, 376, 377, 378, 379, 383, 384, 385, 386, 387, 388, 390, 391, 392, 393, 394, 395, 396, 397, 398, 399, 400, 401, 406, 434
Marich Man Singh Shrestha 319, 409, 413, 418
Mashal 43, 44, 119, 168, 193, 351, 388
Master General of Ordnance 349
Mathura Prasad Shrestha 358
Matrika Yadav 290, 316, 327, 330, 339, 358
Matthew Hodes 386
Mayo College 33
Mayor 83, 164, 187, 260, 261, 269, 328
McKnoll Foundation 385, 386
Medal Award Committee 328
MI-17 132, 154, 304, 309
Michael E. Malinowski 195, 214, 351
Midwestern 112
Military Attaché 124, 127, 136, 179, 186, 187, 193, 229, 255, 287, 321, 322, 328, 337, 369, 370, 375, 376, 377
Military Secretariat XXI, 3, 4, 10, 11, 48, 211, 212, 230, 236, 238, 266, 268, 298, 302, 303, 326, 346, 356, 407, 415, 432, 433, 434
Military Secretary 7, 8, 9, 11, XXI, XXII, XXIII, XXIV, XXV, 7, 9, 10, 11, 12, 13, 25, 27, 31, 36, 40, 48, 86, 93, 97, 99, 103, 113, 117, 122, 143, 147, 151, 180, 211, 212, 228, 234, 265, 266, 298, 301, 302, 316, 324, 329, 384, 400, 403, 404, 408, 412, 413, 414, 424, 427, 428, 432, 433, 434, 435, 436, 437
Mine Clearing Vehicle 204
Minendra Rijal 387, 388
Minister 7, 9, 12, XVII, XXII, 1, 3, 4, 5, 9, 10, 12, 13, 17, 18, 19, 20, 21, 22, 23, 24, 25, 28, 29, 44, 45, 46, 47, 48, 49, 50, 51, 53, 54, 57, 58, 63, 67, 68, 69, 70, 71, 72, 73, 74, 75, 78, 79, 80, 82, 83, 84, 85, 86, 87, 88, 89, 90, 91, 93, 99, 107, 117, 120, 121, 122, 123, 124, 125, 129, 130, 131, 133, 135, 137, 138, 139, 140, 141, 143, 145, 148, 158, 161, 166, 169, 170, 172, 189, 190, 195, 198, 205, 208, 210, 212, 213, 214, 215, 216, 222, 225, 229, 235, 236, 242, 244, 245, 246, 247, 250, 251, 252, 253, 254, 255, 256, 257, 264, 266, 268, 269, 277, 278, 279, 280, 282, 284, 285, 286, 287, 288, 289, 295, 296, 298, 300, 302, 304, 305, 309, 310, 313, 314, 317, 318, 333, 335, 337, 338, 339, 343, 347, 352, 357, 358, 362, 363, 366, 368, 369, 371, 372, 374, 375, 376, 379, 386, 388, 389, 391, 394, 397, 398, 402, 409, 413, 418, 431, 432, 439
Minister of Labor and Transportation 268, 338
Minister of Physical Planning and Construction 85, 130, 269, 310, 337, 339
Ministry of Communication 253
Ministry of External Affairs 190, 387
Ministry of Finance 146, 163, 259, 278, 340
Ministry of Foreign Affairs 140, 188, 204, 227, 277, 339, 354, 389, 395, 427
Ministry of Home Affairs 4, 5, 21, 27, 29, 46, 54, 83, 85, 109, 118, 121, 123, 126, 128, 152, 161, 173, 186, 198, 248, 249, 259, 260, 315, 339, 355, 360, 364, 377, 434
Ministry of Information and Communication 362
MOD 117
Mohammad Mohsin 17, 192, 309
Mohan Bahadur Pandey 4, 7, 14, 17, 21, 268
Mohan Baidhya 99, 157, 168, 316
Mohan Bikram Singh 43, 44, 351
Mohan Prasad Lohani 289, 332
Mohan Pratap Shah 426, 427
monarchy 8, 9, 11, XVII, XIX, XXIII, 9, 10, 11, 13, 19, 20, 21, 23, 24, 27, 28, 29, 30, 38, 39, 43, 45, 52, 58, 59, 60, 62, 64, 65, 67, 81, 84, 86, 89, 94, 99, 114, 116, 132, 145, 147, 152, 156, 163, 165, 167, 168, 169, 171, 173, 174, 176, 180, 182, 192, 212, 217, 218, 228, 231, 232, 234, 236, 238, 262, 264, 275, 279, 281, 282, 283, 285, 289, 295, 301, 302, 306, 316, 317, 318, 319, 320, 324, 325, 326, 327, 335, 336, 340, 341, 343, 345, 348, 349, 363, 368, 376, 392, 393, 395, 396, 397, 399, 406, 409, 416, 437, 438, 440, 441, 442
Mons 421, 422
Moon Drops Overseas 326
Moti Krishna Shakya 180
Moti Lal Bohara 69, 103
Motorola Company 140
Mountain Warfare 149
MP5 submachine gun 16
MRMG 385
Muktinath 148, 178, 222, 288, 322
Mul Bhatta 148, 406
Mulyankan Monthly 167
Muma Ram Khanal 328, 352
Murari Raj Sharma 354

Mustang 172, 222, 225, 232, 237, 371

N

Nagarjun 37, 127
Nagarkot 125, 149, 402, 421
Nanda Kishor Pun 378
Nara Bahadur Khatri 77
Narahari Acharya 284, 359
Narayanhiti Royal Palace 11, XIX, XX, 152, 192
Narayan Khadka 182
Narayan Man Bijukchhe 121
Narayan Prasad Shrestha 410, 412
Narayan Singh Pun 79, 85, 90, 93, 100, 112, 130, 165, 166, 212, 227, 255, 264, 269, 277, 278, 279, 284, 286, 287, 288, 290, 294, 296, 297, 298, 302, 304, 305, 306, 307, 308, 310, 313, 315, 316, 318, 323, 324, 325, 329, 330, 335, 338, 339, 341, 343, 344, 347, 352, 357, 358, 369, 373
Nar Bahadur Gurung 427
Narendra Bikram Shah 264
Nasal Chowk 24
National Accountable Bureau 312
National Assembly 142, 150, 154, 156, 166, 174, 190, 192, 238, 245, 248, 287, 306, 309, 337
National Defense College 310, 320
National Defense University 106
National Investigation Department XXV, 16, 57
National Security Council (NSC) XXII, 47, 75, 102
National Wildlife Conservation Department 206
Nature Conservation (KMTNC) 18
Naya Karyasampadan Niyamawali 1
Nayan Raj Pandey 13, 151
NC President 173, 190, 323, 372
NDC 101, 106, 154, 265
Neer Shah 372, 373
Negotiation of Coalition Building 388
Nepal Army XXII, 3, 5, 42, 47, 75, 76, 103, 127, 132, 139, 146, 186, 187, 194, 197, 222, 253, 332, 333, 397, 398, 400, 401, 402, 403
Nepal Bank 123, 154, 226, 359
Nepal Bhushan Chand 230, 356
Nepali Congress 10, XXII, 43, 44, 45, 51, 52, 58, 67, 68, 69, 70, 71, 75, 76, 77, 78, 79, 80, 82, 83, 92, 93, 97, 98, 100, 108, 111, 115, 116, 125, 126, 133, 134, 145, 146, 147, 149, 150, 155, 156, 157, 159, 160, 161, 166, 169, 182, 186, 195, 196, 197, 198, 199, 200, 201, 203, 207, 208, 209, 210, 213, 214, 216, 218, 220, 223, 224, 225, 227, 229, 232, 234, 235, 236, 242, 245, 248, 249, 252, 254, 255, 256, 257, 259, 260, 265, 271, 275, 276, 279, 284, 285, 287, 293, 299, 306, 307, 308, 314, 316, 320, 322, 328, 329, 330, 332, 333, 337, 340, 341, 342, 346, 349, 351, 353, 359, 361, 362, 365, 366, 368, 372, 374, 388, 396, 399
Nepal Peasants and Worker's Party 346
Nepal Police 1, 5, 12, 16, 23, 44, 46, 47, 49, 51, 54, 55, 56, 73, 93, 98, 106, 109, 118, 122, 126, 147, 151, 152, 161, 172, 176, 177, 184, 187, 189, 193, 195, 196, 198, 199, 213, 216, 237, 240, 241, 256, 267, 269, 279, 283, 289, 292, 304, 309, 311, 312, 317, 331, 347, 358, 359, 364, 365, 366, 377, 401, 428
Nepal Rastra Bank 142, 164, 240, 279
Nepal Sadbhavana Party (NSP) 107, 335
Nepal Samacharpatra 7, 179, 289
New Delhi 58, 59, 62, 105, 114, 128, 170, 174, 186, 193, 208, 210, 216, 217, 278, 296, 349, 368, 369, 371, 387, 390, 394, 395, 396, 410
Newsweek 69
New Year 11, 140, 202, 211, 314, 414
New York 124, 127, 150, 175, 349, 377, 387, 389
NGOs 128, 405, 406
NID XXV, 21, 23, 44, 46, 49, 57, 58, 61, 87, 111, 128, 132, 144, 156, 157, 158, 163, 177, 193, 196, 197, 198, 203, 204, 206, 207, 216, 226, 230, 234, 235, 236, 238, 243, 244, 249, 270, 273, 278, 286, 288, 300, 307, 311, 312, 325, 337, 341, 345, 349, 355, 440
Niranjan Karki 3
Nirendra Aryal 394
Nirmal Chandar Vij 309, 320
Nirmal Lama 43
Nirmal Niwas 24, 25, 81, 99, 113, 153, 169, 212, 433
Nir SJB Rana 153, 216
Nischal Nath Pandey 150, 388, 389
No 2 Brigade 185
No 3 Brigade 18, 197, 215, 329
No 4 Brigade 13, 77, 157, 327
No 5 Brigade 201, 439
No 6 Brigade 206
No 11 Brigade 305
No 14 Brigade 344
Non-Aligned Nations' Summit 430
Norodom Sihanouk of Cambodia 348
NSP 107, 197, 248, 306, 308, 335, 365, 388
NTV (Nepal Television) 172
Nudup Shrestha 286
Nuwagaon 78
Nuwakot 37, 181

O

Om Bahadur Pun 258
Om Sharma 268, 269
operation 16, 18, 19, 54, 62, 77, 102, 105, 120, 125, 127, 131, 144, 145, 155, 156, 159, 186, 187, 191, 215, 218, 221, 238, 239, 258, 263, 320, 325, 326, 366, 367, 370, 377, 378, 395

P

Padam Kumar Acharya 182, 188, 306, 367
Padmanavan 222
Padma Ratna Tuladhar 83, 167, 176, 181, 272, 273, 292, 307, 324, 386, 388

PADT 411
Pakistan 59, 102, 105, 107, 119, 178, 189, 190, 192, 193, 222, 225, 234, 259, 312, 328, 337, 348, 371, 378
Pakistani Ambassador 312
Panchas 132, 411
Panchayat 7, XVII, 30, 34, 35, 38, 43, 44, 45, 56, 57, 66, 67, 68, 132, 152, 168, 275, 324, 342, 363, 405, 406, 409, 410, 411, 412, 413, 437
Panchayat Policy and Evaluation Committee 34, 168
Panchayat system XVII
Panchkhal 102, 118, 131, 139, 215, 268
parachute 100, 424
Paras Prakash Nepal XXV
Paras Shah 113
Pari Thapa 388
Pashupati 6, 17, 20, 22, 26, 33, 77, 87, 94, 117, 143, 147, 148, 156, 162, 183, 211, 220, 251, 252, 269, 280, 282, 283, 287, 294, 295, 307, 309, 320, 357, 363, 397, 411, 413, 432
Pashupati Bhakta Maharjan 6, 17, 87, 94, 143, 220, 251, 252, 269, 280, 282, 283, 294, 363, 413
Pashupatinath 235, 269, 328, 368, 406
Pashupati SJB Rana 33, 156
Pawan Jung Thapa 258, 416
Peace Keeping Mission 102, 268
pension 62, 105, 106, 124, 131, 233, 239, 320
People's Republic of China 137
People's War XVIII, XIX, XXI, 45, 50, 54, 59, 64, 66, 143, 176, 199, 200, 323, 339, 351, 373, 390, 396, 397
People's War Group (PWG) 59
Pervez Musharraf 192, 229
Peter Fowler 322, 370
Philip Trousedell 373
Piyush Bahadur Amatya 318
platoons 78, 105
PM 9, 18, 44, 45, 49, 57, 69, 70, 74, 75, 76, 77, 78, 80, 82, 85, 86, 87, 88, 90, 91, 92, 93, 94, 101, 102, 104, 106, 107, 109, 110, 111, 112, 114, 115, 116, 118, 119, 121, 123, 125, 126, 128, 129, 130, 131, 132, 133, 134, 135, 138, 139, 140, 141, 143, 145, 146, 148, 149, 150, 153, 155, 156, 157, 158, 159, 160, 161, 162, 163, 164, 165, 166, 167, 169, 170, 171, 172, 173, 174, 175, 177, 178, 179, 181, 182, 183, 184, 185, 186, 187, 189, 190, 192, 193, 197, 198, 200, 201, 202, 203, 204, 205, 206, 207, 208, 209, 210, 212, 213, 214, 216, 218, 219, 223, 224, 227, 228, 233, 234, 237, 239, 241, 242, 243, 244, 245, 246, 247, 248, 250, 251, 252, 253, 255, 259, 260, 261, 263, 265, 280, 282, 283, 284, 285, 286, 287, 288, 289, 290, 292, 293, 295, 297, 299, 302, 306, 307, 308, 309, 314, 315, 318, 320, 321, 322, 323, 324, 327, 329, 330, 331, 332, 333, 334, 335, 336, 337, 338, 339, 341, 342, 343, 345, 347, 348, 349, 351, 353, 355, 359, 360, 361, 364, 365, 367, 368, 370, 371, 372, 373, 374, 377, 378, 379, 386, 387, 396, 403, 404, 413, 414, 415, 430
PM Koirala 45, 57, 69, 70, 74, 75, 76, 77, 78, 109, 111, 112, 114, 115, 126, 129, 131, 133, 138, 139, 140, 141, 143, 145, 146, 148, 149, 155, 156, 157, 161, 165, 174, 200, 396, 414
Pokhara 17, 18, 101, 133, 200, 215, 221, 275, 329, 383, 412, 440
Police Headquarters 1, 113, 118, 122, 123, 198, 358, 370
Post Bahadur Bogati 356
PPEC 34, 66, 168, 409, 410, 412
Prabhu SJB Rana 18, 81, 310, 347
Prachanda 43, 58, 63, 88, 99, 105, 121, 162, 168, 171, 174, 184, 204, 253, 291, 293, 297, 298, 316, 322, 349, 353, 358, 360, 398
Prachanda Jung Shah 253
Pradip Giri 162, 171, 208, 223, 245
Pradip Nepal 331
Pradip Pratap Bom Malla 25
Pradip SJB Rana 21, 46, 83, 123, 126, 144, 155, 172, 189, 196, 213, 240, 241, 256, 269
Prague 36
Prahlad Singh Gurung 110
Prajwalla SJB Rana 15, 109, 115, 120, 144, 148, 179, 180, 212, 220, 222, 232, 234, 242, 403, 405, 415, 435
Prakash Bahadur Basnyat 18, 439
Prakash Bikram Shah 114, 340
Prakash Chandra Lohani 77, 93, 138, 327, 335, 339, 366, 388
Prakash Gautam 260
Prakash Koirala 161, 241
Prakash Raj Pandey 9
Pravin Gurung 113, 115, 150, 237
Prem Raj Karki 331
Prem Singh Dhami 46
Press Secretary 4, 7, 14, 17, 21, 268, 391, 408, 409, 412
Prime Minister 7, 9, 12, XVII, XXII, 1, 3, 4, 9, 10, 12, 13, 17, 18, 19, 20, 21, 22, 24, 25, 28, 29, 44, 45, 46, 47, 49, 50, 51, 53, 57, 58, 63, 67, 68, 70, 71, 72, 73, 74, 75, 78, 79, 80, 82, 83, 84, 85, 91, 93, 99, 107, 117, 129, 145, 161, 169, 205, 214, 215, 236, 244, 245, 246, 247, 250, 251, 255, 257, 277, 280, 285, 304, 314, 335, 352, 363, 368, 379, 386, 397, 398, 402, 409, 413, 418, 432, 439
Prince Dipendra 8, 9, XVIII, XIX, XX, 15, 16, 17, 19, 20, 30, 31, 33, 34, 35, 36, 37, 38, 39, 64, 66, 99, 142, 176, 196, 314, 344, 402, 438, 440
Prince Himalaya Shah 436
Prince Hridayendra 282
Prince Nirajan XIX, 16, 37, 38, 131, 344, 438
Prince Paras XX, 8, 25, 28, 99, 113, 115, 116, 200, 209, 227, 261, 376
Princep Shah 436
Princess Komal XIX, 25
Princess Pooja 261
Princess Prekshya XX, 183
Princess Prerana 88, 293, 311
Princess Royal, Anne 128
Princess Royal Shruti XX

Princess Shanti XIX
Princess Shobha XIX
Prince Surendra XVII
Prince William 321
Principal Military Secretary XXI, 10, 11, 12, 48, 99, 103, 113, 122, 147, 151, 180, 212, 234, 265, 301, 302, 403, 404, 408, 412, 413, 414, 424, 428, 433, 434
Principal Personal Secretary 160, 168
Principal Private Secretary 17, 410, 411
Principal Secretary 1, 6, 17, 20, 21, 87, 88, 94, 107, 117, 134, 135, 143, 212, 220, 238, 251, 252, 254, 269, 280, 282, 283, 294, 297, 300, 305, 310, 336, 340, 348, 363, 409, 412, 413, 414, 433
Principal Staff Officer 303
Prithuladhish 179, 233, 336
private staff 16, 31
Property Investigation Commission 233
PSOs 124, 127, 156, 205, 207, 209, 274, 303, 357, 375
puja 40, 90, 206, 282, 410, 411
Purna Chandra Thapa 3
Purnagiri Dam 134
Putali Bagaincha 206
Putali Sadak 146
Pyar Jung Thapa 3, 90, 147, 148, 156, 175, 180, 191, 234, 237, 243, 250, 300, 312, 320, 338, 371, 395, 441
Pyuthan 143, 238, 284, 313, 376

Q

queen XVI, 32, 37, 340
Queen Aishwarya 13, XVIII, XIX, 15, 16, 19, 30, 31, 32, 33, 35, 37, 38, 41, 175, 305, 405, 410, 432, 438
Queen Mother Ratna 14, 18, 20, 24, 32, 33, 37, 38, 416
Queen of Britain 128

R

Rabi Bhakta Shrestha 309
Rabindra Nath Sharma 87, 375
Rabindra Shrestha 320, 328, 351, 356
Rabi Raj Thapa 35
Radha Krishna Mainali 213, 214, 271, 272, 295, 303, 305, 311, 315, 316, 319, 331, 348, 363, 373
Radio Nepal 177, 224, 431
Raghu Chandra Bahadur Singh 254
Raghu Shah 332
Raish Khan 249
Raj Bahadur Singh 293
Rajeev Shahi 16
Rajendra Bahadur Singh 121, 123, 256
Rajendra Shrestha 274
Rajeshwor Devkota 319
Rajiv Bikram Shah 340
Rajsabha 19
Raju Dangol 179
Raju Karki 17
Ram Bahadur Thapa 88, 89, 216, 290, 316, 319, 324, 326, 336, 349
Ram Chandra Poudel 20, 28, 77, 130, 158, 160, 226, 332, 369
Ramesh Nath Pandey 87, 90, 91, 142, 150, 172, 190, 193, 199, 223, 242, 282, 283, 287, 288, 289, 298, 310, 318, 321, 324, 352, 388, 389, 391, 392, 393, 394, 442
Ramesh Raj Kunwar 338
Ramesh Tamrakar 260
Rameshwaram 438
Ramindra Chhetri 74, 94, 217, 360, 376
Ram Kaji Bantawa 57
Ram Narayan Mishra 133
Ram Preet Paswan 238
Ram Raja Prasad Singh 312, 315, 323
Ram Sharan Mahat 190, 261, 328, 388
Rana Dhoj Limbu 103
Ranjan Raj Khanal 409
Ranjan Singh Shah 373
Rara 13, 14, 151, 183, 438
Rashtriya Banijya Bank 199, 359
Rashtriya Prajantra Party (RPP) 67
Rasuwagadi 225
Ratna Bahadur Bista 421
Ratna Bahadur Gurung 427
Ravi Charan Shrestha 409
Ravi SJB Rana XX
RAW 58, 60, 63, 86, 87, 109, 152, 168, 171, 181, 203, 207, 217, 233, 235, 259, 273, 288, 301, 318, 339, 356, 358, 368, 372, 394
R&D Directorate 401
Rekha Sharma 215, 313, 346
Reporters' Club 120, 308
Revolutionary 99, 130, 163, 199, 243, 309, 316, 346
revolutionary forces 283
Revolutionary International Movement (RIM) 99
Rewati Raman Khanal 412, 414
Richard Blum 386
Richard Boucher 195
Ring Road 26, 138
Rishi Dhamala 308
Rishi Khatri 183
Rishi Kumar Pande 412, 424
RJ-100 108, 404
RMA 423, 430
RNA 11, 23, 25, 27, 41, 42, 47, 50, 52, 53, 54, 55, 56, 75, 77, 82, 94, 117, 129, 131, 132, 136, 137, 138, 139, 141, 147, 149, 153, 154, 161, 162, 164, 165, 167, 172, 177, 180, 181, 182, 184, 186, 194, 195, 198, 200, 205, 206, 207, 209, 212, 215, 216, 217, 218, 219, 221, 222, 223, 225, 226, 229, 234, 242, 255, 257, 258, 259, 260, 261, 263, 264, 266, 268, 270, 272, 277, 278, 283, 286, 291, 303, 305, 309, 314, 315, 316, 319, 320, 321, 322, 327, 336, 340, 344, 346, 350, 353, 356, 358, 360, 364, 366, 369, 373, 377, 379, 383, 384, 393,

394, 395, 397, 398, 400, 401, 403, 404, 407, 414, 415, 416, 417, 421, 424, 427, 433, 440
RNAC 410, 431
RNA Headquarters 77, 82, 117, 137, 147, 161, 186, 194, 195, 200, 206, 212, 223, 261, 264, 266, 272, 278, 303, 314, 319, 340, 346, 356, 358, 377, 393, 403, 404, 414, 415, 416, 421, 427, 433
Robin Cook 99
Rohini Thapaliya 133
Rolpa 49, 53, 77, 104, 108, 119, 143, 159, 165, 176, 187, 199, 218, 221, 233, 304, 330, 333, 367, 378, 391
Romeo 47, 55
Rookmangud Katawal 91, 127, 147, 151, 192, 198, 202, 204, 219, 227, 230, 254, 270, 287, 293, 338, 341, 356, 367, 369, 376, 391, 397, 398, 399, 415, 427
Royal Guard Military Police 118, 237
Royal Guards 7, 116, 154, 189, 204, 209, 233, 303, 346, 424, 433
Royal Guards Brigade 7, 189, 209, 346, 424, 433
Royal Highness Himani 25
Royal Nepal XXII, 3, 5, 75, 103, 120, 127, 139, 146, 187, 194, 197, 222, 253, 346, 397, 404, 410
Royal Nepalese Army (RNA) 11, 41, 47, 52
Royal Nepalese Embassy 36, 186, 193, 216, 231, 410
Royal Nepal Military Academy 120
Royal Palace 7, 8, 9, 10, 11, XVIII, XIX, XX, XXI, XXII, XXIV, XXV, 1, 2, 3, 4, 5, 7, 8, 9, 10, 12, 13, 14, 15, 17, 18, 21, 24, 25, 26, 27, 28, 29, 31, 33, 34, 35, 38, 39, 40, 41, 43, 45, 47, 48, 49, 50, 51, 52, 53, 57, 60, 63, 64, 66, 67, 68, 69, 70, 71, 72, 73, 74, 75, 76, 77, 78, 79, 80, 81, 86, 87, 91, 92, 93, 94, 97, 98, 100, 102, 103, 104, 107, 111, 112, 113, 114, 116, 117, 119, 121, 124, 128, 130, 134, 136, 139, 142, 143, 145, 146, 147, 150, 151, 152, 153, 155, 156, 157, 159, 160, 161, 162, 164, 165, 166, 168, 169, 170, 172, 173, 176, 177, 178, 179, 182, 184, 186, 187, 192, 195, 198, 200, 203, 209, 211, 212, 213, 215, 216, 219, 221, 222, 224, 228, 229, 230, 231, 232, 235, 237, 238, 241, 242, 246, 247, 253, 254, 255, 257, 258, 259, 265, 266, 270, 274, 278, 280, 281, 288, 289, 290, 292, 294, 296, 298, 302, 303, 305, 310, 313, 314, 318, 326, 327, 330, 331, 337, 345, 346, 347, 348, 350, 352, 353, 356, 360, 362, 363, 372, 373, 375, 379, 383, 384, 385, 390, 391, 392, 396, 398, 400, 403, 405, 406, 407, 408, 409, 410, 412, 414, 415, 416, 417, 418, 424, 425, 427, 428, 432, 433, 434, 435, 436, 437, 438, 440, 441
Royal Palace Massacre 7, 8, 9, 10, 11, XVIII, XX, XXII, 13, 29, 38, 39, 40, 43, 49, 64, 75, 77, 78, 81, 145, 151, 153, 155, 157, 160, 161, 172, 176, 182, 195, 198, 228, 231, 281, 302, 327, 350, 391, 403, 432, 435
Royal Seal 41, 129, 134, 136, 159, 177, 212, 433, 434
RPP 67, 71, 77, 93, 114, 115, 120, 138, 149, 156, 157, 158, 197, 225, 251, 261, 275, 276, 287, 295, 306, 307, 308, 309, 314, 327, 332, 353, 357, 364, 375, 388
RP Pradhan 201, 355
Rudra Bahadur Poudel 77
Rudra Prasad Dahal 102
Rukum 2, 11, 44, 53, 119, 142, 143, 149, 233, 304, 355, 378, 418, 419, 420, 430
Rumjatar 257, 258
Rupendra Raj Pandey 319

S

SAARC 60, 102, 181, 184, 188, 189, 190, 191, 192, 193, 309, 394, 410
Sadip Bahadur Shah 27, 154, 176, 234, 258, 302, 427, 435
Sagar Prasad Timilsina 7, 17
Sahabir Thapa 61, 311, 366, 441
Salyan 78, 101, 143, 157, 229, 233, 329, 330, 418
Samacharpatra 7, 179, 289
Sam 'Bahadur' Manekshaw 429
Sam Cowan 112, 180, 242, 255
Samuel Tamrat 352, 353
Sandhikharka 239
Sandhurst 7, 321, 420, 421, 422, 423, 424, 425, 430, 437
Sand Model 60
Sanduk Ruit 271
Sanghu Weekly 204
Sanphe Bagar 200
Sanskrit 105, 113, 129, 440
Saptahik Bimarsha 130, 165
Saraswoti Choudhary 182
Satbaria 60, 214
Satchit SJB Rana 81, 123, 221, 369, 374, 377
Scotland 422
SD Muni 58, 296, 323, 335, 368
Secretary 7, 8, 9, 11, XXI, XXII, XXIII, XXIV, XXV, 1, 4, 6, 7, 9, 10, 11, 12, 13, 14, 17, 20, 21, 23, 25, 27, 29, 31, 36, 40, 46, 48, 62, 68, 69, 83, 86, 87, 88, 91, 93, 94, 97, 99, 100, 101, 102, 103, 104, 105, 107, 109, 111, 113, 117, 118, 119, 121, 122, 123, 130, 132, 134, 135, 136, 139, 140, 141, 143, 146, 147, 151, 160, 161, 163, 168, 171, 173, 180, 182, 183, 188, 190, 192, 193, 194, 195, 196, 198, 199, 200, 210, 211, 212, 216, 220, 223, 227, 228, 229, 234, 238, 243, 244, 245, 248, 249, 251, 252, 254, 257, 259, 263, 265, 266, 268, 269, 275, 280, 282, 283, 287, 289, 290, 293, 294, 295, 297, 298, 300, 301, 302, 305, 306, 307, 309, 310, 315, 316, 318, 322, 324, 325, 329, 331, 334, 336, 340, 344, 345, 347, 348, 349, 350, 351, 352, 353, 361, 363, 364, 365, 367, 368, 370, 372, 374, 377, 384, 387, 388, 389, 391, 395, 400, 403, 404, 406, 408, 409, 410, 411, 412, 413, 414, 419, 420, 424, 427, 428, 432, 433, 434, 435, 436, 437
Shailaja Acharya 156, 190, 196, 198, 208, 235, 315, 319, 322, 329, 333, 334, 337, 359
Shailendra Kumar Upadhyaya 324
Shakti Amatya 377
Shakti Lamsal 209, 210, 306
Shambhu Thapa 370

Shangri-La XV, 83, 267, 273, 299, 352
Shankaracharya 438
Shankar Hotel 90
Shanta Kumar Malla XXI, 10, 35, 40, 48, 151, 403, 404, 412, 413, 416, 428, 435, 440
Shanti XIX
Sharad Adhikari XXV
Sharada Prasad Pradhan 6, 7, 17, 211, 432
Sharada Singh 183
Sharad Bhattarai 273
Sharad Chandra Shah 81, 119, 129, 252, 261, 274, 280, 281, 283, 329, 333, 392, 442
Sharad Chandra Wosti 8, XXV
Sharat Singh Bhandari 316, 386
Sharvendra Nath Shukla 376
Shashi Bikram Singh 435
Shashi Pratap Bom Malla XXV
Shashi Pratap KC 407, 427, 434, 440
Sheela Dixit 278
Shekhar Koirala 334
Sher Bahadur Deuba 9, 10, XXII, 9, 45, 49, 50, 51, 69, 72, 76, 78, 79, 82, 94, 125, 126, 130, 146, 160, 161, 214, 226, 229, 234, 237, 247, 250, 256, 287, 308, 332, 367, 386, 387, 404
Sher Bahadur Malla 428
Sheshadri 252
Shishir Chandra Shah 425, 427
Shivapuri Army Staff College 334
Shiva Ram Khatri 221
Shova Bhagawati 22
Shree Kanta Regmi 21, 23, 198
Shree Panch 285
Shridhar Khatri 227
Shri Ram KC 432
Shri Sadan 416, 428
Shyam Bhakta Thapa 5, 292, 304, 358, 364
Shyam Sharan 302, 345, 366, 395
Shyam Shrestha 167, 306, 358, 369
Siddhi Lal Singh 274
Sierra Leone 131, 141, 152
Sikkim 65, 135, 259
Siliguri 52, 59, 99, 168, 177, 178, 279
Sinamangal 89, 323
Sindhunath Pyakurel 283, 362
Singapore 58, 140, 190, 319, 364, 421
Singha Durbar 190, 202, 206, 337
Singha Pratap Shah 35, 426
Singh Bahadur Basnyat 429
Siraha 112, 200, 342, 360
Sir Geoffrey James 338, 369, 394
Sir Michael Bois 225
Sir Michael Jackson 378
Sir Samuel Cowan 124
Sita Nandan Roy 274
Sitashma 8, 9, 295
Situation Center 326, 356
SLR 152, 401
Soaltee Hotel 191, 386
Solukhumbu 185
Sonam 128
South Asia 57, 59, 102, 138, 195, 304
South Block 58, 114, 276, 301, 318, 342, 353
South Korea 188, 326, 350
Speaker of House 92
squad 105, 121
S. Ramesh 234
Sri Lanka 59, 61, 101, 108, 111, 221
SSNCC 406
SSP 122, 128, 147, 193, 256, 370
Staff College 36, 115, 130, 142, 147, 149, 334, 429, 437
Standing Committee 17, 19, 182, 224, 245, 373
Star News 221
State Council 17, 19, 20, 21, 23, 170, 182, 245, 373, 374
Subash Ghising 368
Sudan Prasad Pokharel 4, 7, 17, 160, 168, 229
Sudan Thapa 209
Sudheer Sharma XXIV, XXV
Sudhir Shah 85
Sujata Koirala 134, 186, 199, 223, 242, 284, 372
Sundar Pratap Rana 13, 15, 151, 156, 180
Sun Heping 371
Sunsari 154, 327
Supreme Court 17, 41, 45, 69, 70, 72, 82, 124, 134, 142, 196, 227, 230, 233, 255, 337, 365, 394, 395, 399
Supriya Shah 32
Suraj Shumsher Rana 420
Surendra Nepal 159
Surendra Prasad Choudhary 196, 198
Suresh Hamal 188
Suresh Raj Sharma 8, 164
Surya Bahadur Thapa 3, 10, 12, 53, 72, 73, 74, 77, 87, 92, 93, 186, 197, 210, 213, 230, 244, 245, 248, 250, 252, 273, 295, 314, 322, 323, 327, 329, 334, 335, 341, 345, 352, 355, 356, 360, 363, 365, 372
Surya Nath Upadhyaya 220
Surya Regmi 286
Surya Subedi 317, 344
Sushil Koirala 79, 160
Sushil Kumar 146
Swami 168, 178
Swayambhu 21, 430
Sweden 343
Switzerland XVII, 84, 85
Sworoopananda Sarswoti 438

T

talk facilitator 284, 285, 288, 295, 300, 302, 310, 325, 334, 339

Tamil Separatist 61
Tanakpur 69, 72, 73, 299
Tapa Bahadur Magar 332
Tara Bahadur Thapa 24, 35, 37, 111, 114, 121, 133, 141, 155, 165, 255, 363, 404, 407, 412, 428, 434, 440
Taranath Ranabhat 17, 69, 92, 94
Tarun Dal 147, 362
Tej Bikram Shah 340
Terai 118, 135, 285, 372, 378
Thakur Prasad Choudhary 200
Thapathali Palace XIX
Tharuwan Liberation Front 349
Three-Naught-Three 240
Tibet 59, 61, 118, 127, 232, 315, 378
Tiger Mountain 128
Tihar 62, 260, 370, 371, 374, 375
Tika Datta Niroula 253, 315, 344, 349, 350
Tika Dhamala 3, 4, 7, 8, 9, 15, 36, 37, 151, 168, 178, 265, 300, 301, 310, 320, 390, 392, 393, 399, 414
Tilak Bahadur Kunwar 404
Tilak Rawal 142, 195, 240
Time Magazine 204
Tirtha Man Shakya 315
Tirupati 410, 411
Top Bahadur Rayamajhi 170
TOR 125, 142, 254
Toran Jung Bahadur Singh 258
Trailokya Nath Uprety 9, 437
Treaty of Commerce and Transit 304
Tribhuvan Army Officers' Club 192, 298, 314, 357
Tribhuvan International Airport 89, 111, 117, 305, 323, 352
Tulsi Giri 119, 123, 126, 250, 393

U

Uddhav Poudel 253
Ujjwal Bikram Rana 109, 110
UN 12, 50, 102, 113, 124, 127, 131, 137, 139, 141, 150, 151, 152, 188, 194, 195, 215, 221, 240, 263, 264, 268, 295, 336, 340, 349, 352, 353, 354, 355, 358, 368, 371, 377, 387, 388, 389, 390, 401, 415, 416
UNDP 352
UNHCR 264, 344
Unified Command 5, 54, 206, 373
UNIFIL 139
United Nations Mission 63, 118
University Campus 36
UNMIN 63
UN Peacekeeping Mission 12, 131, 215, 264, 355
UN Peacekeeping Training Center 131, 215
Upendra Devkota 16, 85, 90, 333
Upendra Kanta Aryal 374
Upendra Yadav 182
Upper House 17, 19, 150

V

Vaaj (hawk) 43
VDCs 104
Victory SJB Rana 180, 234, 302, 338, 384, 415, 435
Vienna 184
Vijaya Bikram Khadka 300
Vijay Kumar [Pandey] 236
Village Defense Forces 378
VIP 27
Vishwo Hindu Parishad 368
Vishwo Nath Prasad Upadhyaya 72, 230, 338, 341
VP Singh 196
VVIP 111, 148, 194, 195, 428

W

weapons 11, 12, XXII, 16, 41, 47, 51, 52, 53, 55, 56, 62, 63, 64, 65, 66, 67, 79, 84, 98, 103, 106, 107, 108, 109, 110, 114, 117, 118, 120, 122, 124, 126, 129, 131, 132, 138, 142, 148, 150, 152, 153, 154, 167, 175, 184, 185, 186, 187, 189, 193, 194, 195, 199, 201, 205, 207, 209, 214, 217, 219, 220, 221, 222, 223, 226, 228, 231, 234, 237, 238, 239, 240, 261, 272, 276, 277, 288, 300, 304, 343, 370, 401, 402, 403
Wei Hwang Kwong 350
Western Development Region 278, 301, 310
Western Division 234
William 321, 386
William Wiry 386
World War II 131

Y

Yagya Bikram Rana 217
Yankila Sherpa 154
Yogendra Pratap Jung Rana 414
Yogesh Bhattarai 46
Yogi Narahari Nath 117, 233, 285
Yoshiro Mori 117
Yubaraj Ghimire 28, 167
Yugoslavia 34
Yu Yung Kong 264

Z

Zee TV 181
Zenji Kaminaga 344
Zhou Rongji 148
Zonal Police Chief 121
Zone of Peace 42, 62

PHOTOGRAPHS

Royal Military Academy, Sandhurst, UK
Officer Cadet, 1969

During the visit of the then Military Secretary Kshetra Bikram Rana to Royal Military Academy, Sandhurst

Boxing Team of RMA, Sandhurst, 1971

Officer Cadet Training

On top of the highest mountain in the UK, Ben Nevis, 1969

Graduation Ceremony, Royal Military Academy, Sandhurst, 1972

2nd Lieutenant in Gorakh Bahadur Battalion

3" and 81 mm Mortar Training

Royal family's visit during the marriage ceremony of my eldest son, 2054 BS

King Birendra receiving the Guard of Honor (along with the Chinese President) during his visit to China

Prince Nirajan's initiation ceremony

Receiving an Honor

CGSC Graduation, Fort Leavenworth, USA, 1982

On the occasion the President of Maldives, Maumoon Abdul Gayoom's visit to Nepal

On the occasion of Indian Prime Minister Atal Bihari Vajpayee's visit to Nepal

During a Royal Visit and the dinner programme at the Rastrapati Bhavan, India

Meeting with General Tan of the PLA during China visit

With Shree 1008 Sachchidananda Saraswoti (Khaptad Baba)

With wife and sons in my Major Rank

With wife at the Staff College, USA

Military Secretary, Lieutenant General Vivek Kumar Shah

In a formal program at the Royal Palace

On the day of promotion to the rank of Lieutenant General (14 April,2003)

During an informal visit to UK, 2036 BS

Prince Dipendra watching training event at the UN Peace Keeping Training Center, Panchkhal

King Gyanendra appoints Pyar Jung Thapa as the Chief of the Army Staff

King Gyanendra confers the rank of Honorary General of the Nepal Army on Indian Chief of the Army Staff, General Nirmal Chander Vij

The Crowning Ceremony of King Gyanendra, Hanuman Dhoka, 2001

Procession on its way to Royal Palace after King Gyanendra's Crowing Ceremony

The royal family's visit to Mayo College, Ajmer, India, where Princess Shruti was studying

Receiving tika from King Birendra

His Majesty King Birendra receiving felicitation on the occasion of 25th anniversary of his ascension to the throne

The royal family conducting pooja at Haridwar

With the former president of the United States, Jimmy Carter

King Gyanendra in consultation

The Roya Palace

While a scout , with the Defense Minister of India, Mr Swarna Singh in New Delhi

Family photo, on the occasion of the marriage of the youngest son, 2063 BS